T0174684

Space-time

"[T]he gravitational field is space…it is a real entity that undulates, fluctuates, bends, and contorts"

(Rovelli, 2017)

Space-time
An Introduction to Einstein's Theory of Gravity

Jonathan Allday

CRC Press
Taylor & Francis Group
Boca Raton London New York

CRC Press is an imprint of the
Taylor & Francis Group, an **informa** business

Cover image courtesy of Event Horizon Telescope Collaboration.

CRC Press
Taylor & Francis Group
6000 Broken Sound Parkway NW, Suite 300
Boca Raton, FL 33487-2742

First issued in paperback 2020

© 2019 by Taylor & Francis Group, LLC
CRC Press is an imprint of Taylor & Francis Group, an Informa business

No claim to original U.S. Government works

ISBN-13: 978-1-138-05668-8 (hbk)
ISBN-13: 978-0-367-77969-6 (pbk)

This book contains information obtained from authentic and highly regarded sources. Reasonable efforts have been made to publish reliable data and information, but the author and publisher cannot assume responsibility for the validity of all materials or the consequences of their use. The authors and publishers have attempted to trace the copyright holders of all material reproduced in this publication and apologize to copyright holders if permission to publish in this form has not been obtained. If any copyright material has not been acknowledged please write and let us know so we may rectify in any future reprint.

Except as permitted under U.S. Copyright Law, no part of this book may be reprinted, reproduced, transmitted, or utilized in any form by any electronic, mechanical, or other means, now known or hereafter invented, including photocopying, microfilming, and recording, or in any information storage or retrieval system, without written permission from the publishers.

For permission to photocopy or use material electronically from this work, please access www.copyright.com (http://www.copyright.com/) or contact the Copyright Clearance Center, Inc. (CCC), 222 Rosewood Drive, Danvers, MA 01923, 978-750-8400. CCC is a not-for-profit organization that provides licenses and registration for a variety of users. For organizations that have been granted a photocopy license by the CCC, a separate system of payment has been arranged.

Trademark Notice: Product or corporate names may be trademarks or registered trademarks, and are used only for identification and explanation without intent to infringe.

Library of Congress Cataloging-in-Publication Data

Names: Allday, Jonathan, author.
Title: Space-time: an introduction to Einstein's theory of gravity / by Jonathan Allday.
Description: Boca Raton, FL : CRC Press, Taylor & Francis Group, [2019] |
Includes bibliographical references and index.
Identifiers: LCCN 2019003925 | ISBN 9781138056688 (hbk) |
ISBN 1138056685 (hbk) | ISBN 9781315165141 (ebk) | ISBN 1315165147 (ebk)
Subjects: LCSH: General relativity (Physics) | Space and time. | Quantum gravity.
Classification: LCC QC173.6 .A45 2019 | DDC 530.11—dc23
LC record available at https://lccn.loc.gov/2019003925

Visit the Taylor & Francis Web site at
http://www.taylorandfrancis.com

and the CRC Press Web site at
http://www.crcpress.com

Contents

Preface

Until I started my research prior to writing, I was not aware of some of the philosophically contentious issues surrounding general relativity's foundations and formulation. Indeed, there is a case for referring to the theory as *Einstein's theory of gravity*, as many dispute that the theory does 'generalise' the principles of relativity. While I am sensitive to philosophical niceties of this type, I love the name *general relativity* so much I can't bring myself to refer to it any other way.

Authors always, I believe, aim high and compromise to finish. Hopefully, the necessary compromises needed to keep this book in bounds will still give the reader sufficient interest and understanding to pursue the subject further.

Notes on Mathematical Terminology Used in This Book

Various symbols are used in this book, some of which may not be familiar to the reader, at least not with my specific lexography:

$\sim$	$x \sim y$	Similar sort of size, 10 and 20 are of the same order
$\approx$	$x \approx y$	Approximately equal to, $3.01 \approx 3$
$\sum$	$\sum_i x_i$	Sum: add up a sequence of terms: $x_1 + x_2 + x_3$ etc.

I have not followed an especially rigorous mathematical path in some aspects of this development. The aim has been to give the interested person with a background in physics and mathematics, appropriate to someone who was involved in the subject right up to the end of school, an understanding of the physics. I suspect that some of the liberties that I have taken would be frowned upon by the experts. In particular, I have not been at all careful about the difference between infinitesimal and very small. Differentiation, partial and ordinary, has been replaced by 'rates of change'. All I can say is that, aside from any typos that may exist, all the results are correct, even if the means of getting to them is a bit 'hand wavy'.

Various conventions that are widely used among professionals in this field have not been followed here. The most conspicuous are the following.

The Einstein summation convention: implied summation over a repeated index:

$$T_{abc} g_{cd} = \sum_c T_{abc} g_{cd}$$

while this does clean up some otherwise rather inelegant algebraic expressions, I think it makes it harder for the beginner to follow 'by eye'.

Contravariant and covariant indices being indicated by superscripts and subscripts: T^u – contravariant and T_u – covariant. Instead I have placed all indices in the subscript (which I believe is where people are more used to seeing them) and indicated covariance with a bar over the index.

Appendix

The book's appendix is available for downloading from this book's CRC Press website: https://www.crc-press.com/Space-time-An-Introduction-to-Einsteins-Theory-of-Gravity/Allday/p/book/9781138056688.

Jonathan Allday
14 August 2017, Madeira
17 December 2018, Yorkshire
jallday40r@me.com

Thanks

Of course thanks go to the Taylor & Francis team and especially Francesca McGowan, Rebecca Davies and Kirsten Barr. It is well known that Douglas Adams, author of *The Hitch Hiker's Guide to the Galaxy*, referred to deadlines in the following way: "I love deadlines. I love the whooshing noise they make as they go by...". In this case, life got in the way of starting this book so it finished somewhat past deadline. The patience and understanding of the team was much appreciated, as was their gentle pressure in the closing struggles. Anyone else who was involved in the production process, please accept my apologies for not directly mentioning you.

I also want to thank the classes of 1979/1980 at the Liverpool Blue Coat School. For whatever reason, and I am sure it had a fair amount to do with the exceptional teaching, those years produced a decent crop of scientists. I had the pleasure of friendships, discussions and flat arguments with them about physics and other matters, Professors Simon Hands (Swansea) and Roger Jones (Lancaster) principally among them. I dedicate this book to them.

A variety of people were kind enough to read chapters and make suggestions, principally Simon Hands and Alex Dalton (6th form student at Woodhouse Grove School). Any mistakes or confusions, of course, remain my responsibility.

A project like this can't take place without the tacit support of my family. Once again thanks go to them, although fortunately as the children are now grown up, their father's mental absences did not scar them too badly this time. Carolyn remains a baffled onlooker, but without her support and suffrage of no nonsense, I would not have been able to do this.

And finally, Jack Richards did not get to see this book completed. He always liked looking at fonts, etc., in my other books. As a professional typesetter he hated having to do the fiddly bits in mathematical expressions. I suspect this book may well have driven him mad. I will think of him whenever I see it on the shelf.

Author

Jonathan Allday teaches physics at Woodhouse Grove School, UK. After taking his first degree in physics at Cambridge, he moved to Liverpool University, where he gained a PhD in particle physics in 1989. While carrying out his research, he joined a group of academics and teachers working on an optional syllabus to be incorporated into A-level physics. This new option was designed to bring students up to date on advances in particle physics and cosmology. An examining board accepted the syllabus in 1993 and now similar components appear on most advanced courses and some aimed at intermediate level.

Jonathan has authored *Quarks, Leptons and the Big Bang* (1998, 2001, 2017), published by CRC Press and now in its 3rd edition, which was intended as a rigorous but accessible introduction to these topics. Since then he has also written *Apollo in Perspective* (1999) and *Quantum Reality* (2009), published by CRC Press.

Introduction

What We Know

Modern thinking about gravity is dominated by Einstein's theory, which is now well over 100 years old. In that time, the *general theory of relativity* has not lost its power to baffle, astound and beguile in equal measure. Mastery of the theory represents a challenge for physicists, partly due to its mathematical complexity (it is centred around a collection of ten non-linear differential equations, which is *exactly* as bad as it sounds) and partly as it is a theory involving the geometry of four dimensions (three spatial dimensions and time), which stretches our ability to visualise what is happening in a given situation. Many of us rely on such visualisations to help us calculate and to provide reassurance that we have developed an understanding of what is happening. We are left to work with half-pictures, simulations and analogies, which can lead to dangerous oversimplifications.

Despite its fearsome reputation, general relativity has been a remarkably successful theory. From its earliest triumphs in explaining an odd quirk of Mercury's orbit (1916) and predicting the extent to which light would be deflected as it passed by our own Sun (confirmed in 1919), through to the successful detection of gravity waves announced on 11 February 2016, general relativity has met every experimental challenge. It has also opened up new avenues of research related to neutron stars and black holes and without general relativity our extraordinary level of understanding of the processes at work in the Big Bang would not be possible.

More prosaically, the accurate timekeeping that lies behind the operation of the GPS satellite network would not be possible without general relativity.

Yet we also know that general relativity is not infallible. Buried inside its diamond-like mathematical structure there is a flaw, demonstrating that it cannot successfully model reality in all situations. In the right circumstances, the theory predicts a physical impossibility: the *singularity*. Given a sufficiently concentrated mass, gravity will engender irreversible collapse. The mass will compress under the action of its own gravity, and as it collapses so that gravity will increase, hastening the collapse. Despite being the weakest of nature's forces, nothing can insulate the effects of gravity, so a sufficiently compressed mass cannot withstand this collapse: the very structure of the matter from which it is constructed will break down. The theory predicts that the collapse will continue until the mass reaches a point of zero size and hence infinite density. We have no idea what such a mathematical concept could possibly mean physically; hence the term singularity is more of a label flagging ignorance than a name ascribed to a physically existing structure.

To see such an outcome emerge from an otherwise plausible set of calculations is embarrassing enough; to compound the issue we believe that the correct physical situation exists in nature to trigger the onset of such collapse. This is how black holes form from the death of sufficiently massive stars. There is now plenty of astronomical evidence confirming the existence of black holes, and general relativity works beautifully well with regard to their gross physical properties. However, inside the *event horizon* of the black hole, the theory predicts the on-going collapse to the singularity.

We are also faced with the even more disturbing prospect that the universe, according to general relativity, must have started with a singularity, and may possibly collapse back to one in the future.[1]

When a theory malfunctions in this manner, it is a signal that something in its structure does not fully correspond to nature. Clearly general relativity is not hopelessly wrong, or none of its predictions would have been confirmed. In truth its predictions must be *approximately* correct; so close to being right that we cannot currently detect the difference between what it predicts to be the case and what we can measure in reality. That is until we get to an extreme case, where the flaw gets prised wide open.

1

So, we confidently work with general relativity while at the same time searching for the successor theory that will somehow show why the singularity does not form in reality and explain the undoubtedly esoteric structure that will take its place.

Beyond General Relativity

The picture of gravity built by general relativity is quite unlike that portrayed by our theories of the other fundamental forces of nature. Considerable progress has been made in that area over the last 50 years as our understanding of the fundamental particles of nature and the forces of interaction between them has developed. The other great theory of the 20th century, *quantum theory*, has grown into *quantum field theory* and then the *gauge theory* of fundamental interactions. This progression of ideas has taken us a long way from our everyday understanding of forces as the pushes and pulls that we exert on objects, and which we occasionally suffer the consequences of. Yet, the line of ideas is unbroken. We may think in terms of interactions being mediated by quantum disturbances in fundamental fields, but ultimately these ideas can be used to explain the pushes and pulls of everyday life. If we think in terms of games, then as we have progressed from amateur status towards Grand Mastery, our understanding of chess has expanded and transformed, but we are still playing the same game.

General relativity dissolves the notion of gravity being any sort of pulling force that influences the way in which objects move by erecting a picture of objects being guided, as if on invisible rails, by the fundamental geometry of the space-time they are passing through. In the words of one of our most revered experts on general relativity, John Wheeler: "Spacetime tells matter how to move; matter tells spacetime how to curve".[2]

At school we are taught that when we drop a ball, the pull of gravity exerted by the Earth's mass accelerates the ball towards the centre of the Earth, a fall that is only broken by contact with the ground or some otherwise solid object. General relativity tells us that the Earth's mass does not extend its influence through space, reaching out via a gravitational field to exert a force on the ball. Rather the very structure of space and time in the vicinity of Earth is distorted and the ball is moving, un-deflected by any forces,[3] along the straightest (laziest) path that it can in this warped geometry. As the distortion affects both time and space, the path looks to us like acceleration towards the centre of the Earth.

Given this radically different posture towards the 'force' of gravity, it is perhaps not surprising that general relativity does not easily mesh with the quantum field theories we use for the other fundamental forces of nature. It is possible to create a gauge theory of gravity, but only in the limit of very weak gravitational effects. As a toy theory it helps frame the issues, but it is not the ultimate theory that we need.

Quantum theory arose out of the need to provide a framework for explaining the physics of the subatomic world. Its natural home lies in the smallest dimensions of matter, which is just where we need it to operate in order to save us from the physical impossibility of the singularity. Unfortunately, current efforts to combine the principles of quantum theory with the maxims of general relativity have the tenor of a forced marriage rather than an elopement. The quantum revolution has forced us to confront many counter-intuitive aspects to reality, not the least of which is the strange wave/particle duality whereby particles thought to be little lumps of matter (like tiny marbles) experimentally display wave-like features, and phenomena safely categorised as wave in nature (e.g. light) can be shown to have particle aspects. In dealing with this we have had to transcend the primitive categories of wave and particle and adopt a perspective whereby reality is described by quantum fields which can have a 'granular' nature. Accordingly, gravitation should be described by such a quantum field, which is hard to reconcile with the curved geometrical picture of the general theory.

Progress is being made, but there is still the sense that something fundamental is missing. Is it the quantum picture that needs expanding to accommodate the requirements inherent in the geometry of space-time, or does our understanding of space-time need radically transforming in the light of quantum ideas? Most likely, some combination of the two will emerge as our ultimate candidate theory.

Work in this area is already opening up new avenues of thinking. Talk of *strings*, *branes*, higher dimensions and *the bulk* is rippling down to a less professional audience, albeit partly through the works of science fiction.

We are living in a curious period for physics, which in some ways calls to mind a type of thinking last seen among the ancient Greeks. Theorists are exploring highly abstract and speculative avenues and experimentally testable predictions are hard to come by. As a result, theories are rated and ranked according to their mathematical utility and, in a very real sense, their aesthetic appeal, rather than their experimental validity. This not meant as a criticism, we have to make progress how and when we can, but theorists' claims to certainty need to be judged in this context.

Notes

1. All current cosmological measurements related to the structure of the universe indicate that there is not sufficient mass/energy in the universe to trigger such ultimately catastrophic collapse.
2. John Archibald Wheeler, 1911–2008, *Geons, Black Holes, and Quantum Foam* (2000), p. 235.
3. Let's agree to ignore air resistance and the like…

1

Four Keystones

1.1 Starting Points

In developing our understanding of gravitation, we are going to be faced with some tricky concepts and fiddly mathematical ideas. Consequently, it is important to ground ourselves as and where we can in experiment and observation; otherwise, the subject becomes somewhat abstract.

There is little point in dwelling on ideas that have been rejected or replaced, but a lot of what we learn about gravity in school, or at college, needs a spring clean and setting afresh in a context that allows us to move forward. For this reason, we are going to start by focussing in on four keystones that have helped support conceptual development in the area of gravitation. I am not saying that these are the most important keystones, still less that they are the only ones. I selected them partly out of personal interest, but also as I believe there is an interlinking thread through them that illustrates our advancing experimental sophistication alongside a crucial conceptual development. The selection is as follows:

- Galileo's supposed observation, involving the leaning tower of Pisa, that all objects fall with the same acceleration in a gravitational field, provided that air resistance can be ignored.
- Newton's fabled observation of an apple falling that led to his theory of universal gravitation, from which he could prove that Galileo's assertion for falling objects was correct.
- Eddington's careful measurements during an eclipse, which helped establish Einstein's general theory of relativity and in the process transform the conceptual foundations for our understanding of gravity. One crucial step in the development of this theory took place when Einstein was meditating on the relevance of Galileo's observation.
- The results of the LIGO experiment, which for the first time directly detected *gravitational waves*, a prediction of the general theory with no Newtonian counterpart, hence showing that the general theory's conceptual core is better suited to the description of nature.

1.2 Galileo and the Tower of Pisa

Philosophy is written in this grand book—I mean the Universe—which stands continually open to our gaze, but it cannot be understood unless one first learns to comprehend the language and interpret the characters in which it is written. It is written in the language of mathematics, and its characters are triangles, circles, and other geometrical figures....

G Galilei[1]

As one of the key figures in the history of science, Galileo (1564–1642) is justly admired, not only for his discoveries and insights but also for his communication skills and his insistence on the value of experiment and observation in discovering natural law. He was also one of the earliest people to declare that the laws of nature are inherently mathematical, something that we take for granted now, despite the deep puzzles that it suggests. Fascinatingly, as the quote above illustrates, Galileo was alluding to the role played by geometry in the laws of nature way back in the 1600s. Given that the general theory of relativity places geometry front and centre as one of the fundamental aspects of our universe, his views turned out to be somewhat prophetic.

Galileo's research ranged widely over topics in astronomy (Sunspots, the moons of Jupiter, phases of Venus and more), engineering (compasses, thermometers, telescopes and microscopes) and physics (the movement of objects, pendulum swings, the frequency of sound waves and of course, gravity). Most significant for us, however, are two of his key ideas:

- that the laws of physics for any observer moving at a constant speed in a straight line are the same as for any other stationary observer[2];
- that objects dropped from the same height will hit the ground after the same time of flight, irrespective of their masses.

The first of these thoughts (known as *Galileo's principle of relativity*) connects with one of Einstein's central aims: to show how the laws of physics could take the same mathematical form, irrespective of the choice of co-ordinate system selected for their expression. Galileo, on the other hand, was concerned to dissolve any objections to the Copernican view that the Earth orbited the Sun. His relativity principle ensured that the Earth's orbital motion (and its rotation for that matter) did not directly impact on people's daily lives. In one of his writings, Galileo discusses the fate of a collection of butterflies trapped below deck on a ship in gentle (unaccelerated) motion "nor will it ever happen that they [the butterflies] are concentrated towards the stern, as if tired out from keeping up with the course of the ship".[3]

The second thought nudged Einstein towards the radical conceptual jump that replaced the notion of gravitational force with that of curved space-time.

In Galileo's day, people generally accepted that our world is formed from four elements: *earth, air, fire* and *water* – a philosophical view that dated back to the ancient Greeks. In a perfect world, these elements would exist in pure form as layers, like an onion, from the earth element at the centre through water, air and fire at the outside (Figure 1.1).

FIGURE 1.1 A 1524 representation of the universe, according to Aristotle's cosmology. The water and earth elements (drawn as continents and oceans) are at the centre, immediately surrounded by regions of air and fire. Then, in their respective spheres come the Moon, Mercury, Venus, The Sun, Mars, Jupiter and Saturn. Finally, we have the firmament with its stars and astrological groupings. (Image credit: Public Domain, and Edward Grant, 'Celestial Orbs in the Latin Middle Ages', Isis, Vol. 78, No. 2. [June 1987], pp. 152–173 [PD-US].)

However, due to some event (which Christian theology later associated with the Fall), the elements had become mixed up so that none of the objects we actually come across exist in pure elemental form. However, manipulating things, for example, dropping them, can show their innate nature. Equally, setting fire to something releases the fire element to rise to its outer layer. The impermanence and decay that we see in the world, according to this philosophy, also arises due to the mixing of the elements.

Working in this context, Aristotle proposed that objects fall at a speed related to their mass, so that heavier objects, imbued with more of the earth element, fell faster due to their greater impulsion to return to their natural position in the hierarchy of elements.

Galileo is supposed to have refuted this by dropping different masses from the top of the leaning tower at Pisa (Figure 1.2): a story that seems to have grown from a biography by one of Galileo's pupils, Vincenzo Viviani.[4]

As Galileo himself never wrote about this demonstration, most historians doubt that it actually took place. Most likely, Galileo came to his conclusion via an ingenious thought experiment.[5]

Imagine two objects of somewhat different mass tethered together by a light cord that is resistant to being extended (Figure 1.3).

According to Aristotle, the object with greater mass should fall faster, so shortly after being dropped, the tethered pair should arrange themselves vertically with the heavier object towards the bottom. The lower mass object at the top will be falling at a lower speed and so will tend to hold back the motion of the heavier one and pull the cord taut. Yet, if we consider the collection of objects and cord together, the group has more mass than any of the separate parts and so should fall with a combined speed that is greater than that of the pieces.

The only way to resolve this apparent contradiction is to assume that all the objects fall at the same rate, irrespective of their mass (neglecting air resistance).

Even if Galileo never actually lobbed anything off the tower at Pisa, such an experiment did take place in 1586 when Simon Stevin and Jan Cornets de Groot dropped lead balls of considerably different masses from the church tower in Delft,[6] a fall of 9 m to the ground (Figure 1.4).

FIGURE 1.2 The leaning tower at Pisa where Galileo is supposed to have demonstrated that objects of different masses fall to the ground in the same time. (Free stock image.)

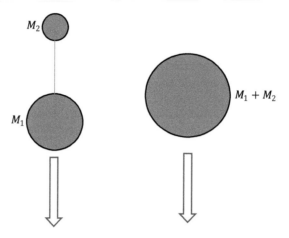

FIGURE 1.3 Galileo's thought experiment on falling bodies compares two masses tethered together with a single object of mass equal to the combined mass of the separate objects.

FIGURE 1.4 The tower of New Church in Delft (Netherlands) where Stevin and De Groot carried out experiments with falling objects. (Image: W Bulach, CC-BY-SA 4.0.)

Possibly the most amusing public demonstration of Galileo's idea took place on 7 August 1971 when astronaut Dave Scott dropped a geological hammer and a falcon feather from the same height (~1.6 m) on the surface of the Moon. This was right at the end of the Apollo 15 mission's last Moonwalk (all of which were televised from the Moon) and shortly before Scott re-entered the lunar module. In his commentary at the time, Scott stressed the relevance of Galileo's work to spaceflight (Figure 1.5).

FIGURE 1.5 Astronaut David Scott on the surface of the Moon demonstrating that objects accelerate under gravity at the same rate irrespective of their mass. The mission transcript for this period reads: 167:22:06 Scott: Well, in my left hand, I have a feather; in my right hand, a hammer. And I guess one of the reasons we got here today was because of a gentleman named Galileo, a long time ago, who made a rather significant discovery about falling objects in gravity fields. And we thought where would be a better place to confirm his findings than on the Moon. 167:22:28 Scott: And so we thought we'd try it here for you. The feather happens to be, appropriately, a falcon feather for our Falcon. And I'll drop the two of them here, and hopefully, they'll hit the ground at the same time. (Pause) 167:22:43 Scott: How about that! 167:22:45 Allen: How about that! (Applause in Houston) 167:22:46 Scott: Which proves that Mr. Galileo was correct in his findings. (Image and transcript credit: NSSDCA/NASA.)

1.2.1 Galileo and the Laws of Motion

Galileo's understanding of motion in general, and falling objects specifically, was undoubtedly bolstered by his experiments with balls rolling down inclined planes. This helped him establish the relationship between the distance travelled by an object accelerating from rest, x, and the time duration of the acceleration, t:

$$x - \frac{1}{2}at^2$$

although he expressed this in geometrical, rather than algebraic, terms. His observations also lead him to the conclusion that "… any particle projected along a horizontal plane without friction… will move along this same plane with a motion which is uniform and perpetual, provided the plane has no limits",[7] which Newton later folded into his first law of motion.

So far, we have not discussed Galileo's views regarding the laws of nature for moving and stationary observers. Galilean relativity develops into one of the key principles of Einstein's special theory of relativity, so we will pick it up again in Chapter 2. For now, we move forward a generation, to Newton and the law of universal gravitation.

1.3 Newton and the Apple

On 15 April 1726 I paid a visit to Sir Isaac, at his lodgings in Orbels buildings, Kensington: din'd with him… after dinner, the weather being warm, we went into the garden, & drank thea under the shade of some appletrees, only he, & myself. amidst other discourse, he told me, he was just in the same situation, as when formerly, the notion of gravitation came into his mind. "why should that apple always descend perpendicularly to the ground," thought he to himself: occasion'd by the fall of an apple, as he sat in a contemplative mood: "why should it not go sideways, or upwards? but constantly to the Earths centre? assuredly, the reason is, that the Earth draws it. there must be a drawing power in matter."

W Stukeley after an account by I Newton[8]

It seems very likely that Newton did, at some time, see an apple fall from a tree (Figure 1.6), which triggered a train of thought leading to the theory of universal gravitation. It seems much less likely that the apple hit Newton on the head, or any of the other embellished versions of the tale that you sometimes come across.[9]

The deep significance of Newton's theory of universal gravitation was not so much his formulation of the equations that determine how gravity influences the motion of objects, but his incorporation of the heavens and the Earth under the same remit for his law.

As I mentioned when we were discussing Galileo, Greek philosophy saw our world as formed from four elements: earth, air, fire and water that ought to exist in elemental layers, but for some process that had mixed them together, leading to our imperfect world of change and decay.

However, this is not the complete story. One aspect of the world contradicts this neat system: the stars, the Moon and the Sun are observably unchanging and eternal, at least to the naked eye and for periods measured in years. Hence, Greek thought divided the cosmos into two realms: the portion below the Moon contained the four elements in their mixed state, whereas the Moon and above were constructed from *aether*, the fifth element in its pure unchanging form.[10] As a consequence, different laws of nature were in operation in different parts of the cosmos. With historical hindsight, we can see how the development of science was constrained by this way of thinking.

Imagination is an under-rated and under-advertised trait in science. Yet the act of jumping from a casual observation to conceiving the underlying laws of nature is as powerful a form of imagination as any found in other human activities more commonly thought of as creative. For Newton, the apple fell because something was pulling it towards the ground. His imagination built a conceptual bridge between the apple and the Moon. Perhaps the same effect could be pulling the Moon towards the Earth, thus holding it in orbit. Although the idea took root and developed gradually in

FIGURE 1.6 The apple tree at Woolsthorpe Manor in Lincolnshire thought to be the 400-year-old tree from which Newton observed an apple falling and contemplated the existence of a universal gravitational force. (Image credit: M Pettitt licensed under CC-SA2, https://creativecommons.org/licenses/by/2.0/.)

Newton's mind, he transformed the approach to natural philosophy by drawing the Earth and the cosmos together into one overarching physical system. Einstein shared this gift of a powerful physical intuition.

By Newton's time, the geocentric universe had been supplanted through the work of Copernicus and Kepler, so the basic facts about the solar system, that the Earth and other planets orbited the Sun and the Moon orbited the Earth, were not in dispute. As Newton had incorporated Galileo's observation that objects moving without friction will tend to continue in a straight line at constant speed, he realised that in order to remain in orbit, the Moon had to be acted upon by a force that was pulling it towards the centre of the Earth. From the known period of the Moon's monthly cycle from wax to wane and the estimated distance to the Moon, he was able to calculate how far the Moon falls towards the Earth every second.

In essence, Newton's calculation went as follows.

Suppose the Moon orbits the Earth in a circular path of radius R_M. In Figure 1.7, the Moon starts at position A and, projecting forward in time, should subsequently coast to position B. However, as the force of gravity pulls the Moon towards the Earth, it ends up at C instead. Hence, we can argue that the Moon has 'fallen' towards the Earth through a distance x.

From the diagram, we can see that:

$$\cos(\theta) = \frac{R_M}{R_M + x}$$

producing, after some re-arrangement:

$$x = R_M \frac{1 - \cos(\theta)}{\cos(\theta)}. \tag{1.1}$$

As the calculation will be more accurate the smaller the angle involved, we will take the time between A and C to be 1 s (so the diagram is grossly out of scale...).

Given that the period of the Moon is roughly 27.3 days, the angle per second comes out to be:

$$\frac{2\pi}{27.3 \times 24 \times 60 \times 60} = 2.66 \times 10^{-6} \, \text{rad}.$$

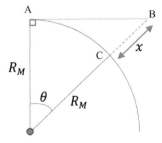

FIGURE 1.7 A portion of the Moon's orbit around the Earth. From a starting position such as A, the Moon's motion ought to carry it to B. However, the pull of gravity towards the Earth curves the path around so that the Moon ends up at C instead.

using the radian measure of angle.[11] Inserting this into Equation 1.1 and using the modern value for the radius of the Moon's orbit gives

$$x_M = R_M \frac{1 - \cos(\theta)}{\cos(\theta)} = 3.85 \times 10^8 \times \frac{1 - \cos(2.66 \times 10^{-6})}{\cos(2.66 \times 10^{-6})} = 1.36 \times 10^{-3} \, \text{m}.$$

In other words, the Moon falls towards the Earth by about 1.36 mm/s.

If we drop an object, such as an apple, while we are standing on the surface of the Earth, it will fall a distance of ~4.91 m in the first second. This is obtained from Galileo's formula for the distance travelled by an accelerating object and the known acceleration due to gravity on Earth, which is 9.81 m/s^2:

$$x_A = \frac{1}{2} g t^2 = \frac{1}{2} \times 9.81 \times 1^2 = 4.905 \, \text{m}.$$

In other words, the apple falls towards the Earth by about 4.91 m in the first second.

Newton could account for the very different distances fallen by the apple and the Moon if the strength (size) of the force acting on the Moon was much smaller than that on the apple. Even though he supposed that the two forces had, at root, the same physical origin, this did not imply that they had to have the same magnitude. In fact, he was coming to believe that this force decreased inversely with the square of the distance. Putting it mathematically, if the force is inversely proportional to distance from the centre of the Earth squared, R^2:

$$F_G \propto \frac{1}{R^2}$$

and, from Newton's second law of motion, the resulting acceleration is proportional to the force:

$$F_G \propto a$$

then the acceleration must be inversely proportional to R^2:

$$a \propto \frac{1}{R^2}.$$

Finally, as we have just seen, the distance fallen from rest in 1 s is given by the acceleration:

$$x = \frac{1}{2} a t^2,$$

so that this distance must in turn be proportional to $1/R^2$

$$x \propto \frac{1}{R^2}.$$

Applying this to the distance fallen by the apple, x_A, and by the Moon, x_M, we get:

$$\frac{x_M}{x_E} = \left(\frac{R_E}{R_M}\right)^2 = \left(\frac{6.37 \times 10^6}{3.85 \times 10^8}\right)^2 = 2.74 \times 10^{-4}$$

dividing one by the other to eliminate the constant in the proportionality.[12]

The ratio of our calculated distances is:

$$\frac{x_{\text{M}}}{x_{\text{E}}} = \frac{1.36 \times 10^{-3}}{4.905} = 2.77 \times 10^{-4}.$$

The closeness of these two figures convincingly suggests that the apple and the Moon are under the influence of the same universal force, corrected for distance.

1.3.1 Universal Gravitation

According to Newton's own accounts, he devised the idea of an inverse square dependence from combining his own work on circular motion with Kepler's discoveries regarding planetary orbits. With his meticulously careful work, Kepler[13] had uncovered three laws of planetary motion, the third of which states that:

$$\left(\text{period of planet's orbit}\right)^2 \propto \left(\text{radius of orbit}\right)^3$$

or

$$T^2 \propto r^3.$$

Newton's own work on circular motion included an independent calculation of an object's acceleration while following a circular path[14]:

$$a \propto \frac{v^2}{r} = \frac{1}{r} \times \left(\frac{2\pi r}{T}\right)^2,$$

so that

$$a \propto \frac{r^2}{rT^2}.$$

Now, if you pull in Kepler's relationship for the period you get:

$$a \propto \frac{r^2}{r \times r^3} = \frac{1}{r^2}.$$

Newton's work on his three laws of motion clarified the relationship between force, mass and acceleration, and in turn constrained the form of the law of gravitation:

- Newton appreciated that when two objects interact, the forces that they exert on each other must be equal in size, albeit opposite in direction. So, the law of gravitation must respect this symmetry.
- As the force required to accelerate an object is proportional to its mass, the force of gravity must be proportional to the mass of the object doing the pulling. After all, from the perspective of the apple, it is stationary and the force that it exerts on the Earth is responsible for pulling the Earth towards it. The distance the Earth rises towards the apple from the apple's perspective has to be the same as the distance the apple falls towards the Earth from the Earth's perspective. In one case, the Earth's mass is doing the pulling and the apple's mass is accelerating. In the other, the roles are switched. For the same result, the pulling and the response must both be proportional to the mass.

Putting these two thoughts together, we conclude that the gravitational force must be proportional to the *product* of the masses involved.

In modern terms, Newton's law of universal gravitation is presented in terms of an equation:

$$F = -\frac{GM_1M_2}{r^2} \qquad (1.2)$$

which allows us to calculate the force between two point-masses, M_1 and M_2 separated by a distance r. In this equation, G is the universal gravitational constant,[15] which was first determined with any precision by Cavendish[16] in 1797–1798. The minus sign in the equation is a signal that the force is attractive. This law is given the tag 'universal' to emphasise that it applies to all objects with mass, wherever they find themselves in the cosmos. The Earth and the Heavens have been brought under the same physical authority.

Newton was able to show, with some clever mathematics, that spherical masses (and planets are close to being spherical) exert gravity in the same pattern as a point-mass located at the centre of the sphere. This enabled him to do a lot of work on celestial mechanics, including establishing the orbit of Halley's comet.

1.3.2 Inertial and Gravitational Masses

Newton's second law of motion relates the force applied to a body to the resulting acceleration and can be expressed in a simple equation, provided the mass of the body does not change:

$$f = ma. \qquad (1.3)$$

We can then link Equations 1.2 and 1.3 together to calculate the rate at which a body will accelerate under gravity:

$$M_1 a = -\frac{GM_1M_2}{r^2},$$

so that

$$a = -\frac{GM_2}{r^2}$$

with M_1 cancelling out, showing that in the presence of a mass acting as the source of gravity, M_2, all other masses fall with the same acceleration, as proposed by Galileo.

However, this simple calculation hides a subtle point. Here we are dealing with two separate definitions of what we mean by mass, and it is too hasty to *assume* that they produce the same value.

Newton's second law of motion defines a property known as *inertial mass*: a bodies 'reluctance' to accelerate when a force is applied to it. Using this idea, we can build a template for determining inertial mass. We simply have to have some mechanism for producing a calibrated and constant force, apply that force to a range of objects in turn and measure their resulting acceleration. The inertial mass of each object is then:

$$m_{\text{inertial}}\left(\text{kg}\right) = \frac{\text{Force applied}\left(\text{N}\right)}{\text{resulting acceleration}\left(\text{m/s}^2\right)}.$$

On the other hand, we can determine the *gravitational mass*, $m_{\text{gravitational}}$, by picking some (preferably very large) test mass which acts as a static source of gravity, placing a range of smaller masses a defined distance from the test mass (1 m would be good) and measuring the force exerted on the test masses. The gravitational mass is then:

$$m_{\text{gravitational}} = \frac{-Fr^2}{GM_2}.$$

To make the calculation exact, we need the value of M_2, which is awkward as we can't *assume* that this is the same as the value obtained from the inertial mass.

In practice, we define a given object as being 1 kg and use that as the test mass and value for M_2. We then get gravitational masses for other objects as fractions/multiples of the test mass.

When we put Newton's second law and his law of gravity together, the result should actually read something like the following:

$$M_{1\,\text{inertial}} a = -\frac{GM_{1\,\text{gravitational}} M_2}{r^2},$$

and the cancellation we applied earlier will only work if the two varieties of mass have the same value.

Galileo's assertion that all objects accelerate at the same rate suggests that the two masses do have the same value (or rather that we are determining the same physical property in two different contexts), but this is not a given. Physicists can imagine a sensible world where this was not the case. With the electrostatic force, the strength of the force experienced by a charge is based on the magnitude of the electrical field and the size of the charge: nothing to do with the inertial mass of the charge. However, the acceleration produced by this force is related to the charge's inertial mass, but not its electrical charge. So, on a sauce for the goose type of argument, one could imagine a theory where the gravitational force on an object was determined by the gravitational field present and the object's gravitational 'charge', while its resulting acceleration was determined by its inertial mass.

Our world is striking: the gravitational 'charge' is either the same as the inertial mass, or has the same value as that mass (which many would argue is tantamount to saying the same thing…).

The equality of inertial and gravitational mass was absorbed as the basis for an important principle (the *principle of equivalence*) by Einstein and marked a crucial stepping stone on the way to his view of gravity.

As an important tenant of the general theory of relativity, the equality of inertial and gravitational mass has been checked to high precision and is true for all known substances, independent of their chemical composition. It has also been checked for particles of anti-matter, although at present, that experimental evidence is not so stringent.

Incorporating the principle of equivalence, Einstein generalised his special theory of relativity and in the process devised a new picture of gravitational effects, based not on forces but on the geometry of space and time surrounding masses. From this work, he was able to calculate the extent to which light paths would be deflected in passing close to large masses, an effect that had its first tentative experimental confirmation during an eclipse in 1919.

1.4 Eddington and the 1919 Eclipse

One thing is certain, and the rest debate—Light-rays, when near the Sun, DO NOT GO STRAIGHT.

A Eddington[17]

As the problem then presented itself to us, there were three possibilities. There might be no deflection at all; that is to say, light might not be subject to gravitation. There might be a 'half-deflection', signifying that light was subject to gravitation, as Newton had suggested, and obeyed the simple

Newtonian law. Or there might be a 'full deflection', confirming Einstein's instead of Newton's law. I remember Dyson explaining all of this to my companion Cottingham, who gathered the main idea that the bigger the result, the more exciting it would be. 'What will it mean if we get double the deflection?' 'Then,' said Dyson, 'Eddington will go mad, and you will have to come home alone'.

A Eddington, as quoted by Chandrasekhar[18]

On the morning of 29 May 1919, it was raining on the island of Principe, in the Gulf of Guinea off the west coast of Africa. Two astronomers were camped on the island at the time, having set sail from the UK in March of that year in order to be on that island that day. Their purpose was to take photographs showing the positions of stars as the Sun tracked across the sky and passed in front of certain constellations, something not normally possible due to the tremendous glare of the sunlight. However, on this day there was to be a total solar eclipse. For six brief minutes, as the Moon's shadow passed over the island (Figure 1.8), the glare of the Sun would be obscured, allowing the pictures to be taken.

The expedition was led by Arthur Stanley Eddington who was on the island with Edwin Cottingham.[19] They were the Cambridge team. As with any astronomical observation, there has to be independent confirmation (and a back-up for bad weather), so Charles Davidson and Andrew Crommelin from Greenwich observatory had been sent to the city of Sobral in Brazil, which was also on the eclipse's track.

Eddington was nearly prevented from taking part in the expedition due to compulsory conscription having been introduced in Britain by 1916 as part of the war effort. He had intended to apply for conscientious objector status,[20] but Cambridge University, where he was Plumian Professor of Astronomy and Director of the Observatories, had been granted an exemption for him on the grounds that Eddington's work was in the national interest. That judgement was appealed in 1918 and tribunal hearings set up. Thanks in part to a written statement from Sir Frank Dyson (who was Astronomer Royal, Director of Greenwich Observatory and in overall charge of the expedition), in which he stressed Eddington's

FIGURE 1.8 The track of the 1919 eclipse across the Southern Hemisphere. (Image credit: Eclipse Predictions by Fred Espenak, NASA's GSFC—http://eclipse.gsfc.nasa.gov/.)

vital role in the eclipse measurements, the tribunal granted a further 12 months exemption, provided Eddington continued his astronomical work. The war ended before the term of this second exemption expired.

During the war, Eddington was secretary to the Royal Astronomical Society and as such invited the Dutch physicist Willem de Sitter to write a series of papers for the RAS Monthly Notices[21] on Einstein's work. Direct communication across Europe was understandably difficult at the time, and it is one of those fortunate historical quirks that Eddington was mathematically well equipped to understand Einstein's ideas and temperamentally amenable to considering the work of a scientist from an enemy country.

Eddington grasped the import of Einstein's ideas and used de Sitter's papers as the basis of his own report on the subject, in 1918.[22] Later he expanded this into a book[23] (1920) and then a mathematical text in 1923,[24] which Einstein described as the finest presentation of the subject in any language.

When, in 1915, Einstein brought his work on gravitation to a culmination, he also proposed three experimental tests that could be performed to confirm his theory. One of them involved a direct measurement of the deflection of light passing near to a large mass. Eddington was extremely enthusiastic about Einstein's ideas, while Dyson remained cautious and sceptical. However, it was clear to both that the 1919 eclipse provided an ideal opportunity to try and settle the matter.

Accounts of the eclipse expedition sometimes give the impression that Einstein predicted the deflection of light in the presence of a mass, but a deflection can be calculated in the Newtonian paradigm.[25]

In 1911, Einstein made his first attempt at such a calculation based on his published theory of special relativity and his new equivalence principle. He obtained a value equal to the Newtonian one. In 1915, Einstein's general relativity had reached its full published maturity and a new calculation of the deflection of starlight showed twice the Newtonian value: 8.48×10^{-6} rad. It was this value and hence full general relativity that Eddington and Dyson determined to check.

As the greatest deflection would occur for light skirting the edge of the Sun on the way to Earth, it was clear that the best measurements could be obtained from an image of distant stars taken as the Sun was passing across their part of the sky. Comparing that image to one taken at another time of the year should show a shift in the star's positions, allowing the gravitational deflection to be calculated. The problem would be seeing the stars. As I suggested earlier, the Sun's glare would make it impossible to expose an image that showed the stars in that part of the sky. Hence, the importance of the eclipse: the Sun's mass would still exert its influence on the passing starlight, but the Moon would block out the glare of the sunlight.

1919 was not the first opportunity to make such a test. A German astronomer had led an expedition to the Crimea in 1914, but was unfortunately arrested as a spy when war broke out.

A team from America did make it to that eclipse, but were not able to make any measurements due to rain.[26] To make matters worse, their instruments were impounded by the Russians and not returned in time for next eclipse in 1916. Indeed, they still had not been returned when an ideal opportunity presented itself in 1918 with an eclipse visible in the USA. Unfortunately, the improvised equipment used instead did not work out and the images were too fuzzy to be of use. So, 1919 was shaping up to being the confluence of history. It certainly helped that this eclipse was due to pass by the *Hyades cluster* (Figure 1.9), a region dense with bright stars ideal for imaging. Indeed, the haste evident in getting this expedition ready can be explained by their being no other star field along the Sun's annual track that had such a collection of bright stars.

Through January and February 1919, Eddington arranged for comparison images to be obtained using instruments in Oxford for both the Hyades cluster and, as a check, images from another part of the sky.[27] In contrast, the Greenwich team stayed on for 2 months after the eclipse in order to take their comparison images on-site, something that would have required a wait nearer to 6 months on Principe.[28] While on the island, however, Eddington did image the check star field previously captured in Oxford.

The rainy morning had dampened hopes on Principe, but fortunately during the eclipse a few images could be obtained through the clouds.

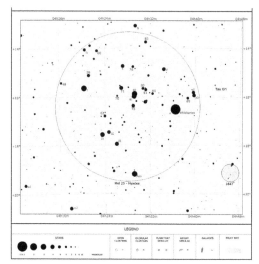

FIGURE 1.9 The Hyades cluster and a map of its most conspicuous stars. (Image credit [left]: NASA, ESA, and STScI CC-BY-SA 4.0, [right] Roberto Mura CC-BY-SA 3.0.)

Over in Brazil, the weather had been much kinder, and although cloud coverage was 9/10, at the start of the eclipse, matters steadily improved and by totality, a large clear space presented itself. As the Greenwich team had two instruments at their disposal and less cloud cover to contend with, one might have guessed that their data would be much better than that obtained on Principe (Figure 1.10). Things are rarely that simple...

Of the two instruments available at Sobral, the technically superior (based on the *Astrographic lens*) suffered a loss of focus during the eclipse. It seems likely that the rapidly changing temperatures on-site distorted the optics, resulting in the focus shift.[29] As a result, measurements from these images were much harder to extract. In the original data analysis, the angular shift established from this instrument gave a result much closer to Einstein's 1911 (Newtonian) calculation. However, in their published paper, Dyson and Eddington argued that this result was not reliable due to the focus shift. The other Sobral instrument (based on the *4-inch lens*) performed better, but had a smaller field of view, giving fewer stars to measure. It obtained results in line with Einstein's 1915 prediction (Table 1.1).

Eddington had much less data to work with from Principe. The clouds had only cleared in the last few minutes of the eclipse, and so there were two usable images with five stars on each. However, his Oxford check images turned out to be crucial. Extracting an angular shift in star positions is a tricky business. In essence, the two images have to be placed on top of each other and the shift in star location measured. However, there are many factors that can produce an apparent shift. A change in image scale will do it.

The Greenwich Astrographic images suffered focus change during the eclipse, which would certainly contribute to a shift in scale between eclipse and comparison images. As they did not have any check images for a different part of the sky, they could not calculate an independent scale determination.

Eddington had two sets: Hyades comparison images (on-site and Oxford) and check images (on-site and Oxford). He argued that the temperature at Principe had not varied by more than 1° throughout his visit[30]; hence, there was no focus shift in his instrument. As a result, he could calculate the scale change between the check images, brought about by other optical variations, and apply it to the eclipse and comparison images. His result for the deflection was comparable to Einstein's prediction (Table 1.1). In the end, combined data from both wings of the expedition were published by Dyson, Eddington and Davidson.

FIGURE 1.10 One of the images from the 1919 eclipse expedition. This image was taken using the 4-inch lens at Sobral and is the only image shown on the paper published by Dyson, Eddington and Davidson. This reproduction (as shown on the paper) is in negative. The black 'fuzzy' area surrounding the Sun is the visible corona with the Sun's main body obscured by the Moon. The short lines, like dashes, are the star traces. (Image credit: Dyson, F.W., Eddington, A. S, and Davidson, C. 1920. A determination of the deflection of light by the Sun's gravitational field, from observations made at the total eclipse of May 29, 1919. *Philosophical Transactions of the Royal Society A*, **220**, pp. 291–333.)

TABLE 1.1 The Deflection Data Obtained from the Instruments at Sobral and Principe

Instrument	Location	Value/10^{-6} rad	Comment
4-inch	Sobral	9.60 ± 0.87	Based on seven stars
		(9.21 ± 0.53)	1979 result
Astrographic	Sobral	4.51	Excluded from final results
		7.56	No scale change
		(7.51 ± 1.65)	1979 result
Astrographic	Principe	7.81 ± 1.45	Based on two stars
			No surviving images to reanalyse in 1979
Einstein's prediction		8.49	

The deflection from the Astrographic lens at Sobral was not used in the complete analysis, and its value was quoted without an accompanying error (due to the apparent shift in focus during the eclipse). A 1979 re-evaluation of the data produced the values quoted in brackets. In this table, the deflections have been converted from arc seconds to the radian measure which will be more familiar to a modern audience.

Source: Data from Dyson, F.W., Eddington, A. S, and Davidson, C. 1920. A determination of the deflection of light by the Sun's gravitational field, from observations made at the total eclipse of May 29, 1919. *Philosophical Transactions of the Royal Society A*, **220**, pp. 291–333. doi:10.1098/rsta.1920.0009.

Eddington's evident enthusiasm for Einstein's theory, coupled with the exclusion of the Astrographic data from Sobral, has led some modern commentators to suggest that Eddington was biased. However, in my mind at least, the work of D Kennefick[31] has settled the issue in favour of Eddington. As part of his argument, Kennefick refers to a modern re-analysis of some surviving plates from Sobral which obtained a result for the

Astrographic lens that is more in line with those from the other instruments[32] (Table 1.1). This chimes with a later paper[33] by Dyson, in which he shows how assuming that the Astrographic lens suffered only a loss of focus and not a scale change, produces a deflection comparable to that found via the other instruments.

The expedition's results were announced at a joint meeting of the Royal Society and the Royal Astronomical Society held in London on 6 November 1919. For whatever reason, this caught the public's imagination. The next day a headline in the *Times* read[34]:

Revolution in science

NEW THEORY OF THE UNIVERSE: NEWTONIAN IDEAS OVERTHROWN.

On 10 November, the *New York Times* had a slightly less temperate headline (Figure 1.11), and later that week, the *Illustrated London News* (22 November) had a detailed diagram explaining the eclipse measurements (Figure 1.11).

Many factors may have contributed to this explosion of public interest. That the ideas of a German scientist had overthrown the work of Newton and been verified, in part, by a British expedition shortly after the war may well have played a substantial part in the narrative. Many physicists at the time freely confessed that they did not fully understand Einstein's theory, which certainly helped the headlines. As reported by the *Times*,[35]

> Even the President of the Royal Society, in stating that they had just listened to 'One of the most momentous, if not the most momentous, pronouncements of human thought', had to confess that no one had yet succeeded in stating in clear language what the theory of Einstein really was.

Einstein's life changed markedly. Almost overnight he was catapulted into being one of the most famous scientists in the world. As for Eddington, as one of the most accomplished theoretical physicists in

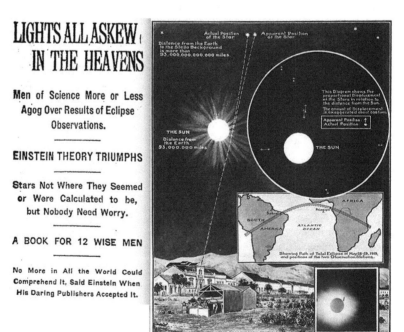

FIGURE 1.11 The headline in the *New York Times*, 10 November 1919 (left) and the article from the *Illustrated London News*, 22 November 1919 (right).

the UK, he was one of the few with a grasp of the meaning and implications of the general theory of relativity. He quickly found himself in demand as a lecturer on the subject, to both technical and lay audiences. It turned out that Eddington had a gift as a communicator, and he became a respected populariser of science. As a result, his name is most linked with the 1919 eclipse expedition to this day.

While Dyson et al.'s results were a persuasive demonstration that the Newton deflection was wrong, they did not completely clinch the case for Einstein. Work on his other two proposed experimental tests played a part (to be discussed later), and subsequent eclipse expeditions (notably 1922) further refined the data. However, it was not until the advent of radio astronomy, with its ability to measure the deflection of radio waves from powerful extra-galactic sources called *quasars*, that the data on gravitational light deflection became completely convincing.[36]

The general theory of relativity has now been part of the scientific vocabulary for over 100 years. It has been used to explore the creation of the universe and the properties of black holes. Undoubtedly, the theory is not the final word as it predicts the existence of space-time singularities which have no clear physical interpretation, but it is widely accepted as a valid description of nature, within its bounds of relevance. Like all great theories, it has opened our eyes to new possibilities most recently in the detection of gravitational waves using technology which promises to provide an entirely new window on the universe.

1.5 LIGO and the Search for Gravitational Waves

Gravitational waves will bring us exquisitely accurate maps of black holes—maps of their space-time. Those maps will make it crystal clear whether or not what we're dealing with are black holes as described by general relativity.

K Thorne, by permission

About 1.3 billion years ago, something cataclysmic happened in an otherwise quiet corner of the universe. Two black holes in orbit around their common centre of mass reached a decisive tipping point, and shortly after, their orbits started to collapse as they spiralled towards each other and merged into one even larger black hole. Before the merger, one of the black holes was approximately 36 solar masses and the other about 29 solar masses. Afterwards, the resulting object amounted to 62 solar masses. Somewhere in the process, a mass roughly equal to 3 times that of our Sun was lost. If that were not astonishing enough, the entire process from tipping point to complete merger took a mere 0.45 s.

A black hole is itself an extraordinary object, representing a region where the space-time is so distorted, it forms a one-way street, with any matter in the vicinity being trapped on a path that falls hopelessly into the black hole. Not even light can avoid such a fate.

Although the possibility of a star so massive that light could not escape its gravitational grip had been suggested in the 18th century by John Michell and Pierre-Simon Laplace,[37] black holes fall decidedly within the scope of the general theory of relativity.

Aside from black holes, one of the more remarkable consequences of the general theory is the existence of *gravitational waves*: distortions of space-time that are able to move through space-time. Gravitational waves are emitted whenever non-spherical or non-cylindrically symmetric masses undergo acceleration. During the black hole merger, powerful gravitational waves propagated out from the event at the speed of light and carried away the energy equivalent of the missing mass[38] (remember—3 times that of our Sun in this case).

1.3 billion years later, these waves were still moving through the universe, and on 14 September 2015 at 09:50:45 UTC,[39] they passed through the Earth and were detected.

The instrument that made this ground-breaking discovery, LIGO, is an extraordinary example of precision engineering, design brilliance and patient development over decades (Figure 1.12).

FIGURE 1.12 The LIGO facility at Hanford, Washington State, USA. The two 4-km detector arms can be clearly seen. A second station is at Livingston in Louisiana. (Image credit: LIGO Laboratory.)

The LIGO project has a complex funding and management history behind it, dating back to the 1980s, never mind the technical and theoretical challenges it has faced. Between 2002 and 2010, the initial version of the Laser Interferometer Gravitational-Wave Observatory (or iLIGO) ran without detecting any gravitational waves. After a 5-year $200 million upgrade programme (which brought the total cost of the project to $620 million), advanced LIGO (aLIGO) was completed. On 18 September 2015, aLIGO began formal science data gathering at 4 times the sensitivity of iLIGO.

The discovery, which in 2017 was celebrated with the Nobel Prize, was made 4 days *before* the official start of data taking, during the final testing stages of the upgraded instrument. In fact, the observations were at one stage suspected of being an artificial signal injected to the system[40] to the test whether the operation of the detectors was correct and if the operators on duty were able to identify a real signal. Given the potential importance of the result, it is not surprising that months of patient analysis was done before it was felt that the discovery could be announced on 11 February 2016.

Detecting the presence of a gravitational wave is an extremely delicate business. Such a wave passing perpendicularly through a detector would expand its length along one axis and compress it along another (during half a wave cycle – see Section 11.3.4), but only by microscopic fractions. The discovery event shifted the detector's 4 km length by a strain factor $\sim 10^{-21}$. In real terms, that is a distance shift of $\sim 10^{-18}$ m. To put that into context, the nucleus of an atom $\sim 10^{-15}$ m across…

To achieve such extraordinary sensitivity, LIGO uses two light rays split from a single laser beam and directed to travel along perpendicular arms before they are reflected back to the splitter and combined again. Any shifts in the distances travelled by the two rays will be detectable after the combination (Figure 1.13).

To ensure that spurious signals caused by local vibrations or seismic events could be eliminated, a second detector was built with the two stations (one at Hanford and the other at Livingston) separated by ~1,900 km, a 'flight time' of around 7 ms at light speed.

One of the things that makes the 14th September observation so impressive, aside from the beautiful match to the computer calculated theoretical model, is the correlation between the signal recorded at Hanford and that at Livingston (Figure 1.14).

Modern physicists working with the general theory of relativity have one distinct advantage over their predecessors: the availability of significant computing power. Einstein's equations are complex and subtle, and very few situations exist where exact mathematical solutions can be found. Generally, simplifying assumptions have to be made and approximations found in order to make the situation

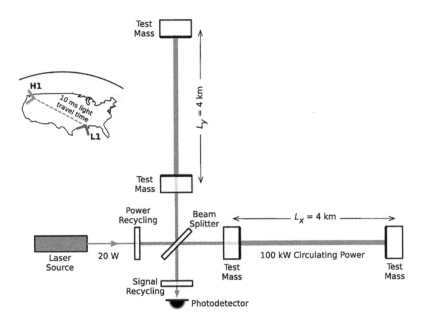

FIGURE 1.13 A schematic diagram of the aLIGO detector (not to scale) and its two 4-km detecting arms. Also shown, top left, are the locations of the two duplicate detectors: one at Hanford, in Washington State and the other located at Livingston in Louisiana and the relative orientations of their detection arm. The test masses are suspended partial or fully silvered mirrors. (Image credit: Abbott, B. P., et al. [LIGO Scientific Collaboration and Virgo Collaboration]. Observation of Gravitational Waves from a Binary Black Hole Merger. *Physical Review Letters*, **116**, p. 061102. doi:10.1103/PhysRevLett.116.061102, CC BY 3.0, https://commons.wikimedia.org/w/index.php?curid=46922746.)

in any way tractable. By using computers, physicists can construct mathematical models and make predictions where they would not be able to make progress with pen and paper. Using such computer models, the LIGO team was able to predict what happens when two black holes merge and compare the expected form of the gravitational waves produced with that detected by the instruments (Figure 1.15). In turn, the parameters that go into the model (such as the black holes' masses) can be refined to fit the profile of the data, which is why we can suggest the values of the masses concerned and the energy lost to gravitational waves.

The LIGO result is of such profound importance to physics that the founding members of the team, Rainer Weiss, Kip Thorne and Barry Barish, were rightly and promptly awarded the Nobel Prize in physics for 2017. However, as the prize winners point out, the group behind the development of the detector and analysis of the data numbers close to 1,000 physicists across multiple universities and countries, each with an important part to play. This is typical of how much of modern physics is done, especially as the equipment required to push the frontiers of knowledge is often so expensive it cannot be afforded by a single institution or country.

This Nobel Prize is not the first to be awarded for the study of gravitational waves. In 1993, Russell Hulse and Joseph Taylor, Jr. received the prize for the discovery and analysis of two pulsars in common orbit. Their decades-long study of that system provided convincing evidence of orbital changes due to energy being lost by gravitational waves emitted by the neutron stars. Hence, the physics community was confident that these waves existed, but their direct detection is still of fundamental significance.

The theoretical story of gravitational waves, the work of Hulse and Taylor and the LIGO discovery is a fascinating episode in gravitational physics, so we will spend a Chapter 11 looking into this in more detail.

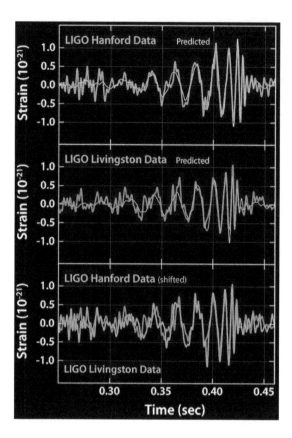

FIGURE 1.14 The LIGO discovery event from 14 September 2015. The top graph shows the signal recorded by the Hanford station (with the theoretical prediction overlay), the middle graph is the signal from the Livingston station with theoretical overlay, and the final lowest graph shows the signals from the two stations superimposed with the Hanford data time shifted by 6.9 ms. The Hanford data have also been inverted for comparison due to the relative orientation of the detection arms at the two stations. (Image credit: Caltech/MIT/LIGO Lab.)

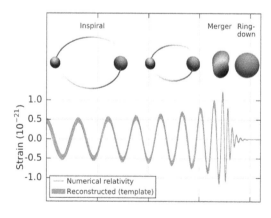

FIGURE 1.15 The computer-calculated gravitational-wave signal produced by the steps in the merging of orbiting black holes, from inspiral through merger and final settling into a single black hole (ring-down). (Image credit: Abbott, B. P., et al. [LIGO Scientific Collaboration and Virgo Collaboration]. Observation of Gravitational Waves from a Binary Black Hole Merger. *Physical Review Letters*, **116**, p. 061102. doi:10.1103/PhysRevLett.116.061102, CC BY 3.0, https://commons.wikimedia.org/w/index.php?curid=46922746.)

1.6 Conclusion

In this chapter, I have introduced concepts that will form a backbone to our understanding of gravity. A few more ideas of equal importance need to be grafted on and those that we have already seen need to be further refined. Where we can, we will refer to fully formed experiments carried out by experts using equipment of relative sophistication. However, as I have already mentioned, one of Einstein's gifts was his ability to construct pictures of the world that enabled him to see through details into underlying physical law. Consequently, we will also make use more informal observation, and imagined situations which cast light on the way that the world is.

There is room for both in a physicist's toolkit.

Notes

1. Galileo G., 1960. The Assayer, English trans. Stillman Drake and C. D. O'Malley. *The Controversy on the Comets of 1618*. Il Saggiatore (in Italian) (Rome, 1623); (University of Pennsylvania Press, 1960). Full text can be viewed at: https://web.stanford.edu/~jsabol/certainty/readings/Galileo-Assayer.pdf.
2. *Dialogue Concerning the Two Chief World Systems*, 1632. Full text translation: http://archimedes.mpiwg-berlin.mpg.de/cgi-bin/toc/toc.cgi?step=thumb&dir=galil_syste_065_en_1661.
3. Drake, S., 1953. *Discoveries and Opinions of Galileo*. New York: Doubleday, pp. 186–187.
4. Vincenzo Viviani Racconto istorico della vita di Galileo Galilei (1717).
5. From *On Motion*, unpublished manuscript: http://galileo.rice.edu/sci/theories/on_motion.html.
6. Stevin, S., De Beghinselen Der Weeghconst, 1586.
7. Galileo G., 1968. *Dialogues Concerning Two New Sciences*. Full text available at http://oll.libertyfund.org/titles/galilei-dialogues-concerning-two-new-sciences.
8. *Memoirs of Sir Isaac Newton's Life*, William Stukeley, (1752). Full text available at http://ttp.royalsociety.org/ttp/ttp.html?id=1807da00-909a-4abf-b9c1-0279a08e4bf2&type=book.
9. Including, sadly, that The Doctor had to drop a sequence of apples on Newton's head, until he got the idea…
10. Not a bad movie as well…
11. The radian is a measure of angle preferred by mathematicians as it is more 'natural'. It is based on the circumference of a circle of unit radius so that 360° is equivalent to 2π radians.
12. Note that this is only true if Galileo is right—that all objects fall under gravity at the same rate at a given distance from the Earth…
13. Kepler, J. 1571–1630, *Harmonices Mundi* 1619 for the third law, the first two having been published earlier.
14. Also discovered by the Dutch physicist Christiaan Huygens (1629–1695).
15. Newton worked in terms of proportionality, so the introduction of G came later.
16. Cavendish, H. 21 June 1798. Experiments to determine the density of the earth. *Philosophical Transactions of the Royal Society of London* (part II), **88**, pp. 469–526.
17. Douglas, A.V., 1956. *The Life of Arthur Eddington*. Thomas Nelson and Sons. p. 44.
18. Subramanian, C., 1976. Verifying the theory of relativity. *Notes and Records of the Royal Society of London*, 30, pp. 249–260.
19. Cottingham was a clockmaker who maintained many of the astronomical instruments at Cambridge.
20. Eddington was a devout Quaker and pacifist.
21. On Einstein's Theory of Gravitation and its Astronomical Consequences, 14 July 1916. First Paper de Sitter, W., *Monthly Notices of the Royal Astronomical Society*, **76**(9), pp. 699–728, https://doi.org/10.1093/mnras/76.9.699 and subsequent.
22. *Report on the Relativity Theory of Gravitation*. London, Fleetway Press, Ltd.

23. *Space, Time and Gravitation: An Outline of the General Relativity Theory.* Cambridge University Press. ISBN 0-521-33709-7.

24. *The Mathematical Theory of Relativity.* Cambridge University Press. ISBN 978-0282596781.

25. Soldner, J.G.V., 1804. On the deflection of a light ray from its rectilinear motion, by the attraction of a celestial body at which it nearly passes by. *Berliner Astronomisches Jahrbuch,* pp. 161–172. Also Henry Cavendish (1784) unpublished manuscript.

26. This actually worked in Einstein's favour. At the time, his calculation of the deflection was wrong, so measurements in 1914 would not have confirmed that work. By the time the measurement was done in 1919, he had the right value.

27. In the vicinity of *Arcturus.*

28. The eclipse passed over Sobral relatively early the morning. Two months later, Hyades was rising at roughly the same angle above the horizon at night so comparison images were easy to obtain. At Principe, much further to the East, the eclipse started in the afternoon, and so it needed longer for the Sun to have moved sufficiently far through the sky for Hyades to be visible at the correct angle at night. As it happened, a strike of the steamship company forced them to return by the first boat, if they were not to be marooned on the island for several months.

29. Indeed, when they used the same instrument again 2 months later to take the comparison images, the focus had been restored without any adjustments having been made.

30. Even during the eclipse itself, as the cloud cover had evened out the change in Sunlight due the Moon's shadow.

31. Kennefick, D., 2007. https://arxiv.org/pdf/0709.0685.pdf.

32. Harvey, G. M., 1979. Gravitational deflection of light: a re-examination of the observations of the solar eclipse of 1919. *The Observatory,* **99**, pp. 195–198.

33. Dyson, F., 1921. Relativity and the eclipse observations of May, 1919. *Nature,* **106**, pp. 786–787. doi: 10.1038/106786a0.

34. *The Times,* 7 November 1919.

35. As 34.

36. As the Sun does not emit much in the way of radio waves, there is much less of an issue with 'glare' and so the measurements can be done with greater precision.

37. Montgomery, C., Orchiston, W. and Whittingham, I., 2009. Michell, Laplace and the origin of the black hole concept. *Journal of Astronomical History and Heritage,* **12**(2), pp. 90–96. Bibcode:2009JAHH...12...90M.

38. Here I am making an indirect reference to the Einstein equation $E = mc^2$ which ties mass and energy together. It is a topic we will pick up in more detail when we get to Chapter 5.

39. *Coordinated Universal Time (UTC):* a time standard adopted across the globe which allows timings to be compared without having to convert to different time zones.

40. Four members of the LIGO team had the authority to do that for testing purposes.

2

The Road to Relativity

2.1 How Science Works

There is a traditional story about the process of doing science. Young physicists absorb this account at school, either because it is specifically taught in the curriculum, or because it is implicit in the way that science is presented to them.

The story goes something like this: the world contains a collection of facts; these facts tell us about how things look, feel, sound, taste and smell. If we are curious enough and ingenious enough, we can extend this collection of basic *sensory* facts by the use of *instruments* that allow us to look further, see smaller and measure more accurately.

Armed with a suitable collection of facts, a scientist will ponder them (generally with coffee, although this vital aspect is not stressed in the story) until they have thought up a *theory* that explains the facts, predicts possible new facts and is consistent with what we know already (other theories and facts).

The next step is to devise experiments that test the theory by looking for the predicted facts. If they check out, the theory is successful and is used more and more to predict and explain, as well as inspire new work and possibly new technologies.

This is a simple and straightforward story that many professional scientists would recognise and even describe themselves, if pushed. However, it does not stand up to more than surface analysis. Science is more complicated and needs more subtlety than this, for a variety of reasons:

- Fact and theory do not cleave into separate categories in a clean and distinct manner. Every fact gets its intelligibility (its factuality) from being embedded in some theory, or descriptive model about the world. Even an apparently simple fact, such as the observation that grass is green, is implicitly loaded with reference to the nature of colour, the way in which colour is perceived and even the distinction between grass and other plant matter…
- Over time, successful theories become part of the basic context in which science is carried out. As such, they become relatively immune to contradiction by new facts. An experimental test that fails to conform with one of the standard theories does not trigger a rejection of the *theory*. Instead, there is a close inspection of the *experiment* and its data in order to find out why the *experiment* has gone wrong, or why the *circumstances* have produced a special case that exempts it from some aspect of the theory. In practice, this is a sensible and pragmatic approach (see later for an example of this: the Michelson–Morley experiment).
- Creating theories is a highly skilled endeavour. Only rarely does a new theory of significance occur to one worker contemplating a collection of bald facts. Modern science inches forward as experts, often working in teams, produce scientific papers that are reviewed and checked by the community. Experiments are repeated by others, perhaps using slightly different techniques and certainly different equipment, in order to confirm and extend results. Conferences give opportunities for new ideas to be presented, debated and refined. The history of science is punctuated with great revolutions that transform subject areas, but history tends to compress perspective and the

years of quiet development that fill most people's careers, are forgotten. Equally, revolutions are not always recognised as such at the time and what, with hindsight, can be seen as a turning point is seldom recognised as such until the dust has settled more evenly, some years later.

Much more could be said about how science works in practice, but we have done enough to show that it is not as straightforward as we have been led to believe. Einstein's theories of relativity, both the special and general versions, although primarily authored by one man, came about in very different scientific and historical contexts and certainly provide interesting counter cases to the simple picture.

2.2 The Origins of Special Relativity

At the turn of the 20th century, science was enjoying a broadly triumphant period. Confidence was high, and there was a growing view that all major discoveries had been made, and so quietly progressive work was now the way forward.

The accepted picture of the world was based on a combination of Newtonian mechanics, which had been successfully refined and developed since the 1600s, burgeoning atomic theory (although that would not be cemented in place until the first decade was nearly complete) and the field theories of gravity (Newton again) and electromagnetism (James Clerk Maxwell). Accordingly, the world was thought to be composed of *particles* (atoms) moving through space under the action of forces brought about by collisions between particles, or via the indirect influence of *masses*, *charges* and *currents*, mediated by *fields*.

Despite the confidence at the time, our modern perspective shows that cracks were already appearing in this picture. One of them, relating to the radiation emitted by hot, black objects, was to usher in the profoundly counter-intuitive *quantum mechanics* (which we will touch upon in Chapter 13). The other centred on reconciling the historically well-proven Newtonian mechanics with Maxwell's more recent theory of electromagnetism.

2.2.1 Einstein and Electromagnetism

Einstein's first paper on the special theory is titled *On the Electrodynamics of Moving Bodies* which, at first glance, seems curious for a document that transformed our view of space and time, but it makes sense when you consider the problem that Einstein was aiming to solve.

Maxwell had shown that light was an electromagnetic phenomenon, which was a brilliant advance. However, his wave equation that specified the speed of light did not square with Newtonian mechanics (Section 2.2.7). Einstein's bold move was to use this as a reason to reject the *grand damme* of physics,[1] Newtonian mechanics, and erect a new mechanics in its place.

At the start of his paper, Einstein sets out two assumptions as the basis for the rest of his analysis:

1. *The principle of relativity: the laws of nature should be the same for all observers in relative inertial motion.*
2. *The speed of light is the same no matter what the state of motion of the light source.*
   Although assumption 2 is phrased as per Einstein's paper, in practice he broadened its scope somewhat into 2a:
2a. *The speed of light is the same no matter what the state of motion of the observer*, without actually stating the assumption in this form.

Assumption 1 clearly links back to Galileo's principle of relativity from Chapter 1, although Einstein's true aim is somewhat veiled in this statement. Underlying much of Einstein's work prior to 1918 is a conviction: that it should be impossible for an observer to detect their own state of motion in any *absolute sense*, no matter what experimental data they obtain. All an observer can do is determine their motion *relative* to any objects under observation.

At least that would be the case if the principle of relativity held true.

If not, then observers might come to different conclusions about the laws of nature and, assuming that none had made any mistakes, the differences could only be explained by their respective states of motion. Observers with the 'right' laws of nature would then be in a state of *absolute rest* and other observers could establish their absolute motion by comparing their laws of nature with the 'right' ones.

Einstein had a deep philosophical objection to this possibility (which we discuss in Chapter 6), hence the statement of Galileo's relativity as a defining principle.

The second assumption ties in with the first. Maxwell's theory of electromagnetism predicts the speed of light, but without any reference to the motion of the observer determining that speed. As a result, it does not fit with Newtonian mechanics, according to which a fast-moving observer should see a ray of light moving more slowly relative to them. In a Newtonian world, only someone absolutely at rest should be able to measure a speed of light consistent with the Maxwellian prediction. Given that Maxwell's equations make no reference to the motion of an observer, another interpretation is that *all* observers see the same speed of light, no matter how they are moving. Maxwell's theory is consistent with the first and second assumptions. Given the choice between Newtonian mechanics and Maxwell's theory, Einstein chose Maxwell. Accordingly, a new mechanics was needed to explain how the speed of light can be the same for all observers and to support the principle of relativity. Remarkably, this new mechanics also established a deeper relationship between electric and magnetic fields, dissolving some issues to do with observers viewing these fields from different states of motion. Einstein was acutely aware of these other issues, and resolving them was another of his motivating factors.

Accordingly, the most appropriate place for us to start is with one of Einstein's meditations on electricity and magnetism.

2.2.2 Magnets and Circuits

According to Faraday, when a magnet is in relative motion with respect to a conducting circuit, an electric current is induced in the latter. It is all the same whether the magnet moves or the conductor; only the relative motion counts, according to the Maxwell-Lorentz theory. However, the theoretical interpretation of the phenomenon in these two cases is quite different...*The thought that one is dealing here with two fundamentally different cases was, for me, unbearable. The difference between these two cases could not be a real difference, but rather, in my conviction, could be only a difference in the choice of reference point.*

A Einstein[2] (my emphasis)

At school we learn that a wire carrying an electrical current will be surrounded by a magnetic field, a discovery first made serendipitously by Ørsted,[3] who was lecturing to a group of students about electrical current when he noticed a compass, that happened to be sitting on the lecture bench, twitching whenever he turned the current on. Of course, we utilise this when we wrap wire around a magnetic material, such as iron, to make an electromagnet. However, there is far more to electromagnetism than that. The real magic arises with *electromagnetic induction*: a magnet moving in the vicinity of a closed loop of wire will induce a current in the loop (Figure 2.1). Technologically we exploit this effect in electrical power generation. However, what is beautifully mysterious about this simply observed phenomenon is the way it comes about by two apparently different physical mechanisms.

From the point of view of an observer sitting on the magnet (Figure 2.1, right), the loop is moving towards the magnet and so the charges (protons and electrons) within the wire experience a magnetic force. In accordance with their nature, magnetic forces act at 90° to the plane containing the magnetic field and the direction in which the charge is moving (Figure 2.2).

Ordinarily, the charges within the wire will be moving along at the same speed and in the same direction as the wire.[4] However, a magnetic force will deflect their path away from this shared movement (Figure 2.3).

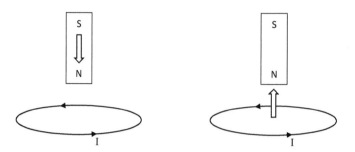

FIGURE 2.1 A magnet moving towards a closed loop of wire, which will induce an electrical current in the wire (left). From the point of view of the magnet (right), it is stationary and the wire loop is moving towards it. This change in perspective cannot alter the physics of the situation, so a current must be present in this case as well. However, a different physical explanation has to be employed to explain its existence.

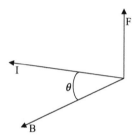

FIGURE 2.2 The magnetic force, *F*, on a current (moving charge) *I* is directed at 90° to the plane containing the magnetic field B and the current.

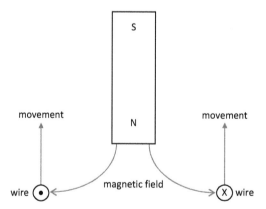

FIGURE 2.3 The magnetic force acting on the free charges within a moving wire loop, which will deflect their path so they circulate around the loop as well as moving vertically with the loop.

If you orient the lines in Figure 2.2 with the movement of the wire (*I*)[5] and the magnetic field of the bar magnet (*B*), you will see that the force (*F*) on a positive charge points into the page on the right-hand side (the X) and out of the page on the left (the dot). The force on a negative charge acts in the opposite direction on each side. As a result, free charges will move along the wire as the wire also carries them past the magnet. Protons in the nuclei of atoms will experience the magnetic force, but they are too strongly bound within the nucleus to move. Electrons, on the other hand, are much lighter and within a metallic structure, some of them at least are free to move: a current is created. So, the magnetic forces acting on the charges within the wire bring about a current, as long as the loop is moving through the field of the magnet.

Switching to the point of view of an observer sitting on the loop (Figure 2.1, left), the wire, along with the charges it contains, is stationary and stationary charges cannot experience a magnetic force. However, the magnet is moving. As it approaches the loop, the magnetic flux passing through the wire is steadily increasing. According to Faraday's law of electromagnetic induction,[6] a changing magnetic field induces an electric field into existence, without the normal accoutrements of charges to act as the sources and sinks of the field. Induced electric fields close in loops, rather like magnetic fields naturally do as well. The observer on the loop then sees the current in the wire being driven by the induced electric field.

In both cases, a current appears in the wire loop. Both observers agree on that. So, the physical *observations* are consistent, but the physical *explanation* is different. This is the issue regarding observers and electric and magnetic fields that I referred to at the end of the last section and that Einstein is objecting to in the quote at the start of this section. His intuition is telling him that the physics relating the moving magnet to the moving circuit is not yet complete. Indeed, Einstein refers to this very experimental scenario in the introduction to his paper. To develop the context further, however, we need some more preliminary work on the physics of electromagnetism and light.

2.2.3 Magnetic and Electric Fields

The hinge point in a physical discussion of the magnet and wire loop experiment is the *Lorentz Force*[7]:

$$F = q(E + v \times B)$$

here written in a compact vector notation. The $v \times B$ is not normal multiplication. To break it down, suppose that an electric field, E, has components (E_x, E_y, E_z) along the x-, y- and z-axes, as does a magnetic field $B = (B_x, B_y, B_z)$ and the velocity of a moving object $v = (v_x, v_y, v_z)$. If the object has charge q, then the force, F, acting on that charge will also have three components (F_x, F_y, F_z) where:

$$F_x = q(E_x + v_y B_z - v_z B_y)$$

$$F_y = q(E_y + v_z B_x - v_x B_z)$$

$$F_z = q(E_z + v_x B_y - v_y B_x).$$

Although it was discovered long before Einstein developed his special theory of relativity, this formulation for the Lorentz force holds true in Einstein's wider picture.

We tend to think of electric and magnetic fields as being utterly different, as they are produced in radically dissimilar ways. Our early pictures are filled with bar magnets, current-carrying wires and electrostatically charged rods or balloons. The Lorentz force switches that viewpoint to focus on the *effect* that the fields bring about. Indeed, the Lorentz relationship *defines* what we mean by electric and magnetic fields. A small test charge,[8] δq, experiences a force determined by the size of its charge, its velocity and two other quantities, defined as the electric, E, and magnetic, B, field strengths at the location of the charge at that moment.

The electric field exerts a force dependent only on its field strength and the size of the charge, whereas the magnetic force depends on the magnetic field strength, the size of the charge and the velocity of the charge.

Now this is a very curious thing. If the charge is moving through the magnetic field, it experiences a magnetic force. However, if the charge is stationary and the magnetic field moving, there is no magnetic force acting on the charge. Normally in such situations the physics is *symmetrical*: only the *relative motion* of objects impacts on the physics. Worse, the judgement as to which is in motion, the charge or the field, is based entirely on the perspective of the observer.

Of course, we are rescued by the convenient appearance of an electric force to replace the effect of the magnetic force. But, that suggests (in Einstein's imagination) a deeper link between electric and magnetic forces via the motion of the observer.

If we can make the velocity of a charge disappear (and hence also the magnetic force acting on that charge) via a simple change in our observational perspective, then *perhaps a magnetic field can be changed, at least in part, into an electric field by a change of observational perspective as well.*

If we apply the Lorentz force equation to the moving wire loop from the previous section, then the electric field is zero and the charges experience a magnetic force:

$$F_{\text{moving circuit}} = qvB.$$

Switching to an observer sitting on the wire loop, the velocity of the charges is zero, so the magnetic force disappears, but now induction ensures an electric field is present so:

$$F_{\text{moving magnet}} = qE_{\text{induced}}$$

and if the physical effect is going to be the same for both perspectives, then:

$$E_{\text{induced}} = vB.$$

In fact, $E_y = -v_x B_z$, from the appropriate Lorentz force equation. So, the electric and magnetic fields are linked via the relative velocity between the loop and the magnet.

All of this is *described* perfectly well by electromagnetic theory, but Einstein is after a deeper explanation born from the shift in perspective when we flip between moving magnet and moving circuit. This draws us into investigating exactly what is involved in a flip between one observer and other in relative motion.

2.2.4 Galilean Transformations

To do any formal work, an observer needs a system of co-ordinates so that the position of any object can be recorded at any time. We imagine a combination similar to that shown in Figure 2.4, where a clock has been placed, for convenience, at the origin of the system.

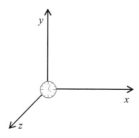

FIGURE 2.4   A system of co-ordinates used to record the location of an object at a given time.

The location of the origin is purely arbitrary, although in a given case the situation might suggest a convenient choice; perhaps in the centre of the loop of wire, in the example we have been considering. Equally, the direction of the axes is not fixed, although they should be maintained at 90° to each other. Once again, convenience generally suggests the best choice for this. It would certainly make sense to have one axis pointing along the direction that the magnet/wire loop is moving, for example.

In principle, the three axes can stretch out to infinity in their respective directions, but this will start to cause us problems for two reasons:

- If we are only equipped with one clock at the origin, we need to consider the time it takes for light to travel to the origin from the object we are trying to record; otherwise, we will get the time it was at a certain position wrong. We will always be viewing the object as it was and where it was in the past. While this is a small error in many cases, as the speed of light is a brisk 3×10^8 m/s, it is no good for precision work and certainly fails if the object is a substantial distance away.
- As we develop Einstein's general theory of relativity, we will appreciate that the curvature of space-time means that a 'flat' co-ordinate system such as Figure 2.4, where the axes are straight lines, cannot be conveniently imposed on real space over arbitrary distances. Space-time turns out to be *locally* flat, but *globally* anything but... However, that comes later.

Given our co-ordinate system, it now becomes pertinent to consider how another observer might view this and how to compare their readings to ours. For example, if one system is set up to record positions and times for an observer based with the wire loop, then we will need another to record the observations from the perspective of the magnet. These two systems will be moving with respect to each other, as shown in Figure 2.5, where moving system has dashes over the co-ordinate axes.

Note that the clocks record times t and t', respectively.

Now we need to relate the co-ordinates of one system to another. Some aspects of the relationship are simple. If the dashed system is sliding exactly along the x-axis of the other system, and provided the same scale is used on all the axes in both systems, then:

$$y' = y$$

$$z' = z.$$

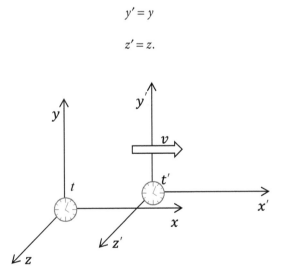

FIGURE 2.5 Two systems of co-ordinates where the dashed system is moving along the x-axis at a speed v. Note that for clarity, the dashed system has been drawn slightly displaced so that the x-axis and the x'-axis are separately visible.

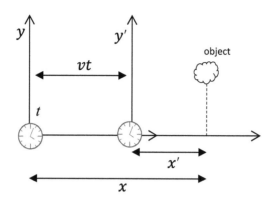

FIGURE 2.6 Relating two co-ordinate systems, the z-axis has not been drawn but should be imagined as pointing straight out of the page.

Also, 'common sense' tells us that provided both clocks are working properly and that they have been synchronised at some point in their past (conventionally they are taken to both read zero the moment the two origins coincide) then:

$$t' = t.$$

Relating the x and x' co-ordinates is slightly trickier, but aided by a diagram such as Figure 2.6, where I have suppressed the z-axis for clarity.

An object is sitting with a co-ordinate equal to x in the stationary system, at time t. It also has co-ordinate x' in the moving system at the same time (as $t' = t$). By time t after the origins crossed, the dashed system has moved a distance vt along the x-axis, so:

$$x' = x - vt.$$

We now have the full transformations between one set of co-ordinates and the other:

$$x' = x - vt$$
$$y' = y$$
$$z' = z$$
$$t' = t$$

which are known as the *Galilean transformations*.

2.2.5 Transforming Velocities

To extend the thinking further, consider an object moving through space in a direction parallel to the x-axis with speed u (Figure 2.7).

From the perspective of the moving system, the object has speed $u - v$ so that the velocity transformation becomes:

$$u' = u - v.$$

Any velocity components in the y- or z-directions will transform like the co-ordinates: without changing.

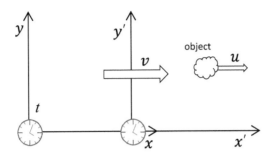

FIGURE 2.7 An object moving at velocity u through the stationary system is recorded at velocity $u' = u - v$ in the dashed system.

Clearly the two systems disagree about the velocity of any object moving through their shared space, but this hardly matters. Within any one system, measurements will be entirely consistent. Take a simple example to do with *momentum*.

In the context of Newtonian mechanics, momentum is defined as the product of mass and velocity for an object:

$$p = mv$$

and gets its importance by being a *conserved quantity*: the total momentum in the universe is always the same.

Now imagine our moving object, of mass m, collides with a stationary mass M (Figure 2.8).

As a result, the two objects merge to become one of mass $M + m$ which then proceeds along at a reduced velocity (Figure 2.9).

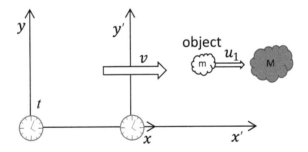

FIGURE 2.8 An object moving through the co-ordinate systems collides with a stationary object.

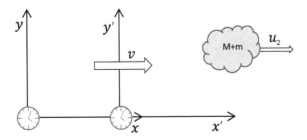

FIGURE 2.9 After the collision, the two objects merge and continue in the same direction with a reduced velocity.

Within the stationary co-ordinate system, the analysis proceeds as follows:

$$\text{initial momentum} = mu_1$$

$$\text{final momentum} = (M+m)u_2$$

$$\text{momentum is conserved, so } mu_1 = (M+m)u_2$$

$$\text{hence } u_2 = \frac{mu_1}{(M+m)}.$$

Switching to the moving system, the analysis is different as now the second object is not stationary, it is moving to the left with speed v:

$$\text{initial momentum} = mu_1' - Mv = m(u_1 - v) - Mv$$

$$\text{final momentum} = (M+m)u_2'$$

$$\text{momentum is conserved, so } m(u_1 - v) - Mv = (M+m)u_2'$$

$$\text{hence } u_2' = \frac{m(u_1 - v) - Mv}{(M+m)} = \frac{mu_1 - mv - Mv}{(M+m)} = \frac{mu_1}{(M+m)} - \frac{(M+m)v}{(M+m)}$$

$$u_2' = \frac{mu_1}{(M+m)} - v = u_2 - v$$

as we would expect.

This simple analysis shows how the laws of physics are operating in an even-handed manner between the two co-ordinate systems. While we have referred to one as stationary and the other as moving, there is no absolute meaning to that phrase. Observers in each system are entitled to view their co-ordinate array as stationary and the other as moving relative to that.

Note that throughout this, we have assumed that the mass is the same in each system. Conservation of momentum would be considerably more complicated if that were not the case.

2.2.6 Acceleration

From school physics, we know that acceleration is defined by:

$$\text{acceleration} = \frac{\text{change in velocity}}{\text{time interval}} = \frac{\Delta u}{\Delta t},$$

where Δu indicates 'the change in u', not some quantity Δ multiplied by u and Δt 'the change in t' or, perhaps more evidently, an interval of time.

So, in order to transform the acceleration, we first need to transform the change in velocity from one system to another. This is a simple extension to the velocity transformation:

$$\Delta u' = \Delta u - \Delta v. \tag{2.1}$$

For the moment, we are going to stick to a situation where the two systems are separating at a constant velocity v so that $\Delta v = 0$ and hence:

$$\Delta u' = \Delta u.$$

Dividing through by Δt and remembering that $\Delta t = \Delta t'$, we can relate the accelerations:

$$\frac{\Delta u'}{\Delta t'} = \frac{\Delta u}{\Delta t} \quad \text{or} \quad a' = a.$$

Now imagine applying a calibrated force to an object of mass m and measuring the resulting acceleration. According to the second law of motion:

$$F = ma$$

in the stationary system. In the dashed system, the acceleration is a' and the force is F'. Provided that the systems are separating at constant velocity, the accelerations in the two systems are equal, so:

$$F = ma \Rightarrow F' = ma'.$$

The mathematical form of the second law of motion is the same in both co-ordinate systems. The equation looks the same, just with dashes over the variables that transform.

In Newtonian physics, we also commonly use the *work done* formula, which relates the energy transferred to an object to its change in kinetic energy:

$$F\Delta x = \frac{1}{2}mu_2^2 - \frac{1}{2}mu_1^2,$$

which also preserves its form under the Galilean transformation:

$$F'\Delta x' = \frac{1}{2}m\left(u_2'\right)^2 - \frac{1}{2}m\left(u_1'\right)^2,$$

and the interested reader is referred to the online Appendix for that proof.

One of Einstein's primary motivations was to ensure that the laws of nature were cast in the same mathematical form for all systems. He believed that achieving this goal removed any basis for finding a system that was 'right' and so guaranteed immunity from absolute states of motion, which he found philosophically objectionable. However, as we will come to see later (Chapter 6), this view came under sharp criticism from his contemporaries and he eventually had to abandon it. However, without the pursuit of this aim, Einstein may not have come to his final theory of gravitation.

2.2.7 Inertial and Non-Inertial Systems

Going back to our transformations, let's see what happens if the two systems are separating at an *accelerating* rate (Figure 2.10).

Starting from Equation 2.1:

$$\Delta u' = \Delta u - \Delta v,$$

and given that $\Delta v \neq 0$ in this case, we must have:

$$\frac{\Delta u'}{\Delta t'} = \frac{\Delta u}{\Delta t} - \frac{\Delta v}{\Delta t} \quad \text{so that} \quad a' = a - \alpha,$$

where $\alpha = \dfrac{\Delta v}{\Delta t}$ is the rate at which one system is accelerating away from the other.

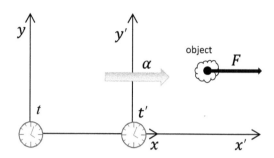

FIGURE 2.10   An object has a force *F* applied to it and is observed from two co-ordinate systems. In this case, the dashed system is accelerating at rate α relative to the stationary system.

Now the transformation of the second law looks like this:

$$F = ma \Rightarrow F' = m(a' + \alpha).$$

The form of the equation has changed in moving from one system to the other. As a result, observers in one system are able to conclude that they accelerating. There is no relativity in this case.

To see why this is, we call on the assistance of two observers, Fred and Ethel. Fred is stationed in a sealed box which we accelerate upwards by means of a rope and machinery (in other words, we build a lift/elevator). Ethel remains behind on the ground. In order to give us something to measure, Fred is instructed to stand on a weighing scale for the duration of his time in the box.

From Ethel's co-ordinate system, she analyses the situation via a combination of forces acting on Fred. There is his weight, *mg*, and an upwards force, *F*, from the weighing scale. Ethel understands the physics: as the box started to accelerate, Fred was momentarily at rest with the floor moving up towards him. As a result, the scale is trapped between a stationary Fred and a moving floor. This compresses the mechanism in the scale, resulting in an upwards force on Fred. In a short time, the compression reaches a maximum, ensuring that the force, *F*, on Fred is sufficient to overcome his weight and accelerate him upwards at the same rate as the lift. Ethel telephones Fred to report her prediction for the reading on the scale, which will be *F*. According to Ethel, Fred is accelerating upwards at rate α, so:

$$m\alpha = F - mg,$$

where *m* is Fred's mass. So, her prediction of the scale's reading is:

$$F = mg + m\alpha = m(g + \alpha).$$

This mystifies Fred. From his point of view, he does not appreciate that the box, with him inside, is accelerating. He agrees that the mechanism in the scale will be compressed, but only because he is standing on it and the scale needs to support his weight. From his point of view:

$$F' = mg'$$

but Fred will assume that stepping into the box has not also moved him to a new planet, so he will be expecting $g' = g$. I will leave the dash in place to make the link with our previous argument.

To his surprise, when he looks at the scale reading, he sees a value larger than he expected by an amount *N*. At this moment of deep existential crisis, Fred is forced to admit that his analysis is wrong.[9] The reading he sees is as follows:

$$F' = mg' + N,$$

which he may as well write, using his knowledge of his own mass, as follows:

$$F' = mg' + m\alpha = m(g' + \alpha),$$

forcing Fred to conclude that he is in accelerated motion at a rate α. At first glance, it seems as if Fred's laws of physics have betrayed him, but by adding in an additional term he is able to explain the discrepancy and establish his own state of motion.

But this is not the only possibility. On further reflection, Fred realises that he need not conclude that he is accelerating. He can retain the belief that he is stationary, provided he slots in a new force acting on him and all other objects in his co-ordinate system. Rather than:

$$F' = mg' + N.$$

Fred can write:

$$F' = mg' + F_{\text{inertial}} = F_{\text{gravitational}} + F_{\text{inertial}}.$$

The *inertial force* that Fred has added enables him to get the right answer for the weighing scale without surrendering his self-imposed status as a stationary observer.

Taking a pen out of his pocket, Fred drops it and notices that it falls to the floor of the box at a faster rate than he is used to. To Ethel, this is simply the effect of the pen falling under gravity while the floor of the box is accelerating up to meet it. Fred can either acknowledge and agree with this argument, or insist that he is stationary and that the pen is falling more rapidly due to the additional inertial force acting.

Inertial forces are sometimes called *fictitious* or *pseudo forces*. They can be used to explain effects that would otherwise be attributed to the acceleration of a reference system. For example, when we sit in an aeroplane that accelerates down the runway, we are pushed back into our seats. We can either see this as a result of the seat accelerating past us, or the action of an inertial force pushing us into the seat. The centrifugal effect we experience on a spinning roundabout can be attributed to a lack of force pulling us in, or an inertial (centrifugal) force throwing us out.

Of course, inertial forces can be seen as something of a contrivance. Inventing new forces that have no obvious physical source cannot completely cover up the broken relativity in the situation. We can always choose the system with the simplest physical explanation of our observations to be the one at rest, and the one with the contrived forces must be accelerating.

Einstein was well aware of this issue, which is why his first assumption only claims that observers in relative *inertial motion* see the same laws of nature. The phrase 'inertial motion' is code for systems that are not accelerating with respect to each other. Accelerating systems are *non-inertial*. The special theory of relativity is restricted to systems that are in inertial motion: that is what is 'special' about it (special in the sense of *restricted*). However, Einstein was not happy with this restriction. He glimpsed a connection between:

$$F' = F_{\text{gravitational}} + F_{\text{inertial}} = m\left(a_{\text{gravitational}} + a_{\text{inertial}}\right)$$

and

$$F = q(E + v \times B).$$

If they operated in a similar way, perhaps it would be possible to switch from gravitational to inertial forces by a shift in reference point, just as one can shift from electric to magnetic forces.

Einstein hoped to extend the special principle of relativity, via the principle of equivalence mentioned in Chapter 1, to include accelerating systems and fold gravity and inertial forces together as well. The laws of physics would contain gravitational and inertial forces on an equal footing, and have the same form in all co-ordinate systems, accelerating or not, with just a sliding scale between gravitational and internal forces depending on the vantage point. Then he could dispose of any absolute state of motion, provided a physical cause for the inertial forces could be found as well. This story of this quest will be the subject of Chapter 6, where we will examine how Einstein ultimately failed in one sense, but succeeded spectacularly in another.

2.2.8 Maxwell's Equations

The precise formulation of the time-space laws was the work of Maxwell. Imagine his feelings when the differential equations he had formulated proved to him that electromagnetic fields spread in the form of polarised waves, and at the speed of light! Too few men in the world has such an experience been vouchsafed... it took physicists some decades to grasp the full significance of Maxwell's discovery, so bold was the leap that his genius forced upon the conceptions of his fellow workers.

A Einstein[10] about Maxwell's equations of electromagnetism

Occasionally the work of a single physicist elevates the subject and sets the agenda for decades to come. Such achievements deserve to be ranked alongside classic works of literature, art and music. Unfortunately, physics often demands a significant level of technical background in order to be understood and appreciated, which is perhaps why it does not always gain the cultural plaudits that it deserves.

James Clerk Maxwell's 1865 synthesis of electromagnetic theory is one of the supreme pieces of individual work done in theoretical physics.[11] Maxwell was able to write down four elegant equations linking electric and magnetic fields with currents and charges. Furthermore, manipulating his equations demonstrated how electric and magnetic fields could sustain a self-contained ripple effect that would propagate through the fields at the speed of light.[12] In other words, he showed that light was an electromagnetic wave, which paved the way for further discoveries across the electromagnetic spectrum and a broad swathe of modern technology.

Unfortunately, exploring these profound ideas would take us too far from our main focus and the mathematical knowledge needed to fully exploit Maxwell's equations (and see them in their true elegance) lies beyond the limits I have set. However, the relevance of Maxwell's equations for the development of relativity is worth our time. I present them here in a somewhat mathematically neutered form...

$$\frac{\Delta E_x}{\Delta x} + \frac{\Delta E_y}{\Delta y} + \frac{\Delta E_z}{\Delta z} = \frac{\rho}{\varepsilon_0} \qquad (2.2)$$

$$\frac{\Delta B_x}{\Delta x} + \frac{\Delta B_y}{\Delta y} + \frac{\Delta B_z}{\Delta z} = 0. \qquad (2.3)$$

Equation 2.2 relates the change in electric field components over a small distance (e.g. $\Delta E_x/\Delta x$) to the *charge density* (ρ) in that region. As a result, it shows how electric field lines either start and end with charges ($\rho \neq 0$), or close in loops (if $\rho = 0$). Equation 2.3 does the same job for the magnetic field, although in this case there are no magnetic charges. Hence, the right-hand side is always zero: magnetic field lines close in loops as a result.

$$c\left(\frac{\Delta B_x}{c\Delta t}\right) = -\left(\frac{\Delta E_z}{\Delta y} - \frac{\Delta E_y}{\Delta z}\right) \qquad (2.4)$$

$$\frac{1}{c}\left(\frac{\Delta E_x}{c\Delta t}\right) + \mu_0 J_x = \frac{\Delta B_z}{\Delta y} - \frac{\Delta B_y}{\Delta z}. \tag{2.5}$$

In the form presented here, Equations 2.4 and 2.5 are not the full story as they only involve time changes of the x components of the fields. Both the y and z components have their time changes as well, which are expressed in separate equations which you could write down by cycling through x, y, z in the same pattern (or all in the one equation using a vector notation that we do not have at our disposal).

In essence, Equation 2.4 shows how the time rate of change of a magnetic field generates an electric field circulating in a loop. This is the physics responsible for the moving magnetic and wire loop situation we have been discussing. Equation 2.5 relates the time rate of change of the electric field to a circulating magnetic field. However, it also contains a term $(\mu_0 J)$ which links the magnetic field to the *current density*, J, (current/area) passing through the magnetic loop. Between them, the equations contain three fundamental constants of nature, ε_0, called the *permittivity of free space* which essentially determines the strength of the electric field, μ_0, which is the *permeability of free space* and does the same role for the magnetic field and finally c, the speed of light which Maxwell discovered was related to the other two by $c^2 = 1/\varepsilon_0\mu_0$.

Here we have the small crack that grew to split classical physics and usher in the relativistic era. Maxwell's equations contain the speed of light and predict the existence of the ripples through the fields that we now know as electromagnetic waves. However, they make no reference to a system of co-ordinates in which light has this speed. As we know from the Galilean transformations (if not common sense), the speed at which an object moves is dependent on the system that it is viewed from. Applying the Galilean transformations to Maxwell's equation for the electromagnetic wave produces a shifted velocity as we would expect. However, this implies a shift from a system in which the 'correct' speed is present. Where is that system? Equally, as the Galilean transformed equation contains a different speed of light, it violates Einstein's second assumption: the constancy of the speed of light. To Einstein, *this was an indication that the Galilean transformations were wrong.*

Before Einstein, physicists assumed that there was a preferred system that set the standard for inertial systems, each of which would be moving at a constant speed with respect to this master system. Maxwell's equations were presumably expressed in that master system, which made it the correct system from which to view electromagnetic waves and obtain the speed of light. It also provided an absolute standard of rest, which was an anathema to Einstein.

This system was thought to reside with the *luminiferous aether*, a posited material that permeated the whole of space and formed the medium in which electromagnetic waves could propagate.

By the 19th century, there was plenty of accumulated evidence to show that all forms of wave motion require a material medium. Light had been shown to be a wave, so a medium was required for its existence. The aether assumed that role.

In truth, Maxwell's discoveries had muddied the issue as his seminal paper had not included any direct reference to the aether, so it was unclear how his ripples through electric and magnetic fields related to that. Indeed, the physical nature of electric and magnetic fields was yet to be determined.

With the aether supporting light waves and forming a master standard for absolute rest, detecting its presence became of fundamental importance to 19th-century physics, something that could potentially be done by sensitive determinations of the speed of light. Such measurements should show the effect of the Earth's orbital motion through the stationary aether: a light wave propagating in the same direction as the Earth should display a different speed to that propagating against the Earth's motion. These ideas formed the basis of the famous Michelson-Morley experiment.

2.2.9 The Michelson-Morley Experiment

The collaboration between Albert Michelson and Edward Morley started in 1885, by which time Michelson was professor of physics at Case School of Applied Science just outside Cleveland and Morley professor of chemistry at Western Reserve University on the same campus. Together they worked towards refining the technique that Michelson had used in his first (1881) attempt to measure the 'aether wind': the movement of the aether across the surface of the Earth, due to the Earth's motion through the stationary material.

The basic principle of this experiment would later be echoed in the LIGO instrument discussed in Chapter 1. Michelson's first *interferometer* took a beam of light and split it at a half-silvered mirror so the two beams travelled along perpendicular arms (Figure 2.11).

At the end of each arm, a fully silvered mirror reflected the light beams back so that they met again at the half-silvered mirror and covered the last distance to an eyepiece in tandem. There the beams combined to form an *interference pattern*: bright lines (or *fringes*) where the light rays arrived in step with each other and dark lines where they were totally out of step (Figure 2.12).

The entire assembly could rotate around a vertical axis, with the fringes being observed at different orientations. At some angle around the 360°, one of the arms, say the longitudinal one, would be aligned along Earth's passage through the aether and so the other, transverse one, would be at 90° to that motion.

The light ray departs the half-silvered mirror along the longitudinal arm at speed c. Additionally, the interferometer is moving through the aether at speed v; hence, the light ray is moving with an aether 'tailwind' and can make progress relative to the instrument at speed $c + v$ (Figure 2.13a). This light ray arrives at the fully silvered mirror after a time t_1, which given the length of the arm, L, is as follows:

$$t_1 = \frac{L}{c+v}.$$

Once reflected, the ray travels back to the half-silvered mirror. Now it has the disadvantage of moving into the aether 'headwind' and so covers the instrumental ground at $c - v$. This travel time is t_2:

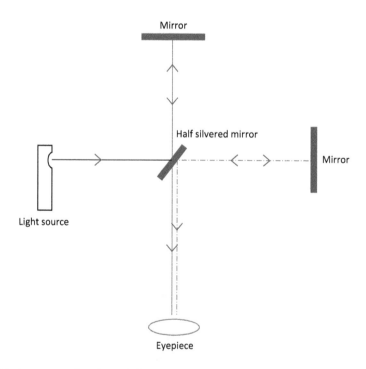

FIGURE 2.11 The basic principles of a Michelson interferometer.

FIGURE 2.12 A typical fringe pattern from a Michelson interferometer.

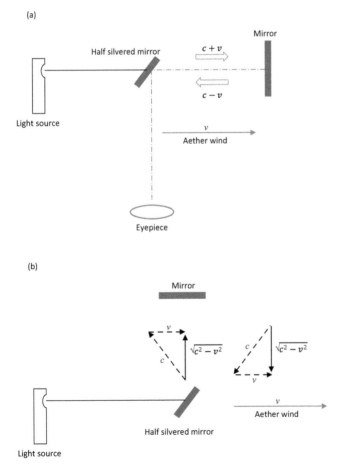

FIGURE 2.13 (a) The aether wind blowing through the Michelson interferometer when the longitudinal arm is aligned along the direction of Earth's motion through the aether. (b) The motion affecting the transverse arm of the interferometer.

$$t_2 = \frac{L}{c-v}.$$

Putting these equations together and with a bit of fiddling, we obtain the total travel time:

$$t_L = t_1 + t_2 = \frac{L}{c+v} + \frac{L}{c-v} = \frac{2Lc}{c^2 - v^2}.$$

Light travelling along the transverse arm of the interferometer is also affected by the aether wind. This case is analogous to the plight of a swimmer trying to cover the shortest distance across a river, but being pushed off course by the current running down the river. In Figure 2.13 (b), we can see how the light ray has to head 'upstream' in order to be blown back into a vertical path towards the half-silvered mirror. On the return journey, the light ray's path is 'downstream' in order to be carried back towards the mirror. In both cases, the net speed across the interferometer is $\sqrt{c^2 - v^2}$.

This gives us travel times to the mirror and back of:

$$t_3 = \frac{L}{\sqrt{c^2 - v^2}} \quad \text{and} \quad t_4 = \frac{L}{\sqrt{c^2 - v^2}},$$

so that the total time for the transverse journey becomes

$$t_T = t_3 + t_4 = \frac{2L}{\sqrt{c^2 - v^2}}.$$

Finally, the time difference between the two legs comes out as[13]:

$$\Delta t = t_L - t_T = \frac{2Lc}{c^2 - v^2} - \frac{2L}{\sqrt{c^2 - v^2}} \approx \frac{Lv^2}{c^3}$$

provided that c is rather larger than v. By multiplying by c, we convert this time difference into an apparent length difference between the arms and dividing by the wavelength of the light, λ we get the expected shift in the fringe positions:

$$\frac{\Delta L}{\lambda} = \frac{Lv^2}{\lambda c^2}.$$

Considering that the entire apparatus can be rotated so that the transverse and longitudinal arms switch places with respect to the aether wind, this will produce a shift in the fringe positions, from one extreme to the other, equal to twice that calculated above. Hence, the detectable shift as the instrument is rotated ought to be:

$$n = \frac{2Lv^2}{\lambda c^2}.$$

From this equation, we can see that the key to ensuring a fringe separation large enough to measure with precision is the length of the light path, L. Michelson's earliest attempt had not succeeded in producing a measurement outside the uncertainties generated by his instrumentation. Hence, this was one area that Michelson and Morley worked on developing.

Their experiment was built in the basement of one of the dormitories on campus and was mounted on a large sandstone slab which was free to rotate on a bath of mercury. By using a system of fully silvered mirrors reflecting the light path back and forth multiple times, they were able to extend L to 11 m, without having to build a massive experimental plinth (LIGO was later to use a similar trick). Measurements took place over 6 days, during which time the Earth moved along its orbit changing the angle of its passage through the purported aether. Additionally, they made several measurements each day rotating the plinth on its mercury bath and charting the fringe separation. Great care was taken to try and isolate the apparatus from thermal variations and vibration (hence the choice of location).

To explain earlier unsuccessful attempts to measure the aether wind, physicists had suggested that the aether was, in part or wholly, 'dragged' around by the Earth in its orbit so that, relative to the surface, it was moving at close to the same speed as the Earth. However, the fringe shift Michelson and Morley detected was far less than smallest predictions (Figure 2.14).

In short, there was no evidence of the speed of light changing, no matter what the direction of the Earth's motion and the orientation of the instrument.

With historical hindsight, Michelson-Morley's failure to detect any motion through the aether can be seen as a turning point,[14] but at the time various attempts were made to explain the results without abandoning the aether hypothesis (we mentioned earlier that this sort of reaction is typical when a theory has gained some credence). Lorentz even suggested that motion through the aether had the effect of compressing the length of objects sufficiently to counterbalance the slowed speed of light. Length contraction was later shown to be a consequence of special relativity, but it follows from the discovery that time is not universally shared by all observers rather than anything to do with the aether.

Interestingly, the extent to which the Michelson-Morley results influenced Einstein is unclear. The experiment is not mentioned in the 1905 paper itself, although it does include a comment about 'unsuccessful attempts to discover any motion of the earth relatively to the "light medium"'.[15] Einstein's later recollections on the matter are inconsistent, if not contradictory. Whatever his motivation at the time, Einstein set down as one of the two defining assumptions of his theory that the speed of light was constant and independent of any motion of the source of light. We can, in the interests of a modern coherent development, point to the Michelson-Morley experiment as justification for this suggestion but in truth this is another example of the complex relationship between theory and experiment. In any case,

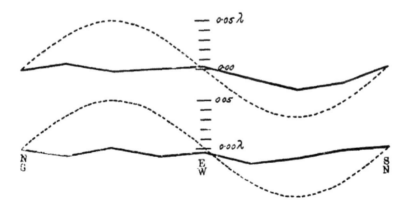

FIGURE 2.14 The results of the Michelson and Morley experiment to measure the difference in the speed of light due to the Earth's motion through the hypothetical aether (On the relative motion of the Earth and the Luminiferous ether. *American Journal of Science*, **34**(203), pp. 333–345). The top solid line represents the results of observations taken at noon, and the lower solid line represents the evening observations. The dashed lines represent the theoretical predictions, but they are plotted to a different scale rendering them only 1/8th of their true vertical size. Clearly the measured shifts are far short of the predictions.

without any shift in the speed of light, the existence of the aether was certainly on shaky ground and with it the whole notion of a master reference system. Einstein dispensed with this idea entirely and replaced the Galilean transformations with a set that preserved the same speed of light for all systems.

2.3 Relativistic Systems of Co-ordinates

If two observers in relative motion at constant speed are to agree on the speed of light, they must disagree about space, or time or both. With this quiet statement, a revolution in our thinking about the universe is acknowledged.

In order to make this plain, Einstein specified a system of co-ordinates developed from the one we used when considering the Galilean transformations. We can no longer blithely assume that a single clock at the origin will serve our purpose. Instead, Einstein proposed that each spatial location within the grid be equipped with its own clock and observer, somewhat like Figure 2.15.

In principle, every co-ordinate point in (x, y, z) should have its own clock, but evidently that would prove challenging to draw... Now, given this plethora of clocks, it becomes imperative to ensure that they are all synchronised.[16] In fact, Einstein takes this as a practical definition of what time *means* in a system of co-ordinates:

> The 'time' of an event is that which is given simultaneously with the event by a stationary clock located at the place of the event, this clock being synchronous, and indeed synchronous for all time determinations, with a specified stationary clock.[17]

His proposed method of synchronisation was to send light beams from the master clock at the origin to all the subservient clocks across the system. Considering any two consecutive clocks on the x-axis, such as shown in Figure 2.16, we specify the time a pulse of light is emitted from clock A as being t_A, recorded on A. That pulse arrives at clock B at time t_B, as recorded on B, and is immediately (without notable delay) reflected back towards A, arriving at t_{A2} as recorded on A.

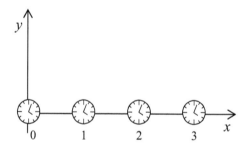

FIGURE 2.15   A co-ordinate system (z-axis omitted for clarity) in which clocks are distributed along the x-axis.

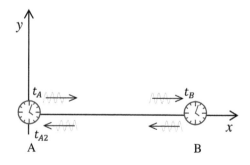

FIGURE 2.16   Synchronising clocks by means of light beams.

The two clocks are synchronised if:

$$t_B - t_A = t_{A2} - t_B.$$

Furthermore, if L_{AB} is the distance between the clocks, then:

$$\frac{L_{AB}}{t_B - t_A} = \frac{L_{AB}}{t_{A2} - t_B} = c.$$

All of which seems terribly straightforward. The problem comes if we wish to set up a second system moving at some speed v with respect to the first. From our stationary vantage point, we can watch the synchronisation process taking place in the moving system, and it seems to go horribly wrong…

From our perspective, the master clock A' is moving with speed v. Nevertheless, the light pulse travels at speed c, not $c + v$ as we might be inclined to think: this is Einstein's assumption about the constancy of the speed of light at work. So, we observe the light pulse travelling at c towards clock B', which is moving away from it at speed v (Figure 2.17).

Eventually, the light pulse will arrive at B' and be reflected. The reflection process has no impact on the speed of the light, aside from reversing its direction, so the pulse is now travelling back towards A' (Figure 2.18).

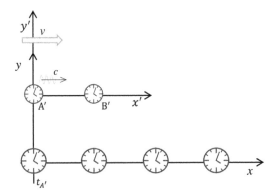

FIGURE 2.17 Synchronising clocks in a moving system. From our perspective in the stationary system, clock A' emits its synchronising light pulse at time $t_{A'}$. However, the pulse is emitted at speed c not $c + v$ as we might expect from classical physics. Hence, the light pulse is making its way towards clock B', gaining ground at $c - v$.

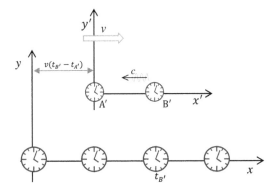

FIGURE 2.18 The light pulse is reflected at clock B' which we believe happens at time $t_{B'}$. The light pulse is now heading back to clock A' closing in at speed $c + v$.

However, A′ is moving towards the light pulse at speed v.

Consequently, the times involved are:

$$t_{B'} - t_{A'} = \frac{L_{A'B'}}{c - v} \qquad t_{A'2} - t_{B'2} = \frac{L_{A'B'}}{c + v},$$

which cannot possibly be the same, so from our perspective the clocks are not synchronised.

Of course, there is nothing to stop us talking to the observers in the other system with the hope of persuading them to change their procedure. However, chances are that we are not going to get anywhere, as from their perspective, they are stationary and we are receding from them at speed v. Hence, from their point of view we are the ones with clocks that are not synchronised.

Alternatively, consider this scenario. The master clock A′ in the moving system sends out a synchronisation pulse at time t'_A. A secondary clock at position x'_1 is programmed to set itself to a time $t'_A + x'_1/c$ on receiving this pulse (this being what the master clock would read when the pulse arrives at the secondary clock). The next clock along is programmed to set its time to $t'_A + x'_2/c$ and so on. However, from our perspective this will set the times wrongly. As the synchronisation pulse travels at speed c and the secondary clocks are moving away from that pulse at speed v, the time taken to arrive at the first clock is actually $x'_1/(c-v)$. Hence, it sets itself slow (compared to the master clock) by an amount:

$$\frac{x'_1}{c-v} - \frac{x'_1}{c} = \frac{vx'_1}{c(c-v)}.$$

Consequently, the further away a clock is from the master clock, the slower it will be set.

From within the system, observers can't see this effect. Information can only get to them from the master clock by travelling at the speed of light, or slower.

It is no use suggesting something less automated, such as synchronising all the clocks in one place and then carrying them to their respective locations. After all, in order to carry them through the system they have to move relative to the system and, as we will see in the next chapter, time dilation would screw up the synchronisation…

So, we are forced to the following conclusion: within a system, it is possible to be satisfied that clocks arrayed across the grid are synchronised with each other; hence, that there is a definable time valid across the whole system (but remember the issues raised earlier with extending these systems out to infinity…), but this time will not be consistent with that of any other system moving with respect to us, and when we observe that other system, the agreed standard of time operating within it does not square with ours.

Not only is there no universal time as we had in the Galilean transformations:

$$t' = t$$

but worse, time has become a function of space! Whatever rule we should now use to go from the time in our system to the moving system has to be different for each clock in the moving system, and their only difference is where they happen to be located in that system.

Of course, all this could be sorted out if there was one aether-based grid that defined an absolute standard of rest and with it the correct vantage point for the speed of light. All other systems could then be synchronised with that. However, the whole thrust of Einstein's argument and his comment about the lack of any ability to find such a thing is that a master system does not exist. Furthermore, the concept is now superfluous. His first assumption, that the laws of physics are the same from the point of view of any inertial system, means that nothing would be gained from having a master system.

2.4 The Lorentz Transformations

It now becomes crucially important to figure out how to transform from one system of co-ordinates into another, given the constancy of the speed of light. In other words, we need a replacement for the Galilean transformations.

Once again, we are going to engage the services of a light pulse. This time, rather than a directed pulse, like a laser beam, we imagine a spherical wave of light spreading out from the origin. As the speed of light is the same in all directions, after a certain time t any point on this spherical wave has co-ordinates (x, y, z, t) where:

$$x^2 + y^2 + z^2 = c^2 t^2. \tag{2.6}$$

We arrange for this spherical wave to be viewed from another system moving at speed v which crosses our origin at the moment the light pulse is emitted. Extraordinarily, when viewed from this second system, the light pulse also appears to spread out from its origin. This is a direct consequence of the speed of light being the same in all systems, via the impact that has on relative time.

Specifying co-ordinates in the moving system by dashed values, a point on the light sphere as viewed in this system, has co-ordinates (x', y', z', t') where:

$$(x')^2 + (y')^2 + (z')^2 = c^2 (t')^2. \tag{2.7}$$

So, we are seeking a set of transformations:

$$(x, y, z, t) \Rightarrow (x', y', z', t')$$

that turn Equation 2.6 into Equation 2.7.

This might seem to be a hopeless quest as, in principle, the transformations could be anything. However, there are some assumptions that help us out:

1. the y and z co-ordinates should be unchanged, as they will be perpendicular to the direction of relative motion between the systems;
2. the transformations should be *linear*, which means no powers of x or t involved;
3. the transformation between t and t' must be a function of x (as suggested earlier), so presumably the transformation between x and x' will be a function of t.

The linearity condition cuts down the possibilities dramatically. It is a sensible assumption as without it, an object travelling in a straight line in one system could well travel along a curved path in another… (a bit of a hint as to how gravity might work here for the future…).

On this basis, we write the transformations as:

$$x' = ax + bt$$

$$t' = At + Bx$$

$$y' = y$$

$$z' = z,$$

where the constants a, b, A, B have to be figured out. We might also note that in this linear form there is the hope that the transformations will be consistent with the Galilean ones we derived earlier, at least at speeds that are not large fractions of light speed.

To obtain the constants, we note that the origin of the moving system $(x' = 0)$ is passing along the x-axis and that it is sensible to set $t' = t = 0$ as the origins coincide. Hence, for times after that crossing, the origin of the moving system must be at a point in the stationary system, x, such that $x = vt$. This tells us that if:

$$x' = ax + bt \quad \text{then} \quad 0 = avt + bt, \quad \text{so that} \quad b = -av,$$

giving:

$$x' = ax + bt = ax - avt = a(x - vt).$$

The next step is to drop our fledgling transformations into Equation 2.7:

$$\left(a(x - vt)\right)^2 + \left(y\right)^2 + \left(z\right)^2 = c^2\left(At + Bx\right)^2.$$

Multiplying things out gets a bit messy, but we are made of stern stuff so:

$$a^2\left(x^2 - 2xvt + v^2t^2\right) + y^2 + z^2 = c^2\left(A^2t^2 + 2ABxt + B^2x^2\right).$$

Gathering terms together gives:

$$\left(a^2 - c^2B^2\right)x^2 + y^2 + z^2 - \left(2a^2v + 2c^2AB\right)xt = \left(c^2A^2 - a^2v^2\right)t^2,$$

which we would like to be identical to Equation 2.6:

$$x^2 + y^2 + z^2 = c^2t^2.$$

Comparing terms tells us that:

$$a^2 - c^2B^2 = 1 \quad 2a^2v + 2c^2AB = 0 \quad c^2A^2 - a^2v^2 = c^2.$$

With a bit of algebra, these can be solved to give all the necessary values.

Re-arranging gives:

$$B^2 = \frac{a^2 - 1}{c^2} \quad A^2 = \frac{c^2 + a^2v^2}{c^2} \quad a^2v = -c^2AB.$$

Squaring and substituting:

$$a^4v^2 = c^4A^2B^2 = \frac{c^4\left(a^2 - 1\right)\left(c^2 + a^2v^2\right)}{c^4} = a^2c^2 + a^4v^2 - c^2 - a^2v^2$$

$$\cancel{a^4v^2} = a^2c^2 + \cancel{a^4v^2} - c^2 - a^2v^2$$

$$a^2\left(c^2 - v^2\right) = c^2$$

$$a^2 = \frac{c^2}{c^2\left(1 - v^2/c^2\right)} = \frac{1}{\left(1 - v^2/c^2\right)}$$

$$a = \pm 1/\sqrt{\left(1 - v^2/c^2\right)}.$$

We select the positive square root in going from a^2 to a, as the transformation will then be:

$$x' = \frac{+1}{\sqrt{1 - v^2/c^2}}(x - vt),$$

ensuring that the x'-axis for $x' > 0$ points in the same direction as the x-axis does for $x > 0$.

After that slightly painful manipulation, the other values follow:

$$B^2 = \frac{a^2 - 1}{c^2} = \frac{1}{c^2}(a^2 - 1) = \frac{1}{c^2}\left(\frac{1}{(1 - v^2/c^2)} - 1\right) = \frac{1}{c^2}\left(\frac{1 - (1 - v^2/c^2)}{(1 - v^2/c^2)}\right)$$

$$= \frac{1}{c^2}\left(\frac{1 - 1 + v^2/c^2}{(1 - v^2/c^2)}\right) = \frac{v^2}{c^4}\frac{1}{(1 - v^2/c^2)}$$

$$B = \pm\left(\frac{v}{c^2}\right)1/\sqrt{\left(1 - v^2/c^2\right)}$$

$$A^2 = 1 + \frac{a^2 v^2}{c^2} = 1 + \left(\frac{v^2}{c^2(1 - v^2/c^2)}\right) = \frac{c^2(1 - v^2/c^2) + v^2}{c^2(1 - v^2/c^2)} = \frac{c^2 - v^2 + v^2}{c^2(1 - v^2/c^2)}$$

$$A = a = 1/\sqrt{\left(1 - v^2/c^2\right)} \quad \text{and} \quad B = \pm\left(\frac{v}{c^2}\right)A.$$

Taking the positive square root for A ensures that the t' and t axes point in the same 'direction' for positive values. Having picked the positive square root for a and A, the relationship:

$$a^2 v = -c^2 AB$$

becomes:

$$\left(\frac{1}{(1 - v^2/c^2)}\right)v = -c^2\left(\frac{+1}{\sqrt{1 - v^2/c^2}}\right)\left(\frac{\pm v}{c^2}\right)\left(\frac{+1}{\sqrt{1 - v^2/c^2}}\right),$$

reducing to:

$$v = -c^2\left(\frac{\pm v}{c^2}\right),$$

forcing us to choose the negative square root for B.

With pride in a job well done, we now present our transformations:

$$t' = \frac{\left(t - xv/c^2\right)}{\sqrt{\left(1 - v^2/c^2\right)}}$$

$$x' = \frac{\left(x - vt\right)}{\sqrt{\left(1 - v^2/c^2\right)}}$$

$$y' = y$$

$$z' = z.$$

It is conventional to package up the $1/\sqrt{1 - v^2/c^2}$ term by calling it γ, resulting in a slightly tidier looking set:

$$t' = \gamma\left(t - xv/c^2\right)$$

$$x' = \gamma\left(x - vt\right)$$

$$y' = y$$

$$z' = z$$

known as the Lorentz transformations.

2.4.1 History of the Lorentz Transformations

While Einstein derived the transformations independently in his paper, they have a history going back before 1905. They first appeared in 1897 in the work of Woldemar Voigt[18] who was looking for transformations that preserved the form of Maxwell's wave equation. His version of the transformations did not have the γ term, but as it cancels out during the transformation this did not affect his result. Similar transformations appealed to Lorentz who used them in his 1904 paper[19] on the electromagnetism of moving objects, and in a later reprint he added a footnote acknowledging that his transformations where the same as those obtained by Voigt. The first person to name the transformations was Poincaré in a 1906 paper,[20] again on electromagnetism. He also included the γ factor and adopted the transformations along with a principle of relativity: that it should be impossible to distinguish between inertial states of motion by means of electromagnetic measurements.

While the Lorentz transformations were 'in the air' at the time, Einstein's relative separation from the community probably meant that he was not that familiar with the zeitgeist, although he adopted the same notation as Voigt, so it is possible that he had read his work.[21]

Einstein was the first to claim that the time transformations were saying something profound about the nature of time itself. Lorentz thought that the time transformation was a mathematical convenience. Poincaré explained them in terms of synchronisation problems in the moving system, but held that time recorded in a system at rest in the aether was the true measure of time. Since Einstein, we accept that within each system there is a valid 'universal' time, but that the times at work in different systems may not be consistent with each other.

This is why the sphere of light used to derive the Lorentz transformations appears centred on each system's origin, even though the two origins are moving away from each other. If you track the progress of the light wave spreading out in all directions from the origin, then you are mapping the 3D shape of the wave from a collection of events all taking place at the same time, t. The same collection of events viewed from the moving system will not be simultaneous, as their different locations mean different values of t'; hence, the shape of the wave will not be spherical in that system. Observers in the moving system will have their own collection of simultaneous events and when they are mapped that will be a sphere centred on their origin. You can see a proof in the online Appendix.

Accepting the Lorentz transformations destroys the Newtonian idea that there is an *absolute time*: a universal standard of time that is accessible to all no matter what their state of motion. Roger Penrose argues that Galilean relativity, if taken completely seriously, *destroys absolute space*, without the intervention of the Lorentz transformations.[22] Given the Galilean relativity principle, Einstein's first assumption, we have to accept that an observer in inertial motion is incapable of discovering their own motion in any absolute sense. So, if such an observer isolates a point in space, by the location of some distinct object, for example, then how are they to be sure that a few moments later, the particular point is the same point as it was earlier? I can colour a dot on the surface of the Earth, but that will not be the same point in space 30 min later, as the Earth rotates, the Earth orbits the Sun, the Sun orbits the galactic core, the Milky way is in motions through the universe, etc. Relative to my system, we can define the point in a fixed manner, but there is no appropriate absolute reference.

Also, can we ascribe any meaning to a point in space, or indeed space as whole, absent of any matter in the universe?

2.5 Next Steps

The most obvious next step is to explore the effect of the Lorentz transformations on Maxwell's equations and see what they have to say about electric and magnetic fields. That, however, needs a more detailed understanding about how transformations work and specifically what the Lorentz transformations have to say about mechanics and the nature of space and time.

Notes

1. Or perhaps the vache sacree, as one reader suggested…
2. The Fundamental Idea of General Relativity in Its Original Form, from an unpublished essay written by Albert Einstein in longhand 1919: 28 March 1972, p. 32. *The New York Times Archives*.
3. Hans Christian Ørsted (1997). Karen Jelved, Andrew D. Jackson, and Ole Knudsen, translators from Danish to English. *Selected Scientific Works of Hans Christian Ørsted*, ISBN 0-691-04334-5, pp. 421–445.
4. Barring some thermally generated wandering for the free charges, which still keeps them within the wire.
5. In this context, I is in the direction of the movement of the wire as the 'current' is the movement of the charges within the wire.
6. Electromagnetic induction was discovered by Michael Faraday in 1831 and separately by Joseph Henry in 1832. As Faraday was the first to publish his results, the effect is named after him.
7. Lorentz, Hendrik Antoon, Versuch einer Theorie der electrischen und optischen Erscheinungen in bewegten Körpern, 1895.
8. We use a small test charge to ensure that the force it exerts on the charges and currents producing the fields does not disturb their location and hence the magnitude of the fields.
9. No observers were philosophically harmed during this thought experiment.
10. *Science*, 24 May 1940.
11. Maxwell, J.C., 1865. A dynamical theory of the electromagnetic field. *Philosophical Transactions of the Royal Society of London*, **155**, pp. 459–512. doi: 10.1098/rstl.1865.0008.

12. Amusingly, Maxwell apparently made this discovery over summer at his Scottish estate, but had to wait several weeks before being able to check the values of what we now call vacuum permittivity and permeability by accessing his books at Kings College. The internet has certainly made some aspects of life easier…

13. Here I have used an approximation that I will cover in more detail in Chapter 5.

14. Supported by astronomical observations of the effect known as *stellar aberration*. Interested readers can follow this point up by consulting the following web sites: Aether drag hypothesis, https://en.wikipedia.org/wiki/Aether_drag_hypothesis (last modified 15 March 2019); Aberration of light, https://en.wikipedia.org/wiki/Aberration_of_light#Aether_drag_models amongst others (last modified 24 February 2019).

15. English translation of Einstein's paper: https://www.fourmilab.ch/etexts/einstein/specrel/specrel.pdf.

16. Or at worst, have a fixed and known time difference with respect to each other.

17. As per 15.

18. Voigt, W., 10. März 1887. *Ueber das Doppler'sche Princip, Nachrichten von der Königlichen Gesellschaft der Wissenschaften und der Georg–Augusts.* Universität zu Göttingen, No. 2.

19. Lorentz, H.A., 1904. Electromagnetic phenomena in a system moving with any velocity smaller than that of light. *Proceedings of the Royal Netherlands Academy of Arts and Science*, **6**, p. 809.

20. Poincaré, H., 1906. Sur la dynamique de l'électron. *Rendiconti del Circolo matematico di Palermo*, **21**, p. 129.

21. In a reprint of papers on relativity, Einstein also added a comment that he was not aware of Lorentz's paper at the time. He also added a footnote to the effect that the Lorentz Transformations could be derived in an easier manner to the way he had done it by using the principle of the constancy of light speed – as Voigt had done.

22. Penrose, R., 2005. *The Road to Reality*, Vintage, New Ed edition, ISBN-10: 0099440687. Penrose also argues, rightly in my view, that this point is not widely enough acknowledged in the physics community.

3

The Theory of Special Relativity

Strangely enough no personal contacts resulted between his teacher of mathematics, Hermann Minkowski, and Einstein. When, later on, Minkowski built up the special theory of relativity into his 'world-geometry', Einstein said on one occasion: 'Since the mathematicians have invaded the theory of relativity, I do not understand it myself any more'. But soon thereafter, at the time of the conception of the general theory of relativity, he readily acknowledged the indispensability of the four-dimensional scheme of Minkowski.

A Sommerfeld[1]

Herman Minkowski (1864–1909) was professor of mathematician at various institutions, culminating in Göttingen where he worked from 1902 until his death. From 1896 to 1902, he taught at the Zurich Federal Institute of Technology where Einstein attended several of his courses. From 1905, Minkowski was interested in electromagnetic theory and, by developing some of Poincaré's ideas, saw how Maxwell's equations could be expressed in a four-dimensional space-time, where the Lorentz transformations became a 'rotation'. It was then quite natural to fold Einstein's special theory of relativity into the same format.

Arguably, Einstein's ideas gained more widespread influence after Minkowski's 1908 lecture in Copenhagen, when he presented the theory in his more 'geometrical' space-time form. Einstein was initially unimpressed by this re-formulation, but as his work on gravity developed, came to see how it was the most natural context in which to express his ideas. Consequently, it makes sense for modern developments of relativity to set up Minkowskian space-time from the start.

3.1 Rotation for Fun and Profit

Given that our eventual aim is to see how relativity can be expressed as a geometrical theory of space-time, we need to start by looking at *rotations*.

Figure 3.1 shows a 2D co-ordinate system (X, Y) along with another co-ordinate system (X', Y') where the axes have been rotated anti-clockwise by the same angle, θ.

Clearly any point in the 2D plane can be given a set of co-ordinates in the (X, Y) system and also in the (X', Y') system. The rules that transform the co-ordinates of one system into another can be worked out via some rather tedious algebra[2]:

$$X' = X \cos \theta + Y \sin \theta \quad Y' = -X \sin \theta + Y \cos \theta.$$

Note that in this transformation, X' is a mix of X and Y co-ordinates and that Y' is a different mix of X and Y. The transformation rules combine the values in one system to make the values in another. A similar set of rules would allow us to go in reverse, from (X', Y') back to (X, Y), in which case X would be a mix of X' and Y', as would Y (but a different mix). We will shortly see an interesting way to figure out the reverse rules.

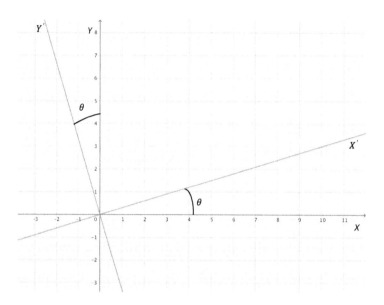

FIGURE 3.1   A pair of 2D co-ordinate systems where one set of axes has been rotated with respect to the other.

While it is possible to work with rotations by manipulating the transformation rules, a much more elegant approach is to use mathematical objects that have been specially designed to work nicely in just this sort of situation: *matrices*.

3.2 Working with Matrices

A matrix is a collection of numbers grouped together into a pattern. For example, $\begin{pmatrix} 4 & 7 \\ 9 & 5 \end{pmatrix}$ is a two-by-two matrix (two rows and two columns). The position of the numbers is of crucial importance as the matrix $\begin{pmatrix} 7 & 9 \\ 5 & 4 \end{pmatrix}$ has the same set of numbers, but is totally different.

The collection can be of any size, for example, $\begin{pmatrix} 4 & 8 & 2 \\ 6 & 3 & 5 \end{pmatrix}$ is a two-by-three matrix (two rows, three columns), whereas $\begin{pmatrix} 4 & 7 \\ 8 & 10 \\ 1 & 6 \end{pmatrix}$ is a three-by-two matrix (three rows, two columns), and $\begin{pmatrix} 1 & 3 \end{pmatrix}$ and $\begin{pmatrix} 12 \\ 13 \end{pmatrix}$ are one-by-two and two-by-one matrices, respectively. Matrices with either one row or one column are often called *vectors*. A vector like $\begin{pmatrix} 12 \\ 13 \end{pmatrix}$ is a *column vector*, whereas $\begin{pmatrix} 1 & 3 \end{pmatrix}$ is a *row vector*.

All these examples employ numerical values, but we can use algebraic symbols in a matrix instead: $\begin{pmatrix} A & B \\ C & D \end{pmatrix}$ is a perfectly good matrix where the symbols A, B, C, D are used to stand for numbers that we can specify or determine. These are the *elements* or *components* of the matrix.

Mathematicians have set up rules that define how matrices are to be combined, either by *adding them* or *multiplying them*.

3.2.1 Adding Matrices

To add two matrices, we add the elements together. So, for example:

$$\begin{pmatrix} 4 & 9 \\ 11 & 6 \end{pmatrix} + \begin{pmatrix} 1 & 7 \\ 3 & 5 \end{pmatrix} = \begin{pmatrix} 4+1 & 9+7 \\ 11+3 & 6+5 \end{pmatrix} = \begin{pmatrix} 5 & 16 \\ 14 & 11 \end{pmatrix}$$

or

$$\begin{pmatrix} A & B \\ C & D \end{pmatrix} + \begin{pmatrix} a & b \\ c & d \end{pmatrix} = \begin{pmatrix} A+a & B+b \\ C+c & D+d \end{pmatrix}.$$

Subtracting matrices works in just the same way:

$$\begin{pmatrix} 4 & 9 \\ 11 & 6 \end{pmatrix} - \begin{pmatrix} 1 & 7 \\ 3 & 5 \end{pmatrix} = \begin{pmatrix} 4-1 & 9-7 \\ 11-3 & 6-5 \end{pmatrix} = \begin{pmatrix} 3 & 2 \\ 8 & 1 \end{pmatrix}.$$

3.2.2 Multiplying Matrices

In practice, we will not have much use for adding and subtracting matrices, but we will often need to multiply them. This is a bit trickier. The rule is as follows:

$$\begin{pmatrix} A & B \\ C & D \end{pmatrix} \times \begin{pmatrix} a & b \\ c & d \end{pmatrix} = \begin{pmatrix} \boxed{Aa+Bc} & Ab+Bd \\ Ca+Dc & Cb+Dd \end{pmatrix}$$

with the elements in the product being made up from the row and column elements of the two multiplied matrices. Importantly, the number of columns in the first matrix has to be equal to the number of rows in the second, or the multiplication rule does not work. You can't multiply the matrices:

$$\begin{pmatrix} A & B & C \\ D & E & F \end{pmatrix} \times \begin{pmatrix} a & b & c \\ d & e & f \end{pmatrix},$$

for example, as the number of columns in the first matrix is not the same as the number of rows in the second. Also, unlike with numbers, the order in which you multiply the matrices is important. For example:

$$\begin{pmatrix} A & B \\ C & D \end{pmatrix} \times \begin{pmatrix} a & b \\ c & d \end{pmatrix} = \begin{pmatrix} Aa+Bc & Ab+Bd \\ Ca+Dc & Cb+Dd \end{pmatrix}$$

but:

$$\begin{pmatrix} a & b \\ c & d \end{pmatrix} \times \begin{pmatrix} A & B \\ C & D \end{pmatrix} = \begin{pmatrix} aA+bC & aB+bD \\ cA+dC & cB+dD \end{pmatrix}$$

is very different. In the vernacular, we say that matrix multiplication does not *commute*.

Just as we can use letters to stand for numbers, we can use letters to stand for matrices. So, we can say $M_1 = \begin{pmatrix} A & B \\ C & D \end{pmatrix}$ and $M_2 = \begin{pmatrix} a & b \\ c & d \end{pmatrix}$, so that:

$$M_1 \times M_2 = \begin{pmatrix} Aa+Bc & Ab+Bd \\ Ca+Dc & Cb+Dd \end{pmatrix}$$

but:

$$M_2 \times M_1 = \begin{pmatrix} aA+bC & aB+bD \\ cA+dC & cB+dD \end{pmatrix}.$$

The matrix $I = \begin{pmatrix} 1 & 0 \\ 0 & 1 \end{pmatrix}$, known as the *unit matrix*, has a special role to play. Multiplying any matrix by I, in any order, produces the same matrix again:

$$M \times I = I \times M = M.$$

You can think of it as the matrix equivalent of the number 1. The I written above is the two-by-two version of the unit matrix, but there are other unit matrices such as:

$$I = \begin{pmatrix} 1 & 0 & 0 \\ 0 & 1 & 0 \\ 0 & 0 & 1 \end{pmatrix},$$

which is the three-by-three version.

3.2.3 The Inverse Matrix

Given a matrix M, its *inverse matrix*, M^{-1}, fits together with M like this:

$$M \times M^{-1} = M^{-1} \times M = I = \begin{pmatrix} 1 & 0 \\ 0 & 1 \end{pmatrix}.$$

Figuring out what the inverse matrix is for any given matrix can be a bit tricky, but in the case of a two-by-two matrix there is a rule that we can use. If

$$M = \begin{pmatrix} a & b \\ c & d \end{pmatrix},$$

then

$$M^{-1} = \frac{1}{ad - bc} \times \begin{pmatrix} d & -b \\ -c & a \end{pmatrix}. \tag{3.1}$$

If you wanted to *divide* matrix A by matrix M, you would multiply A by the inverse of M, so that $A/M = A \times M^{-1}$. Once again, the order of multiplication is important: $M^{-1} \times A$ is a perfectly sensible multiplication, but it is *not* equivalent to A/M and will give a different answer.

3.2.4 Transposes and Symmetry

The *transpose matrix* is what you get if you swap rows for columns. With $M = \begin{pmatrix} a & b & c \\ d & e & f \end{pmatrix}$ its

transpose, $M^T = \begin{pmatrix} a & d \\ b & e \\ c & f \end{pmatrix}$. The transpose is more helpful than it sounds, especially if we add in the

following rule:

$$\{A \times B\}^T = B^T \times A^T \tag{3.2}$$

which we will use shortly.

A *symmetric matrix* has to be 'square' and has equal components either side of the diagonal:

$$M = \begin{pmatrix} a & b \\ b & c \end{pmatrix}$$

in which case, $M^T = M$.

Antisymmetric matrices are defined by $M^T = -M$, in which case they have the form:

$$M = \begin{pmatrix} 0 & b \\ -b & 0 \end{pmatrix}.$$

Note that such matrices *must* have diagonal elements which are zero.

Amusingly, any matrix can be split into symmetric and antisymmetric parts, $M = M_A + M_S$, where $M_A = 1/2(M - M^T)$ and $M_S = 1/2(M + M^T)$.

3.2.5 The Rotation Matrix

Our rules for transforming one co-ordinate system into a rotated one:

$$X' = X \cos\theta + Y \sin\theta \quad Y' = -X \sin\theta + Y \cos\theta$$

can be written as a matrix equation:

$$\begin{pmatrix} X' \\ Y' \end{pmatrix} = \begin{pmatrix} \cos\theta & \sin\theta \\ -\sin\theta & \cos\theta \end{pmatrix} \begin{pmatrix} X \\ Y \end{pmatrix}$$

where $\begin{pmatrix} \cos\theta & \sin\theta \\ -\sin\theta & \cos\theta \end{pmatrix}$ is called the *rotation matrix*, and given the symbol $\mathfrak{R}$ or $\mathfrak{R}(\theta)$ if we wish to

show the angle involved. The rotation matrix is a rather pretty little matrix with some nice properties. For example, the inverse matrix $\mathfrak{R}^{-1}$ is as follows (using Equation 3.1):

$$\mathfrak{R}^{-1} = \frac{1}{\cos^2\theta + \sin^2\theta} \begin{pmatrix} \cos\theta & -\sin\theta \\ \sin\theta & \cos\theta \end{pmatrix}$$

as $\cos^2\theta + \sin^2\theta = 1$, we get:

$$\mathfrak{R}^{-1} = \begin{pmatrix} \cos\theta & -\sin\theta \\ \sin\theta & \cos\theta \end{pmatrix} = \mathfrak{R}^T.$$

The inverse of this matrix is the same as its transpose, which is not a common feature of matrices.

The trigonometric functions sin, cos and tan started their lives as ratios to do with angles and sides in right-angled triangles, but mathematicians have extended their use to include negative angles as well. It works in the following way:

$$\sin(-\theta) = -\sin\theta \quad \cos(-\theta) = \cos\theta \quad \tan(-\theta) = -\tan\theta$$

for negative angles between $0°$ and $-90°$.

Applying this to our rotation matrix gives us:

$$\mathfrak{R}(-\theta) = \begin{pmatrix} \cos(-\theta) & \sin(-\theta) \\ -\sin(-\theta) & \cos(-\theta) \end{pmatrix} = \begin{pmatrix} \cos\theta & -\sin\theta \\ \sin\theta & \cos\theta \end{pmatrix} = \mathfrak{R}^{-1}.$$

A little further thought shows us that this makes complete sense. A *positive* angle turns in an *anti-clockwise* direction, whereas a *negative* angle turns in a *clockwise* direction. Hence, if we rotate our axes *anti-clockwise* through the angle θ and then take the rotated axes and send them *clockwise* through angle θ (which is to say rotating them through angle $-\theta$), this clearly puts the axes back to where they started from. So, $\mathfrak{R}(-\theta) = \mathfrak{R}^{-1}$.

3.3 Invariance

One of our ultimate aims is to see how equations that represent laws of nature can retain their form when transformations are applied to them. Consequently, it is interesting to investigate any quantities that do not change at all when they are transformed. These are known as *invariant quantities* and there is one simple example that is invariant under rotation of co-ordinate axes.

Consider the following combination:

$$\begin{pmatrix} X & Y \end{pmatrix} \begin{pmatrix} X \\ Y \end{pmatrix} = X^2 + Y^2$$

and its partner in the rotated system:

$$\begin{pmatrix} X' & Y' \end{pmatrix} \begin{pmatrix} X' \\ Y' \end{pmatrix} = (X')^2 + (Y')^2. \tag{3.3}$$

The rotation matrix provides the link between the two systems:

$$\begin{pmatrix} X' \\ Y' \end{pmatrix} = \begin{pmatrix} \cos\theta & \sin\theta \\ -\sin\theta & \cos\theta \end{pmatrix} \begin{pmatrix} X \\ Y \end{pmatrix} = \Re \begin{pmatrix} X \\ Y \end{pmatrix}.$$

Now, $\begin{pmatrix} X' & Y' \end{pmatrix}$ is the transpose of the vector $\begin{pmatrix} X' \\ Y' \end{pmatrix}$:

$$\begin{pmatrix} X' & Y' \end{pmatrix} = \begin{pmatrix} X' \\ Y' \end{pmatrix}^T = \left\{ \Re \begin{pmatrix} X \\ Y \end{pmatrix} \right\}^T = \begin{pmatrix} X \\ Y \end{pmatrix}^T \Re^T = \begin{pmatrix} X & Y \end{pmatrix} \Re^T \ \left(\text{using rule 3.2} \right).$$

If we now insert this into Equation 3.3, something rather interesting emerges:

$$\begin{pmatrix} X' & Y' \end{pmatrix} \begin{pmatrix} X' \\ Y' \end{pmatrix} = \begin{pmatrix} X & Y \end{pmatrix} \Re^T \Re \begin{pmatrix} X \\ Y \end{pmatrix}.$$

The properties of the rotation matrix include the uncommon fact that $\Re^T = \Re^{-1}$, so $\Re^T \Re = \Re^{-1}\Re = I$. Hence:

$$\begin{pmatrix} X' & Y' \end{pmatrix} \begin{pmatrix} X' \\ Y' \end{pmatrix} = \begin{pmatrix} X & Y \end{pmatrix} I \begin{pmatrix} X \\ Y \end{pmatrix}$$

and consequently:

$$\left(X'\right)^2 + \left(Y'\right)^2 = X^2 + Y^2.$$

It is easy to see from simple geometry (Figure 3.2) that we have calculated the length of a line connecting the origin to a point: $R^2 = X^2 + Y^2 = \left(X'\right)^2 + \left(Y'\right)^2$.

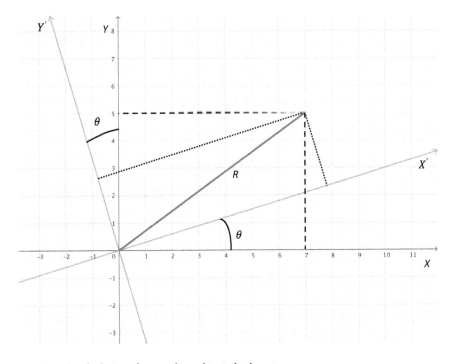

FIGURE 3.2 Length calculation of a co-ordinate line in both systems.

This length is *invariant under rotation*. When we come to look at the Lorentz transformations, we will be looking for quantities that are *invariant under a Lorentz boost*.[3]

3.4 The Lorentz Matrix

We have already met the Lorentz transformations in the form:

$$t' = \gamma\left(t - xv/c^2\right)$$

$$x' = \gamma\left(x - vt\right)$$

$$y' = y$$

$$z' = z$$

remembering that $\gamma = 1/\sqrt{1 - v^2/c^2}$. As they stand, the transformations are not 'balanced' in the way that they treat spatial transformations and temporal transformations. This is hardly surprising as space and time are very different things. We can walk around in space, after all, but nobody has mastered the ability to move through time with the same degree of freedom.[4] The lack of even-handedness in the way that x and t transform comes from their being measured in different units; metres and seconds, respectively.[5] If we introduce the parameter $\beta = v/c$, we can tidy up the transformations. If $\beta = v/c$ then $v = \beta c$ and $\gamma = 1/\sqrt{1 - \beta^2}$, also:

$$t' = \gamma\left(t - x\beta c/c^2\right) = \gamma\left(t - x\beta/c\right)$$

$$x' = \gamma\left(x - \beta ct\right)$$

$$y' = y$$

$$z' = z.$$

Taking the transformation of t, and multiplying through by c, we get:

$$ct' = \gamma\left(ct - \beta x\right)$$

to put alongside:

$$x' = \gamma\left(x - \beta ct\right).$$

By multiplying the time co-ordinate by c, which is a physical constant, we are effectively 'measuring' it in the same units as x and so the transformations now have a nice symmetry to them.

Next, we render these transformations in matrix form, choosing to ignore the y and z transformations for simplicity. Defining the *Lorentz vectors* $X = \begin{pmatrix} ct \\ x \end{pmatrix}$ and $X' = \begin{pmatrix} ct' \\ x' \end{pmatrix}$ along with the *Lorentz matrix* $L = L(\beta) = \begin{pmatrix} \gamma & -\beta\gamma \\ -\beta\gamma & \gamma \end{pmatrix}$, the transformations become:

$$\begin{pmatrix} ct' \\ x' \end{pmatrix} = \begin{pmatrix} \gamma & -\beta\gamma \\ -\beta\gamma & \gamma \end{pmatrix}\begin{pmatrix} ct \\ x \end{pmatrix} \qquad X' = LX.$$

Bringing back our rotation matrix $\Re = \begin{pmatrix} \cos\theta & \sin\theta \\ -\sin\theta & \cos\theta \end{pmatrix}$ we can see that there is some similarity with the Lorentz matrix (elements a and d are the same as each other in both matrices), and a difference (in the rotation matrix elements b and c have the same value but opposite sign, in the Lorentz matrix they are the same value and with the same sign).

As the Lorentz matrix is symmetric, when we transpose:

$$L^T = \begin{pmatrix} \gamma & -\beta\gamma \\ -\beta\gamma & \gamma \end{pmatrix}^T = \begin{pmatrix} \gamma & -\beta\gamma \\ -\beta\gamma & \gamma \end{pmatrix} = L$$

it is unchanged, which was not true of the rotation matrix.

The inverse of L is calculated from Equation 3.3 as follows:

$$L^{-1} = \frac{1}{\gamma^2 - \beta^2\gamma^2} \begin{pmatrix} \gamma & \beta\gamma \\ \beta\gamma & \gamma \end{pmatrix} = \frac{1}{\gamma^2(1-\beta^2)} \begin{pmatrix} \gamma & \beta\gamma \\ \beta\gamma & \gamma \end{pmatrix} = \begin{pmatrix} \gamma & \beta\gamma \\ \beta\gamma & \gamma \end{pmatrix}.$$

So, L^{-1} looks just like L, except for positive values in elements b and c rather than negative ones.

This makes sense. A Lorentz boost, $L(\beta)$, takes us from a system at rest relative to us into another moving at velocity v away from us. Sitting in that new system, the original one appears to be moving with a velocity of $-v$. Transforming back from that system into the first one would use $L(-\beta)$ and must surely be the inverse of the original transformation. In summary:

$$\{L(\beta)\}^{-1} = L(-\beta).$$

3.5 Lorentz Invariants

To find a quantity that is invariant under a Lorentz transformation, we should start by trying the same thing as we did to get a rotational invariant. First, we build:

$$\begin{pmatrix} ct & x \end{pmatrix} \begin{pmatrix} ct \\ x \end{pmatrix} = (ct)^2 + x^2$$

from our vectors and then see if it retains its value when we transform to the new co-ordinate system. As:

$$\begin{pmatrix} ct' \\ x' \end{pmatrix} = L \begin{pmatrix} ct \\ x \end{pmatrix}, \tag{3.4}$$

so:

$$\begin{pmatrix} ct' & x' \end{pmatrix} = \begin{pmatrix} ct & x \end{pmatrix} L^T$$

which we have obtained by transposing Equation 3.4 and applying rule 3.2. Next, we construct the product of our vectors in the transformed system and substitute in the transformations:

$$\begin{pmatrix} ct' & x' \end{pmatrix} \begin{pmatrix} ct' \\ x' \end{pmatrix} = \begin{pmatrix} ct & x \end{pmatrix} L^T L \begin{pmatrix} ct \\ x \end{pmatrix}.$$

When we did this for rotations, the product $\mathfrak{R}^T\mathfrak{R}$ appeared in the centre of the calculation. This time we have $L^T L$. However, $\mathfrak{R}^T = \mathfrak{R}^{-1}$, so that product was the unit vector, I, showing that the line length was invariant. This time, we are not so lucky as $L^T \neq L^{-1}$, so $L^T L \neq 1$ and the form that we have constructed is *not* invariant under the Lorentz transformations. All is not lost however, as we can learn from this and set about constructing something that is invariant.

Comparing $\mathfrak{R} = \begin{pmatrix} \cos\theta & \sin\theta \\ -\sin\theta & \cos\theta \end{pmatrix}$ and $L = \begin{pmatrix} \gamma & -\beta\gamma \\ -\beta\gamma & \gamma \end{pmatrix}$, the most pertinent difference is that

L has negative values in both element b and c, whereas in $\mathfrak{R}$ they are of opposite sign. This is what makes $L^T = L$, when $\mathfrak{R}^T \neq \mathfrak{R}$ and whereas $\mathfrak{R}^T = \mathfrak{R}^{-1}$, this cannot be the case for L. Clearly, to fix this we need to do something with minus signs...

Suppose we use a matrix, η, to construct a product which *is* invariant. In other words:

$$\begin{pmatrix} ct' & x' \end{pmatrix} \eta \begin{pmatrix} ct' \\ x' \end{pmatrix} = \begin{pmatrix} ct & x \end{pmatrix} \eta \begin{pmatrix} ct \\ x \end{pmatrix}. \tag{3.5}$$

This is a rather bold assumption at this juncture, but I will be able to fill in some of the justification for it later on. One good justification is that it will give us the right answer!

If we insert the Lorentz transformations into the left-hand side, Equation 3.5 becomes:

$$\begin{pmatrix} ct & x \end{pmatrix} L^T \eta L \begin{pmatrix} ct \\ x \end{pmatrix} = \begin{pmatrix} ct & x \end{pmatrix} \eta \begin{pmatrix} ct \\ x \end{pmatrix}$$

telling us that $L^T \eta L = \eta$, or as $L^T = L$ then $L\eta L = \eta$.

To make the next stage a little easier, we are going to multiply each side from the right by L^{-1}:

$$(L\eta L)L^{-1} = (\eta)L^{-1}$$

giving $L\eta = \eta L^{-1}$.

Supplying the matrix η with a set of elements and filling in the Lorentz matrix:

$$\begin{pmatrix} \gamma & -\beta\gamma \\ -\beta\gamma & \gamma \end{pmatrix} \begin{pmatrix} a & b \\ c & d \end{pmatrix} = \begin{pmatrix} a & b \\ c & d \end{pmatrix} \begin{pmatrix} \gamma & \beta\gamma \\ \beta\gamma & \gamma \end{pmatrix}$$

or

$$\gamma \begin{pmatrix} 1 & -\beta \\ -\beta & 1 \end{pmatrix} \begin{pmatrix} a & b \\ c & d \end{pmatrix} = \begin{pmatrix} a & b \\ c & d \end{pmatrix} \gamma \begin{pmatrix} 1 & \beta \\ \beta & 1 \end{pmatrix}.$$

Carrying out the matrix multiplication on both sides gives:

$$\gamma \begin{pmatrix} a-\beta c & b-\beta d \\ -\beta a+c & -\beta b+d \end{pmatrix} = \gamma \begin{pmatrix} a+\beta b & a\beta+b \\ c+\beta d & c\beta+d \end{pmatrix}.$$

Comparing the bottom right element in each matrix shows us that:

$$-\beta b+d = c\beta+d$$

which can only work if $c = -b$, or $c = -b = 0$.

Looking at the top right and bottom left elements in each matrix:

$$b - \beta d = a\beta + b \quad -\beta a + c = c + \beta d$$

it is clear that $a = -d$.

So, our matrix takes the form:

$$\eta = \begin{pmatrix} a & b \\ -b & -a \end{pmatrix}$$

making the invariant:

$$\begin{pmatrix} ct & x \end{pmatrix} \eta \begin{pmatrix} ct \\ x \end{pmatrix} = \begin{pmatrix} ct & x \end{pmatrix} \begin{pmatrix} a & b \\ -b & -a \end{pmatrix} \begin{pmatrix} ct \\ x \end{pmatrix}.$$

Expanding this out gives:

$$\begin{pmatrix} ct & x \end{pmatrix} \begin{pmatrix} a & b \\ -b & -a \end{pmatrix} \begin{pmatrix} ct \\ x \end{pmatrix} = \begin{pmatrix} ct & x \end{pmatrix} \begin{pmatrix} act + bx \\ -bct - ax \end{pmatrix}$$

$$= a(ct)^2 + bxct - bxct - ax^2 = a(ct)^2 - ax^2 = a\left\{ (ct)^2 - x^2 \right\}.$$

As the terms involving b cancel out, we may as well take $b = 0$. Also, as the multiple of an invariant must be an invariant, we do not lose any generality by setting $a = 1$.

The matrix η can then take the form $\eta_1 = \begin{pmatrix} 1 & 0 \\ 0 & -1 \end{pmatrix}$ or $\eta_2 = \begin{pmatrix} -1 & 0 \\ 0 & 1 \end{pmatrix}$, either being a perfectly

acceptable choice as there is no difference in the physics produced. Different authors go with different conventions. We are going to use η_1, hereafter simply called η.

The upshot of all this is a belief in the following product's Lorentz invariance:

$$s^2 = \begin{pmatrix} ct' & x' \end{pmatrix} \begin{pmatrix} 1 & 0 \\ 0 & -1 \end{pmatrix} \begin{pmatrix} ct' \\ x' \end{pmatrix} = \begin{pmatrix} ct & x \end{pmatrix} \begin{pmatrix} 1 & 0 \\ 0 & -1 \end{pmatrix} \begin{pmatrix} ct \\ x \end{pmatrix}$$

which is to say:

$$s^2 = (ct')^2 - (x')^2 = (ct)^2 - x^2 \tag{3.6}$$

introducing s^2 as shorthand for this invariant.

Checking the invariance of s^2 is a simple matter of substituting in the Lorentz transformations:

$$s^2 = (ct')^2 - (x')^2 = \gamma^2 (ct - \beta x)^2 - \gamma^2 (x - \beta ct)^2 = \gamma^2 \left\{ (ct)^2 + \beta^2 x^2 - 2ct\beta x - x^2 - \beta^2 (ct)^2 + 2ct\beta x \right\}$$

$$= \gamma^2 (1 - \beta^2) \left\{ (ct)^2 - x^2 \right\} = (ct)^2 - x^2.$$

The other thing that we ought to do is check that η does obey the relationship $L\eta L = \eta$, which I have done in the on-line Appendix. Note that this effectively defines a transformation rule for η, something we will pick up on in Section 3.10.2.

The matrix η is our first example of a crucially important class of objects called *metrics*. Much of what we do in general relativity will revolve around metrics of different form. In special relativity η is the *Minkowski metric* and will be our sole concern.

3.6 The Space-time Interval

While it is one thing having a shiny new Lorentz invariant to play with, it is quite something else to discern its physical meaning. When we found the rotation invariant, working out the physical meaning was trivial when you considered the geometry of the situation. This time, we need a little more subtlety in our approach...

So far, we have been considering the Lorentz transformations as point-to-point mappings, taking us from one set of co-ordinates into another. However, consider the distance between two points along the x-axis, x_1 and x_2, which we represent as $\Delta x = x_2 - x_1$. (Remember that the symbol Δx is taken to mean 'the change in x' not some quantity Δ multiplied by x.)

Each of these points maps into a point on the x'-axis via a Lorentz transformation:

$$x_2' = \gamma(x_1 - \beta c t_1); \quad x_2' = \gamma(x_2 - \beta c t_2).$$

Assuming that we are tracking an object moving from point x_1 at time t_1 to x_2 at time t_2, with $\Delta t = t_2 - t_1$, subtracting tells us that:

$$\Delta x' = \gamma(\Delta x - \beta c \Delta t) \tag{3.7.1}$$

$$c\Delta t' = \gamma(c\Delta t - \beta c \Delta x). \tag{3.7.2}$$

It is clearly possible to build another vector $(c\Delta t, \Delta x)$ which transforms via the Lorentz matrix:

$$\begin{pmatrix} c\Delta t' \\ \Delta x' \end{pmatrix} = L \begin{pmatrix} c\Delta t \\ \Delta x \end{pmatrix} = \begin{pmatrix} \gamma & -\beta\gamma \\ -\beta\gamma & \gamma \end{pmatrix} \begin{pmatrix} c\Delta t \\ \Delta x \end{pmatrix}.$$

In a sense there is nothing new here. After all the vector, (ct, x), is really talking about the x co-ordinate, which is a distance to the origin (Δx with $x_1 = 0$) and a moment in time t, which can be thought of as t seconds after the clock started (Δt with $t_1 = 0$). I could, with a bit of hand waving, have pasted Δ symbols in front of the x and the t and justified this new vector.

Now we proceed to turn our new vector into an invariant, confident that if it worked for our co-ordinate vector, it should work for this as well:

$$(\Delta s)^2 = (c\Delta t \ \ \Delta x)\eta \begin{pmatrix} c\Delta t \\ \Delta x \end{pmatrix} = (c\Delta t)^2 - (\Delta x)^2.$$

If you are so minded, you can check this directly from the Lorentz transformations, but I would start from their (3.9) form. The quantity Δs is known as the *space-time interval*.

3.6.1 Simultaneity

Our new Lorentz vector points us towards one of the 'tent pole' features of relativity. We have seen how transformations mix the components of a vector together: $c\Delta t'$ is built from $c\Delta t$ and Δx; equally $\Delta x'$ is a combination of Δx and $c\Delta t$. Up to now, however, we have not stressed the physical consequences of this mathematical fact. If we observe two events in our system that are *simultaneous* (happening at the same time) they will *not necessarily be simultaneous in another system*. Simultaneity in our system implies that $c\Delta t = 0$, but as $c\Delta t'$ is an admixture of $c\Delta t$ and Δx it will not be zero, unless $\Delta x = 0$ as well. The simultaneity of two spatially separated events is not universal. We skirted this in the previous chapter while discussing clock synchronisation.

Once we have accepted that a universal time, shared by all observers no matter what their state of motion, was an understandable fable, it follows that judgements about the simultaneity of events (or even in some situations the time order in which they happen) should be a matter observational perspective as well. However, this statement rather underplays the significance of a lack of universal simultaneity. It is hard to notice the fundamental way in which we rely on simultaneity until it is pointed out to us.

Consider judging the length of an object. In abstract terms, we discern the length by noting the x co-ordinates of each end. This is straightforward while the object is stationary in our system, but somewhat more of a challenge if the object is moving. Then we have to stipulate that we record the location of the two ends against our co-ordinate axis *at the same moment*. However, as these events are spatially separated, the judgement of the same moment cannot be agreed upon by all potential observers. Consequently, other observers will think that we have not abided by the stipulation that the two x co-ordinates be recorded at the same time and hence will disagree with us about the length of the object. In fact, as we will see in Section 4.4, moving objects are adjudged to be shorter than their stationary doubles[6]:

$$\Delta x' = \frac{\Delta x}{\gamma}.$$

This second 'tent pole' feature of relativity is known as *length contraction*.

The manner in which the Lorentz transformations fold space and time together elevates the notion of space-time as a more helpful concept with a deeper reality than space and time considered separately. Such issues will be explored more thoroughly in the next chapter.

3.6.2 Proper Time

Think back to a co-ordinate system where we are sitting at rest and all the axis are fixed with us. Now imagine that meandering through that system is an object moving along our x axis (for simplicity) at some velocity v. The object passes point x_1 at time t_1 and x_2 at time t_2. As we are sitting watching all this happen, we muse about what it must be like to be sitting 'on' that object. After all, a perfectly sensible choice of co-ordinate system would have an origin nailed to the object and moving along at speed v with it. Experts call this the *co-moving system*. Indeed, any system moving along at the same speed as our object is classed as co-moving, even if the origin is not on the object.

From the point of view of someone either sitting on the object, or comfortably embedded somewhere in a co-moving system, our object is not moving at all. From their perspective, it is us and our system that is moving. Hence, in a co-moving system $\Delta x' = 0$.

If we, rather imaginatively, picture the two systems having their x-axes drawn out, perhaps on gigantic pieces of paper, then our x-axis will be sliding past a co-moving observer and they can note when various points on *our axis pass their position*, which will *not have changed in their system*.

In case you do not quite buy that then consider:

$$\Delta x' = \gamma\left(\Delta x - \beta c\Delta t\right).$$

In our system, the object has moved distance Δx in time Δt which is a velocity $v = \Delta x/\Delta t$. We are then boosting into a system that is also moving at velocity v, so we have:

$$\Delta x' = \gamma\left(\Delta x - \frac{\Delta x}{c\Delta t}c\Delta t\right) = 0, \text{ remembering that } \beta = v/c.$$

Although the co-moving observer is content to feel stationary in their system, they still have a clock that is advancing and recording the time $\Delta t'$ it takes for our system to slide the points x_1 and x_2 past them.

Back to our invariant quantity. In the co-moving system, we have $\Delta x' = 0$, but $\Delta t' \neq 0$, so:

$$(\Delta s)^2 = \begin{pmatrix} c\Delta t' & \Delta x' \end{pmatrix} \eta \begin{pmatrix} c\Delta t' \\ \Delta x' \end{pmatrix} = (c\Delta t')^2 - (\Delta x')^2 = (c\Delta t')^2.$$

The time elapsed in the co-moving system is called the *proper time* and to distinguish it from any other time, t or t', we will give it a unique symbol, τ. With that in place we have $\Delta s^2 = (c\Delta \tau)^2$ or $\Delta s = c\Delta \tau$.

So, in this situation, the space-time interval takes the value of c multiplied by the proper time: the time interval as measured by a co-moving observer.

Let us spell it out to be clear.

An object is moving through space. In different co-ordinate systems, associated with different observers, the object will cover certain distances in certain times. If the observers get around to comparing their measurements, they will find a wide range of different distances measured and elapsed times recorded. The only agreement will be between observers who happen to be at rest relative to each other. The collection of observers who are moving along at the same speed in the same direction as the object (the co-moving observers) will agree with each other, but will not see any distance covered by *them*, only an elapsed time, $\Delta \tau$. However, *all* of the observers will agree on the value of the space-time interval: in each case, they will get $c\Delta \tau$.

Once again, we are prompted to think that the space-time interval has a deeper reality than the separate distances and times measured by different observers.

I must in conscience though warn you against jumping to the conclusion that the space-time interval is *always* related to a proper time. We can potentially pick any two points in space and time and calculate the interval between them, but that will only be a proper time if it is physically possible for an object to move between the two spatial positions in the time given (see Section 4.3.3).

3.7 The Search for Lorentz Vectors

What makes a Lorentz vector distinctive from any other pair of numbers floating around in a bracket is its formation from a pair of physical quantities that transform via the Lorentz matrix:

$$\begin{pmatrix} a' \\ b' \end{pmatrix} = L \begin{pmatrix} a \\ b \end{pmatrix}.$$

It is clearly important to seek out other Lorentz vectors that may exist, not only to ensure that our theory is fully developed, but also as they will build other invariants, as any construction of the form

$\begin{pmatrix} a & b \end{pmatrix} \eta \begin{pmatrix} a \\ b \end{pmatrix}$ will be a Lorentz invariant.

As a further indication of why this is potentially so important, let's consider what happens with a quantity that does *not* transform in a Lorentzian manner (is not part of a Lorentz vector).

For this example, we return to our combination of a stationary co-ordinate system and another moving along the x-axis at some speed v. Also moving in an x direction is another other object travelling with velocity u, which is not the same as the relative velocity of the second system (Figure 3.3). We ask ourselves the question: what is the velocity of the object observed in the moving system?

In our system, $u = \Delta x / \Delta t$, and we need to boost into the other system in order to obtain the velocity measured there. As the top and bottom lines of $u = \Delta x / \Delta t$ transform separately via Lorentz transformations, u itself does not. If we were calculating some quantity that was a multiple of Δx or of $c\Delta t$ then

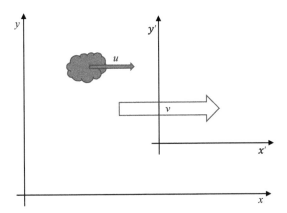

FIGURE 3.3 An object moving through our co-ordinate system with velocity u is also observed from a second system moving at speed v relative to us. In this diagram, I have displaced the x'-axis for clarity; normally, it is taken to be sliding along on top of the x-axis.

that quantity would transform the same way as Δx or $c\Delta t$. But velocity is not like that. The boost is a bit more awkward:

$$u' = \frac{\Delta x'}{\Delta t'} = c\frac{\Delta x'}{c\Delta t'} = c\frac{\gamma_v(\Delta x - \beta c\Delta t)}{\gamma_v(c\Delta t - \beta \Delta x)} = c\frac{\Delta t(u - \beta c)}{\Delta t(c - \beta u)} = \frac{c(u - v)}{c(1 - uv/c^2)}$$

$$= \frac{(u - v)}{(1 - uv/c^2)}.$$

In this calculation, I have introduced a useful convention by writing $\gamma_v = 1/\sqrt{1 - v^2/c^2}$ to distinguish from $\gamma_u = 1/\sqrt{1 - u^2/c^2}$, which I will also need later.

Here we have it then: the object is moving at velocity u in our system, but appears to be moving at velocity $u' = (u - v)/(1 - uv/c^2)$ in the other system.

The top line of this relationship matches with the Galilean velocity transformation we established in Section 2.2.5. Indeed, if u and v are both very much smaller than c, the bottom line becomes irrelevant as $1 - uv/c^2 \approx 1$ and $u' \approx (u - v)$. Of course, this is why we have been fooled into thinking that the 'obvious' relationship $u' = (u - v)$ was right, as we could never detect the difference between that and the true physics, given the context in which we were working. Modern physics deals with much lighter objects, hence they can be accelerated to much greater velocities and the true transformation becomes clear.

This technique for boosting velocities from one system to another also functions as a *velocity addition formula*. Consider a car moving along at 20 m/s and I inside the car (rather recklessly) throw a tennis ball at 5 m/s out of the window in the direction of travel. In a Newtonian/Galilean world, an observer primed to catch the ball would feel it hit the hand at 25 m/s. The ball was already travelling at 20 m/s while I was carrying it; I simply added an extra 5 m/s when it got thrown.

The observer catching the ball sees the car as a co-ordinate system approaching at 20 m/s (v). The ball leaves my hand and becomes an independent object moving at 5 m/s (u') in the car's system. The observer catches the ball moving at velocity u in his system. As we have:

$$u' = \frac{(u - v)}{(1 - uv/c^2)}$$

if follows that:

$$u'\left(1-uv/c^2\right)=(u-v) \quad \text{so} \quad u'-u'uv/c^2=u-v$$

and hence:

$$u'+v=u+u'uv/c^2=u\left(1+u'v/c^2\right)$$

finally giving:

$$u=\frac{(u'+v)}{\left(1+u'v/c^2\right)}=\frac{(5+20)}{\left(1+(5\times20)/9\times10^{16}\right)}$$

$$=24.99999999999997222222222222225308641975308638545953360768 \text{ m/s}$$

which,[7] I think we will all agree, is an awful lot like 25 m/s.

On the other hand, if my 'car' is a particle moving at $c/5=0.2c$ and it decays into other particles, one of which happens to come out moving in the same direction at $c/2=0.5c$ relative to the original particle, then the velocity at which this decay product hits a particle detector right in line with its path is:

$$u=\frac{(u'+v)}{\left(1+u'v/c^2\right)}=\frac{(c/2+c/5)}{\left(1+(c^2/10)/c^2\right)}=\frac{(5c+2c)/10}{1.1}=\frac{7c}{11}=0.64c$$

not $0.7c$ which is what we would have expected in our 'normal' world.

The extreme case comes when we replace the tennis ball, or the particle, with a light beam. Now my aim in the car is to shine a ray of light from a laser onto a photocell. From the co-ordinate system appropriate to the photocell, the ray of light impacts the cell travelling at:

$$u=\frac{(u'+v)}{\left(1+u'v/c^2\right)}=\frac{(c+v)}{\left(1+cv/c^2\right)}=\frac{(c+v)}{(1+v/c)}=\frac{(c+v)}{(c+v)/c}=c\frac{(c+v)}{(c+v)}=c.$$

No matter what speed the car is moving at, if it emits a ray of light moving at c relative to the car, that ray will also be moving at c relative to the photocell. More broadly, any object moving through space at velocity c will be observed from *any co-ordinate system* to be moving at velocity c:

$$u'=\frac{(u-v)}{\left(1-uv/c^2\right)}=\frac{(c-v)}{\left(1-cv/c^2\right)}=\frac{(c-v)}{(c-v)/c}=c\frac{(c-v)}{(c-v)}=c.$$

Space and time adjust in every co-ordinate system to maintain the constant value of the velocity of light for all observers: just as we said in Chapter 2. Of course, we set up the Lorentz transformations on the assumption that the speed of light was the same for all observers, but it is re-assuring to see it work out.

3.7.1 Lorentz Velocity

Important and interesting though all this is, we are no closer to our aim of finding another vector that transforms 'properly', where by properly we mean transforming the same way that X and ΔX do under the Lorentz transformations. The problem with traditional velocity is that it is made by dividing one component of the Lorentz vector ΔX by another: $\Delta x/\Delta t$. As Lorentz transforming a vector mixes its components together, it is not surprising that an object built from $\Delta x/\Delta t$, which is already a mix of two elements, is not going to work.

Earlier I mentioned that any multiple of a Lorentz vector is also going to be a Lorentz vector, which is not especially interesting if we are thinking of multiples as simply numbers like 5, 7, 589, etc. However, if we look for something 'velocity-like' by dividing Δx by a quantity that is 'time-like' yet *is invariant under the transformations*, then that will be a multiple of a Lorentz vector and so will transform 'properly'. We have just the thing already: the *proper time of a moving object*, which is related to the *invariant space-time interval*.

The plan is to construct a Lorentz vector, U, from:

$$U = \frac{1}{\Delta\tau}\begin{pmatrix} c\Delta t \\ \Delta x \end{pmatrix} = \begin{pmatrix} c\Delta t/\Delta\tau \\ \Delta x/\Delta\tau \end{pmatrix}.$$

Using our Lorentz transformations, we can relate the propter time of an object moving at speed u to the elapsed time in the system observing the speed u:

$$c\Delta\tau = \gamma_u(c\Delta t - \beta\Delta x) = \gamma_u c\Delta t\left(1 - \frac{u}{c}\frac{\Delta x}{c\Delta t}\right) = \gamma_u c\Delta t\left(1 - \frac{u^2}{c^2}\right) = \frac{\gamma_u c\Delta t}{\gamma_u^2} = \frac{c\Delta t}{\gamma_u}$$

$$\Delta\tau = \frac{\Delta t}{\gamma_u}.$$

(I have used u as our velocity here as I will be using v as well and I want to avoid confusion.)

With this relationship, I can complete the build of our prototype vector U:

$$U = \frac{1}{\Delta\tau}\begin{pmatrix} c\Delta t \\ \Delta x \end{pmatrix} = \frac{\gamma_u}{\Delta t}\begin{pmatrix} c\Delta t \\ \Delta x \end{pmatrix} = \gamma_u\begin{pmatrix} c\Delta t/\Delta t \\ \Delta x/\Delta t \end{pmatrix} = \gamma_u\begin{pmatrix} c \\ u \end{pmatrix}$$

or, if you prefer:

$$U = \begin{pmatrix} \gamma_u c \\ \gamma_u u \end{pmatrix}.$$

This has to transform via the Lorentz matrix just the way our other vectors do. After all, we built it from a Lorentz vector by dividing by $\Delta\tau$, which is invariant. However, I would not blame you for wishing to see this proven directly. Unfortunately, that is not a straightforward thing to do as we need to see how γ_u transforms when boosted by velocity v. The end result should be:

$$\begin{pmatrix} \gamma_{u'} c \\ \gamma_{u'} u' \end{pmatrix} = L\begin{pmatrix} \gamma_u c \\ \gamma_u u \end{pmatrix} = \begin{pmatrix} \gamma_v & -\gamma_v\beta_v \\ -\gamma_v\beta_v & \gamma_v \end{pmatrix}\begin{pmatrix} \gamma_u c \\ \gamma_u u \end{pmatrix}.$$

If you prefer to see it 'longhand' then the transformations are:

$$\gamma_{u'} c = \gamma_v(\gamma_u c - \beta_v u)$$

$$\gamma_{u'} u' = \gamma_v(\gamma_u u - \beta_v c).$$

Now we can appreciate the value of the convention I brought in earlier: it helps to distinguish between the γ factor that is responsible for the boost (γ_v) and the factor that is *part of the vector we are boosting* (γ_u). Note also that $\gamma_{u'} = 1/\sqrt{1 - (u')^2/c^2}$.

As this demonstration is rather long-winded, I have put it in the on-line Appendix, where interested readers can consult it at leisure.

With our new vector, we gain access to a new invariant:

$$\begin{pmatrix} \gamma_u c & \gamma_u u \end{pmatrix} \eta \begin{pmatrix} \gamma_u c \\ \gamma_u u \end{pmatrix} = \left(\gamma_u c\right)^2 - \left(\gamma_u u\right)^2$$

$$= \frac{c^2}{\left(1-u^2/c^2\right)} - \frac{u^2}{\left(1-u^2/c^2\right)} = \frac{c^2-u^2}{\left(1-u^2/c^2\right)} = \frac{c^2\left(1-u^2/c^2\right)}{\left(1-u^2/c^2\right)} = c^2$$

thus confirming, via another approach, that the speed of light is invariant.

3.8 4-Vectors

In this chapter, we have quietly ignored the y- and z-axes, as they do not change when boosting into a system moving along the x-axis. While this is a very helpful simplification, we must acknowledge that our theory has to include these axes. In order to accommodate that, as a start, we need to expand our Lorentz matrix from its two-by-two form into a four-by-four version:

$$L_4 = \begin{pmatrix} \gamma & -\beta\gamma & 0 & 0 \\ -\beta\gamma & \gamma & 0 & 0 \\ 0 & 0 & 1 & 0 \\ 0 & 0 & 0 & 1 \end{pmatrix}.$$

With our vectors $X = \begin{pmatrix} ct \\ x \\ y \\ z \end{pmatrix}$, $\Delta X = \begin{pmatrix} c\Delta t \\ \Delta x \\ \Delta y \\ \Delta z \end{pmatrix}$, $U = \dfrac{\Delta X}{\Delta \tau} = \begin{pmatrix} \gamma_u c \\ \gamma_u u_x \\ \gamma_u u_y \\ \gamma_u u_z \end{pmatrix}$ and the matrix that we use to cre-

ate invariants, $\eta = \begin{pmatrix} 1 & 0 & 0 & 0 \\ 0 & -1 & 0 & 0 \\ 0 & 0 & -1 & 0 \\ 0 & 0 & 0 & -1 \end{pmatrix}$ our first invariant in its full form becomes:

$$\left(\Delta s\right)^2 = \left(c\Delta t\right)^2 - \left(\Delta x\right)^2 - \left(\Delta y\right)^2 - \left(\Delta z\right)^2.$$

These vectors and others like them are known as *4-vectors* (I have been using the term Lorentz vector up to now). As we proceed further, we will generally use the two-by-two versions when there is no specific need for the other elements, but I shall continue to refer to them 4-vectors, even if I am only showing two components.[8]

It is conventional that the top component in a column (first component in a row) 4-vector is called the *temporal component* (even if it is not time) and the others are *spatial components*. Thus, the temporal component of the 4-velocity is $\gamma_u c$ and the spatial components are $\gamma_u u_x$, etc.

3.9 Transforming Maxwell's Equations

We are now in a position to address the transformation of Maxwell's equations:

$$\frac{\Delta E_x}{\Delta x} + \frac{\Delta E_y}{\Delta y} + \frac{\Delta E_z}{\Delta z} = \frac{\rho}{\varepsilon_0}$$

$$\frac{\Delta B_x}{\Delta x} + \frac{\Delta B_y}{\Delta y} + \frac{\Delta B_z}{\Delta z} = 0$$

$$c\left(\frac{\Delta B_x}{c\Delta t}\right) = -\left(\frac{\Delta E_z}{\Delta y} - \frac{\Delta E_y}{\Delta z}\right)$$

$$\frac{1}{c}\left(\frac{\Delta E_x}{c\Delta t}\right) + \mu_0 J_x = \frac{\Delta B_z}{\Delta y} - \frac{\Delta B_y}{\Delta z}.$$

If we were to do this properly, I would need to develop the important mathematical toolkit of *partial differential calculus*. This would take us outside of the mathematical constraints that I have set and away from focusing on the important physical ideas. So, instead I am going to fudge the issue, although the results I obtain will be completely correct.

Firstly, as per Einstein, we are going to restate Maxwell's equations for empty space, with no charges or currents in the vicinity. This is so we do not have to deal with their transformations; we want to focus on the fields. In other words, we set $\rho = 0$ and $J = 0$ making the equations:

$$\frac{\Delta E_x}{\Delta x} + \frac{\Delta E_y}{\Delta y} + \frac{\Delta E_z}{\Delta z} = 0$$

$$\frac{\Delta B_x}{\Delta x} + \frac{\Delta B_y}{\Delta y} + \frac{\Delta B_z}{\Delta z} = 0$$

$$c\left(\frac{\Delta B_x}{c\Delta t}\right) = -\left(\frac{\Delta E_z}{\Delta y} - \frac{\Delta E_y}{\Delta z}\right)$$

$$\frac{1}{c}\left(\frac{\Delta E_x}{c\Delta t}\right) = \frac{\Delta B_z}{\Delta y} - \frac{\Delta B_y}{\Delta z}.$$

Next, we consider how a mathematical quantity changes when it depends on multiple variables that are themselves changing.

Say we have some quantity f which is obtained from a combination of x and y, which we generally write as $f(x,y)$. This could be something as simple as $f(x,y) = 3x^2 + 2y$, or it could be rather more complicated than that. If we allow x and y to change by small amounts: $x \Rightarrow x + \Delta x$; $y \Rightarrow y + \Delta y$ then the change in f, Δf, is approximately:

$$\Delta f \approx \left(\frac{\Delta f}{\Delta x}\right)_y \Delta x + \left(\frac{\Delta f}{\Delta y}\right)_x \Delta y,$$

where the symbol $(\Delta f/\Delta x)_y$ refers to how much f changes with a change in x, while keeping y constant and $(\Delta f/\Delta y)_x$ tells us about how much f changes with a change in y, while keeping x constant. Both of these are best evaluated via the slope of a line.

Let's take a specific example: $f(x,y) = 3x^2 + 2y$ with $x = 1 \Rightarrow 1.001$ and $y = 3 \Rightarrow 3.002$. First, we calculate the exact change in f to act as a reference:

$$\Delta f = \left\{\left(3 \times 1.001^2\right) + \left(2 \times 3.002\right)\right\} - \left\{3 \times 1^2 + 2 \times 3\right\} = 9.010003 - 9 = 0.010003.$$

To build our approximation, we start by plotting the graph of $f(x,y) = 3x^2 + 2y$, setting $y = 3$. Then, as we are evaluating the change when $x = 1 \Rightarrow 1.001$, we need the gradient (slope) of that curve at the point $x = 1$: that gradient is 6 (Figure 3.4).

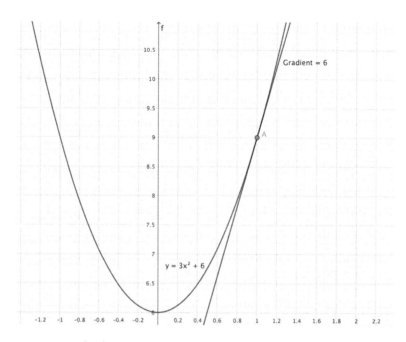

FIGURE 3.4 The graph of $f(x,y) = 3x^2 + 2y$ plotted for a constant y-value of 3. The gradient of the curve has been calculated for $x = 1$. This gradient is equal to that of a line touching the curve at the point $x = 1$.

Next, we plot the graph of $f(x,y) = 3x^2 + 2y$ using the fixed value $x = 1$, which is Figure 3.5. This is a straight line with gradient 2.

Now we can assemble our approximation out of the pieces:

$$\Delta f \approx \left(\frac{\Delta f}{\Delta x} \right)_y \Delta x + \left(\frac{\Delta f}{\Delta y} \right)_x \Delta y = (6) \times 0.001 + (2) \times 0.002 = 0.01.$$

Remember that the exact value we obtained was 0.010003, so this is not a bad approximation.

Clearly, this is a lot of work for an approximation, especially when the exact value is eas(y)(ier) to work out. Well, this simple example was chosen just to illustrate how the approximation works. What we are trying to do with Maxwell's equations needs this approximation, so now we have to go for the full monte…

Imagine we have a quantity that depends on more than just two variables: $f(x, y, z, ct)$. The change in this quantity is as follows:

$$\Delta f = \left(\frac{\Delta f}{\Delta x} \right) \Delta x + \left(\frac{\Delta f}{\Delta y} \right) \Delta y + \left(\frac{\Delta f}{\Delta z} \right) \Delta z + \left(\frac{\Delta f}{c\Delta t} \right) c\Delta t.$$

On an apparent whim, we divide throughout by $c\Delta t'$ to give:

$$\frac{\Delta f}{c\Delta t'} = \left(\frac{\Delta f}{\Delta x} \right) \frac{\Delta x}{c\Delta t'} + \left(\frac{\Delta f}{\Delta y} \right) \frac{\Delta y}{c\Delta t'} + \left(\frac{\Delta f}{\Delta z} \right) \frac{\Delta z}{c\Delta t'} + \left(\frac{\Delta f}{c\Delta t} \right) \frac{c\Delta t}{c\Delta t'}$$

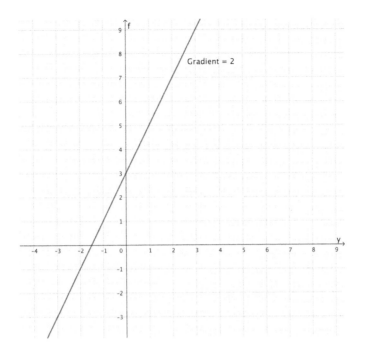

FIGURE 3.5 The graph of $f(x, y) = 3x^2 + 2y$ for a fixed value of $x = 1$. This line has gradient 2.

Now I carry out my promised fudge by re-interpreting terms such as $\Delta x/c\Delta t'$ as 'the way in which Δx changes as we change $c\Delta t''$ i.e. as the gradient of the curve relating Δx and $c\Delta t'$.

The inverse Lorentz transformations give us:

$$c\Delta t = \gamma\left(c\Delta t' + \beta\Delta x'\right)$$

$$\Delta x = \gamma\left(\Delta x' + \beta c\Delta t'\right)$$

$$\Delta y = \Delta y'$$

$$\Delta z = \Delta z'$$

so that the rate at which Δx changes with $c\Delta t'$ keeping $\Delta x'$ constant is:

$$\left(\frac{\Delta x}{c\Delta t'}\right)_{\Delta x'} = \gamma\beta.$$

The transformations also tell us that $\Delta y/c\Delta t' = \Delta z/c\Delta t' = 0$ and $c\Delta t/c\Delta t' = \gamma$.

Substituting into our formula produces:

$$\frac{\Delta f}{c\Delta t'} = \frac{\Delta f}{\Delta x}\gamma\beta + \frac{\Delta f}{c\Delta t}\gamma = \gamma\left(\beta\frac{\Delta f}{\Delta x} + \frac{\Delta f}{c\Delta t}\right)$$

$$\frac{1}{\gamma}\frac{\Delta f}{c\Delta t'} = \beta\frac{\Delta f}{\Delta x} + \frac{\Delta f}{c\Delta t}$$

and re-arranging a little more we obtain:

$$\frac{\Delta f}{c\Delta t} = \frac{1}{\gamma}\frac{\Delta f}{c\Delta t'} - \beta\frac{\Delta f}{\Delta x}.$$ (3.8)

Now we apply this to Maxwell's equations. Taking one of the equations involved with electromagnetic induction:

$$\frac{1}{c}\left(\frac{\Delta E_x}{c\Delta t}\right) = \frac{\Delta B_z}{\Delta y} - \frac{\Delta B_y}{\Delta z},$$

our f is, in this case, E_x, so that the term $\Delta E_x/c\Delta t$ can be converted using Equation 3.8:

$$\frac{1}{c}\left(\frac{\Delta E_x}{c\Delta t}\right) = \frac{1}{c}\left(\frac{1}{\gamma}\frac{\Delta E_x}{c\Delta t'} - \beta\frac{\Delta E_x}{\Delta x}\right),$$

which we can put back into the equation to give:

$$\frac{1}{c}\left(\frac{1}{\gamma}\frac{\Delta E_x}{c\Delta t'} - \beta\frac{\Delta E_x}{\Delta x}\right) = \frac{\Delta B_z}{\Delta y} - \frac{\Delta B_y}{\Delta z}.$$

Now another of Maxwell's equations has:

$$\frac{\Delta E_x}{\Delta x} + \frac{\Delta E_y}{\Delta y} + \frac{\Delta E_z}{\Delta z} = 0, \quad \text{so that} \quad \frac{\Delta E_x}{\Delta x} = -\left(\frac{\Delta E_y}{\Delta y} + \frac{\Delta E_z}{\Delta z}\right),$$

which we can also put back:

$$\frac{1}{c}\left(\frac{1}{\gamma}\frac{\Delta E_x}{c\Delta t'} + \beta\left(\frac{\Delta E_y}{\Delta y} + \frac{\Delta E_z}{\Delta z}\right)\right) = \frac{\Delta B_z}{\Delta y} - \frac{\Delta B_y}{\Delta z}.$$

Gathering terms, we obtain:

$$\frac{1}{c}\left(\frac{1}{\gamma}\frac{\Delta E_x}{c\Delta t'}\right) = \frac{\Delta B_z}{\Delta y} - \frac{\Delta B_y}{\Delta z} - \frac{\beta}{c}\frac{\Delta E_y}{\Delta y} - \frac{\beta}{c}\frac{\Delta E_z}{\Delta z}$$

or:

$$\frac{1}{c}\left(\frac{\Delta E_x}{c\Delta t'}\right) = \gamma\frac{\Delta B_z}{\Delta y} - \frac{\beta\gamma}{c}\frac{\Delta E_y}{\Delta y} - \gamma\frac{\Delta B_y}{\Delta z} - \frac{\beta\gamma}{c}\frac{\Delta E_z}{\Delta z} = \frac{\gamma\Delta\left(B_z - \dfrac{\beta}{c}E_y\right)}{\Delta y} - \frac{\gamma\Delta\left(B_y + \dfrac{\beta}{c}E_z\right)}{\Delta z}.$$

As $\Delta y' = \Delta y$ and $\Delta z' = \Delta z$, we can produce an interesting looking:

$$\frac{1}{c}\left(\frac{\Delta E_x}{c\Delta t'}\right) = \frac{\gamma\Delta\left(B_z - \dfrac{\beta}{c}E_y\right)}{\Delta y'} - \frac{\gamma\Delta\left(B_y + \dfrac{\beta}{c}E_z\right)}{\Delta z'},$$

which suggests that the following set of transformations:

$$E_x' = E_x$$

TABLE 3.1 The Transformation Properties of Electrical and Magnetic Field Components

Electrical Fields	Magnetic Fields
$E'_x = E_x$	$cB'_x = cB_x$
$E'_y = \gamma\left(E_y - \beta c B_z\right)$	$cB'_y = \gamma\left(cB_y + \beta E_z\right)$
$E'_z = \gamma\left(E_z + \beta c B_y\right)$	$cB'_z = \gamma\left(cB_z - \beta E_y\right)$

$$B'_z = \gamma\left(B_z - \frac{\beta}{c}E_y\right)$$

$$B'_y = \gamma\left(B_y + \frac{\beta}{c}E_z\right)$$

will transform Maxwell's equation into:

$$\frac{1}{c}\left(\frac{\Delta E'_x}{c\Delta t'}\right) = \frac{\Delta B'_z}{\Delta y'} - \frac{\Delta B'_y}{\Delta z'}.$$

This new equation has *exactly the same mathematical form as the original one*, except with dashed quantities replacing every variable: *precisely* what we mean by the laws of physics being the same for every observer in relative inertial motion.

I will not subject you to the similar arguments that reveal the transformations for the other field components. It is algebraic manipulation, rather than enlightening physics. The collected set of transformations for the electric and magnetic fields are shown in Table 3.1.

Note that I have 'symmetrised' them somewhat by multiplying the magnetic fields by c. This is similar to the trick I pulled at the start of the chapter to make the Lorentz transformations look slightly nicer.

Also note how an electric field in one system is a mix of electric and magnetic fields in another and vice versa. If there is no electric field in one system ($E_y = 0$ for example) then there can still be an electric field in another ($E'_y = -\gamma\beta c B_z$). Furthermore, if $v \ll c$, $\gamma \approx 1$ and $E'_y \approx -vB_z$ which is what we obtained using the Lorentz force in Section 2.2.3. This is *exactly* the result that Einstein hoped to get, as we discussed in Chapter 2.

3.10 Tensors

The transformations of the electric and magnetic fields pose something of a puzzle as they cannot be collected together into a 4-vector. The electric field has three components, as does the magnetic field. There are too many components around!

What we need is a mathematical object that transforms like a 4-vector, but which can handle more than four components. Fortunately, we can pull such a thing off the shelf in a box marked *tensors*.

At first glance, tensors look exactly like matrices: they are also an array of symbols. A matrix will represent a tensor, if it has the right transformation properties. However, before we can reveal what they are, we need to dig a little more into the mechanism of transformation.

3.10.1 Covariant and Contravariant

Going back to the approximation from the previous section:

$$\Delta f = \left(\frac{\Delta f}{\Delta x}\right)\Delta x + \left(\frac{\Delta f}{\Delta y}\right)\Delta y + \left(\frac{\Delta f}{\Delta z}\right)\Delta z + \left(\frac{\Delta f}{c\Delta t}\right)c\Delta t,$$

at the time, on a whim, I decided to divide through by $c\Delta t'$ to give:

$$\frac{\Delta f}{c\Delta t'} = \left(\frac{\Delta f}{\Delta x}\right)\frac{\Delta x}{c\Delta t'} + \left(\frac{\Delta f}{\Delta y}\right)\frac{\Delta y}{c\Delta t'} + \left(\frac{\Delta f}{\Delta z}\right)\frac{\Delta z}{c\Delta t'} + \left(\frac{\Delta f}{c\Delta t}\right)\frac{c\Delta t}{c\Delta t'},$$

and I then used the inverse Lorentz transformations:

$$c\Delta t = \gamma\left(c\Delta t' + \beta\Delta x'\right)$$

$$\Delta x = \gamma\left(\Delta x' + \beta c\Delta t'\right)$$

$$\Delta y = \Delta y'$$

$$\Delta z = \Delta z'$$

to extract $\Delta x/c\Delta t' = \beta\gamma$ and $c\Delta t/c\Delta t' = \gamma$ keeping $\Delta x'$ constant. As the other terms are zero, this means that:

$$\frac{\Delta f}{c\Delta t'} = \left(\frac{\Delta f}{\Delta x}\right)\beta\gamma + \left(\frac{\Delta f}{c\Delta t}\right)\gamma = \gamma\left(\frac{\Delta f}{c\Delta t} + \beta\frac{\Delta f}{\Delta x}\right)$$

which is a little suggestive…it looks like part of a 4-vector.

Now, on a whim, I have a different whim and divide my approximation through by $\Delta x'$ to give:

$$\frac{\Delta f}{\Delta x'} = \left(\frac{\Delta f}{\Delta x}\right)\frac{\Delta x}{\Delta x'} + \left(\frac{\Delta f}{\Delta y}\right)\frac{\Delta y}{\Delta x'} + \left(\frac{\Delta f}{\Delta z}\right)\frac{\Delta z}{\Delta x'} + \left(\frac{\Delta f}{c\Delta t}\right)\frac{c\Delta t}{\Delta x'}.$$

Looking back at the inverse Lorentz transformation we find, keeping $c\Delta t'$ constant:

$$\frac{\Delta x}{\Delta x'} = \gamma \quad \frac{c\Delta t}{\Delta x'} = \beta\gamma.$$

Using these, we produce:

$$\frac{\Delta f}{\Delta x'} = \left(\frac{\Delta f}{\Delta x}\right)\gamma + \left(\frac{\Delta f}{c\Delta t}\right)\beta\gamma = \gamma\left(\frac{\Delta f}{\Delta x} + \beta\frac{\Delta f}{c\Delta t}\right).$$

Putting these two results alongside each other:

$$\frac{\Delta f}{c\Delta t'} = \gamma\left(\frac{\Delta f}{c\Delta t} + \beta\frac{\Delta f}{\Delta x}\right) \quad \frac{\Delta f}{\Delta x'} = \gamma\left(\frac{\Delta f}{\Delta x} + \beta\frac{\Delta f}{c\Delta t}\right)$$

suggests that we can write:

$$\begin{pmatrix} \Delta f/c\Delta t' \\ \Delta f/\Delta x' \end{pmatrix} = \begin{pmatrix} \gamma & \beta\gamma \\ \beta\gamma & \gamma \end{pmatrix}\begin{pmatrix} \Delta f/c\Delta t \\ \Delta f/\Delta x \end{pmatrix} = L^{-1}\begin{pmatrix} \Delta f/c\Delta t \\ \Delta f/\Delta x \end{pmatrix}$$

or in full:

$$
\begin{pmatrix} \Delta f/c\Delta t' \\ \Delta f/\Delta x' \\ \Delta f/\Delta y' \\ \Delta f/\Delta z' \end{pmatrix} = \begin{pmatrix} \gamma & \beta\gamma & 0 & 0 \\ \beta\gamma & \gamma & 0 & 0 \\ 0 & 0 & 1 & 0 \\ 0 & 0 & 0 & 1 \end{pmatrix} \begin{pmatrix} \Delta f/c\Delta t \\ \Delta f/\Delta x \\ \Delta f/\Delta y \\ \Delta f/\Delta z \end{pmatrix}.
$$

It very much looks as if we have a 4-vector on our hands, but weirdly one that transforms *forwards* by the *inverse* Lorentz transformation.

At this point it would be sensible to ask why f does not transform as well. In part that question arises as I am fudging the maths again here, but there are some quantities (mass would be an example, gravitational potential energy would be another) which do not transform[9]: they are *scalar*. This relationship works out, if f is a scalar.

We have already come across another 4-vector that transforms weirdly, we just didn't point it out at the time. In order to construct invariant quantities, we built a matrix η and formed the structure $\begin{pmatrix} c\Delta t & \Delta x \end{pmatrix}\eta\begin{pmatrix} c\Delta t \\ \Delta x \end{pmatrix}$ but we did not emphasise that:

$$
\eta\begin{pmatrix} c\Delta t \\ \Delta x \end{pmatrix} = \begin{pmatrix} 1 & 0 \\ 0 & -1 \end{pmatrix}\begin{pmatrix} c\Delta t \\ \Delta x \end{pmatrix} = \begin{pmatrix} c\Delta t \\ -\Delta x \end{pmatrix}
$$

is also a vector. If we try and transform this in the normal way, it goes wrong as:

$$
L\begin{pmatrix} c\Delta t \\ -\Delta x \end{pmatrix} = \begin{pmatrix} \gamma & -\beta\gamma \\ -\beta\gamma & \gamma \end{pmatrix}\begin{pmatrix} c\Delta t \\ -\Delta x \end{pmatrix} = \begin{pmatrix} \gamma(c\Delta t + \beta\Delta x) \\ -\gamma(\Delta x + \beta c\Delta t) \end{pmatrix} \ne \begin{pmatrix} c\Delta t' \\ -\Delta x' \end{pmatrix}.
$$

However, it does transform by the inverse transformation:

$$
L^{-1}\begin{pmatrix} c\Delta t \\ -\Delta x \end{pmatrix} = \begin{pmatrix} \gamma & \beta\gamma \\ \beta\gamma & \gamma \end{pmatrix}\begin{pmatrix} c\Delta t \\ -\Delta x \end{pmatrix} = \begin{pmatrix} \gamma(c\Delta t - \beta\Delta x) \\ -\gamma(\Delta x - \beta c\Delta t) \end{pmatrix} = \begin{pmatrix} c\Delta t' \\ -\Delta x' \end{pmatrix}.
$$

Put in these terms, the emergence of an invariant from our structure become clearer. After all, the invariant is constructed from the product of a vector that transforms forwards and a vector that transforms backwards. Unsurprisingly, the net result is to leave the product unchanged.

To formalise this thinking, I'm going to denote our weird backwards transforming vector $\bar{X}$, so that:

$$
X = \begin{pmatrix} c\Delta t \\ \Delta x \end{pmatrix} \quad X' = LX
$$

$$
\bar{X} = \begin{pmatrix} c\Delta t \\ -\Delta x \end{pmatrix} \quad \bar{X}' = L^{-1}\bar{X}.
$$

Our invariant product is then:

$$
(X')^T \eta X' = (X')^T \bar{X}' = X^T L^T L^{-1}\bar{X} = X^T \bar{X}
$$

remembering that $L^T = L$. Equally, we can form the same invariant from $\bar{X}^T X$ as:

$$
\bar{X}^T X = X^T \eta^T X = X^T \eta X = X^T \bar{X}
$$

casual inspection showing us that $\eta^T = \eta$.

As it is not very professional to carry on calling $\bar{X}$, and other objects like it, *weird vectors*, let's adopt the standard terminology, which calls vectors:

- *contravariant* if they transform *forwards* using the *forward* transformation;
- *covariant* if transform *forwards* using the *inverse* transformation.

We have two examples of covariant vectors: $\bar{X} = \begin{pmatrix} c\Delta t \\ -\Delta x \end{pmatrix}$ and $\overline{\nabla f} = \begin{pmatrix} \Delta f/c\Delta t \\ \Delta f/\Delta x \\ \Delta f/\Delta y \\ \Delta f/\Delta z \end{pmatrix}$. Physically, they

encode the same information as contravariant vectors and every contravariant vector has a covariant buddy. We can always switch between them using the matrix, η. After all, we effectively constructed η to generate covariant vectors from contravariant ones, and a simple manipulation shows us that, given $\bar{X} = \eta X$:

$$\eta^{-1}\bar{X} = \eta^{-1}\eta X = X$$

where η^{-1} is the inverse of η. Additionally, as:

$$\eta^2 = \begin{pmatrix} 1 & 0 & 0 & 0 \\ 0 & -1 & 0 & 0 \\ 0 & 0 & -1 & 0 \\ 0 & 0 & 0 & -1 \end{pmatrix}\begin{pmatrix} 1 & 0 & 0 & 0 \\ 0 & -1 & 0 & 0 \\ 0 & 0 & -1 & 0 \\ 0 & 0 & 0 & -1 \end{pmatrix} = \begin{pmatrix} 1 & 0 & 0 & 0 \\ 0 & 1 & 0 & 0 \\ 0 & 0 & 1 & 0 \\ 0 & 0 & 0 & 1 \end{pmatrix} = I$$

it seems that $\eta^{-1} = \eta$, which we will have cause to use in Section 8.1.

USEFUL SUMMARY:

$$\text{Contravariant}: X = \begin{pmatrix} c\Delta t \\ \Delta x \end{pmatrix} \quad X = \eta^{-1}\bar{X} \quad X' = LX$$

$$\text{Covariant}: \bar{X} = \begin{pmatrix} c\Delta t \\ -\Delta x \end{pmatrix} \quad \bar{X} = \eta X \quad \bar{X}' = L^{-1}\bar{X}$$

3.10.2 Transforming Tensors

When we transform a column vector, the rules of matrix multiplication force us to pre-multiply by the transformation:

$$\begin{pmatrix} ct' \\ x' \end{pmatrix} = L\begin{pmatrix} ct \\ x \end{pmatrix} = \begin{pmatrix} \gamma & -\beta\gamma \\ -\beta\gamma & \gamma \end{pmatrix}\begin{pmatrix} ct \\ x \end{pmatrix};$$

whereas for a row vector, we have to post multiply by the transpose:

$$\begin{pmatrix} ct' & x' \end{pmatrix} = \begin{pmatrix} ct & x \end{pmatrix}L^T = \begin{pmatrix} ct & x \end{pmatrix}\begin{pmatrix} \gamma & -\beta\gamma \\ -\beta\gamma & \gamma \end{pmatrix}.$$

A tensor, in the form of a matrix, contains both columns *and* rows. In order to transform a tensor, we need to pre *and* post multiply. This gives us three possible options:

- a *contravariant tensor*, T—which we pre and post multiply by the *forward* Lorentz transformation: $T' = LTL$.
- a *covariant tensor*, $\overline{T}$—which we pre and post multiply by the *inverse* Lorentz transformation: $\overline{T}' = L^{-1}\overline{T}L^{-1}$.
- and a somewhat schizophrenic *mixed tensor*, M, with contravariant and covariant indices: $M' = LML^{-1}$.

As we dig further into general relativity, we will come across tensors that cannot be represented as a matrix as they have more than two indices. Indeed, one very significant tensor, the *Riemann curvature tensor*, will turn out to have four. We will reveal a notation for dealing with such tensors when the time comes. Suffice to say that every index requires a transformation.

3.10.3 The Electromagnetic Field Tensor

As seen from the magnet, there was certainly no electric field; whereas seen from the circuit there certainly was an electric field. Therefore, the existence of the electric field was a relative one, depending on the state of motion of the coordinate system used; and *only the electric and magnetic fields combined, aside from the state of motion of the observer or coordinate system, could be granted a kind of objective reality.*

A Einstein[10] (my emphasis)

We started out our tensorial quest in order to find a way of representing electric and magnetic fields in a single transformable structure. The answer is the *electromagnetic field tensor*:

$$
F = \begin{pmatrix}
0 & -E_x & -E_y & -E_z \\
E_x & 0 & -cB_z & cB_y \\
E_y & cB_z & 0 & -cB_x \\
E_z & -cB_y & cB_x & 0
\end{pmatrix}.
$$

As the $\Delta/\Delta t$, etc. parts of the Maxwell equations transform like $\overline{\nabla}f$, that is to say covariantly, we could guess that the electromagnetism tensor is contravariant. When we transform it according to the appropriate rule we get $F' = LFL$, or fully written out:

$$
F' = \begin{pmatrix}
\gamma & -\beta\gamma & 0 & 0 \\
-\beta\gamma & \gamma & 0 & 0 \\
0 & 0 & 1 & 0 \\
0 & 0 & 0 & 1
\end{pmatrix}
\begin{pmatrix}
0 & -E_x & -E_y & -E_z \\
E_x & 0 & -cB_z & cB_y \\
E_y & cB_z & 0 & -cB_x \\
E_z & -cB_y & cB_x & 0
\end{pmatrix}
\begin{pmatrix}
\gamma & -\beta\gamma & 0 & 0 \\
-\beta\gamma & \gamma & 0 & 0 \\
0 & 0 & 1 & 0 \\
0 & 0 & 0 & 1
\end{pmatrix}
$$

$$
= \begin{pmatrix}
0 & -E'_x & -E'_y & -E'_z \\
E'_x & 0 & -cB'_z & cB'_y \\
E'_y & cB'_z & 0 & -cB'_x \\
E'_z & -cB'_y & cB'_x & 0
\end{pmatrix}.
$$

If you really want to check this, I suggest you use an on-line symbolic matrix multiplier.

3.10.4 and Finally...

This has been a long chapter with some quite dense mathematics in it, but I can't resist the temptation to show one more thing...

A 4-vector version of electrical current can be constructed from our 4-velocity by multiplying by a charge density (the amount of charge in a unit volume). Unfortunately, charge density is not an invariant quantity, as volume is not. Length contraction means that even in a simple rectangular volume $\mathcal{V} = \Delta x \Delta y \Delta z$ the Δx part (assumed to be along the boost direction) will be contracted to $\Delta x/\gamma$ so that the volume is contracted by the same factor $\mathcal{V}' = \mathcal{V}/\gamma$. So, if we measure the charge density in a system that is co-moving with the current and call it ρ_0, then:

$$\rho_0 = Q/\mathcal{V}' = \gamma \, Q/\mathcal{V} = \gamma\rho.$$

We then write our new 4-current as:

$$\bar{J} = \frac{\rho U}{c} = \frac{\rho_0}{\gamma c}\begin{pmatrix} \gamma c \\ \gamma u_x \\ \gamma u_y \\ \gamma u_z \end{pmatrix} = \begin{pmatrix} \rho_0 \\ \rho_0 u_x/c \\ \rho_0 u_y/c \\ \rho_0 u_z/c \end{pmatrix}.$$

If we multiply our contravariant electromagnetism tensor by the covariant version of the 4-current we have (and in my head I can hear Ta Da! as I type this...):

$$F\bar{J} = \begin{pmatrix} 0 & -E_x & -E_y & -E_z \\ E_x & 0 & -cB_z & cB_y \\ E_y & cB_z & 0 & -cB_x \\ E_z & -cB_y & cB_x & 0 \end{pmatrix}\begin{pmatrix} \rho_0 \\ -\rho_0 u_x/c \\ -\rho_0 u_y/c \\ -\rho_0 u_z/c \end{pmatrix} = \rho_0 \begin{pmatrix} \frac{1}{c}\left(E_x u_x + E_y u_y + E_z u_z\right) \\ E_x + u_y B_z - u_z B_y \\ E_y + u_z B_x - u_x B_z \\ E_z + u_x B_y - u_y B_x \end{pmatrix}.$$

The temporal component is made of pieces like $\dfrac{\rho_0}{c} E_x u_x$. Now $\rho_0 E_x$ is the force exerted on the charge by the electric field's x component. Force times distance is work done and force times velocity is power or rate of work done. Only the electric field can do work as the magnetic field exerts a force at 90° to the direction of motion.

The spatial components make up the Lorentz force I showed you in Section 2.2.3.

Notes

1. Sommerfeld, A., 1949. To Albert Einstein's seventieth birthday. In *Albert Einstein, Philosopher-Scientist*, p. 105.
2. For those of you who quite like tedious algebra, or alternatively don't trust authors, the calculation is shown in the on-line Appendix.
3. The term 'boost' is commonly used as an abbreviation for transforming between inertial co-ordinate systems in relativity.
4. Of course, we all advance through time as we age, but that is hardly the same thing...

5. Strictly it is not the *units* that matter, they are human choices, but what physicist refer to as the *dimensionality*. In this context, the term does not refer to physical dimensions, but the fact that x is spatial and t temporal and would have to be measured with different types of unit. It's a way of keeping track of the sort of measurement and unit that is relevant to the quantity, without referring to a specific unit which is an arbitrary human choice or definition in our society.

6. This does not imply that we actually see them as shorter. What we might see when we observe an object moving at a good fraction of the speed of light is a complex issue as our vision relies on light reaching us from parts of the object; light which has to set out at different moments if it is to arrive at our eyes at the same time. Hence what we see is an amalgam of information from the object that set off at different moments—which makes a crucial difference when the object is travelling at such speeds.

7. Thank you, *Wolfram Alpha*…

8. Some habits are hard to break.

9. Invariants transform, but retain the same value. Scalars do not transform.

10. *Fundamental Ideas and Methods of the Theory of Relativity, Presented in Their Development*, A Einstein Collected Papers, Vol. 7, Doc. 31.

4

Space-time

The views of space and time which I wish to lay before you have sprung from the soil of experimental physics, and therein lies their strength. They are radical. Henceforth space by itself, and time by itself, are doomed to fade away into mere shadows, and only a kind of union of the two will preserve an independent reality.

H Minkowski[1]

4.1 Events and World Lines

In relativity an *event* is a point in space-time specified by a set of co-ordinates for position and a single 'co-ordinate' of time: (ct, x, y, z). We do not even require that anything especially distinctive should happen at one of these point events. It is simply a set of co-ordinates that we wish to draw attention to for some purpose. In common usage, events are sports fixtures, concerts, a birthday party or similar. Such events have a *duration* and a *region* of space that they occupy. A sports *event* would contain a vast number of *mathematical events*.

Now consider an object moving through space and observed from the vantage point of a system laid out from and stationary with respect to us. To make things easier, the object is a *rod*, with a one-dimensional length and negligible width. The rod is calibrated in some fashion, so that it can be used for standardised measurements of length. As it moves through our system, we pay particular attention to each end of the rod. From moment to moment, these ends land on particular co-ordinates, each of which we pick out as an event. If you want a visual picture, then think of our system as a collection of grey dots, one for each space-time point. As the ends of the rod pass through the system, they change the colour of each dot that they touch: they turn them dark red. Now, in principle each dot is infinitesimally close to the next one. So, the red dots will draw out two red lines: one line for the progression of each end of the rod. These lines are what specialists call *world lines*. Between these two world lines is the *world sheet* which marks the area swept out by the rod as it moves though our system. Don't forget that these 'lines' and 'sheets' are drawn through space *and* time. This is crucial: even if the rod is sitting stationary in our system, the ends will still have a world line and there will still be a world sheet, as the rod is progressing through time, even if it is not moving through space. Equally, we should not talk about a particle *moving along its world line* as the line is a collection of points occupied by the particle through space and time. The entire past and future of the particle is mapped out along the line.

4.2 Minkowski Diagrams

When we introduced the Lorentz transformations back in Chapter 2, we did so with the aid of a diagram (Figure 2.15) showing two spatial dimensions (x, y) and embedded clocks. In 1908 Minkowski introduced *space-time diagrams* as an easier way to visualise the consequences of the Lorentz transformations.

However, it can be difficult to follow what they represent until you get used to them, so we are going to start off by using them to work with the (simpler) Galilean transformations.

4.2.1 Newtonian Space-time

With Minkowski diagrams you drop another spatial dimension and instead present time on the vertical axis. One of our events from Section 4.1 is a point on this 2D representational plane (having suppressed y and z for ease). To fit with our later relativistic use, the vertical 'time' axis is in units of ct.

In setting up this diagram, we are assuming the perspective of an observer at rest somewhere within the system, most conveniently at the origin (they are sitting at $x = 0$ with a clock that starts from $ct = 0$). In which case, the ct-axis is drawn out by the *world line of the observer sitting at the origin*. Equally, any line rising vertically from a point on the x-axis (and hence parallel to the ct-axis) will be the world line of an object resting at that location in space.[2]

In a full Minkowski diagram, we include the negative axis in both horizontal and vertical directions. This inclusion should not faze the beginner: a negative x co-ordinate simply refers to a point in space somewhere to the left of our chosen origin point; a negative time axis locates those events that took place before our agreed starting of the reference clock.

In this extended representation, the world line of a photon moving through our system and passing through the origin will be a line at 45° to the x-axis (Figure 4.1) – remember that the time axis is in units of ct.

Any other object passing through our system will have a world line if it is a particle, or a world sheet if it is an extended object. Figure 4.2 shows a Minkowski diagram for a point object that starts at $ct = 0$ sitting at point $x = 4$ and subsequently moves through our system at velocity u along the x-axis.[3]

As the object is moving at constant speed, the x co-ordinate is coupled to the ct co-ordinate by $x = \beta_u ct$, where $\beta_u = u/c$. Consequently, the gradient of the world line is related to the velocity at which the object moves:

$$\text{gradient of a straight line} = \frac{\text{change in vertical axis}}{\text{change in horizontal axis}} = \frac{\Delta ct}{\Delta x} = \frac{c\Delta t}{\beta_u c\Delta t} = \frac{1}{\beta_u}.$$

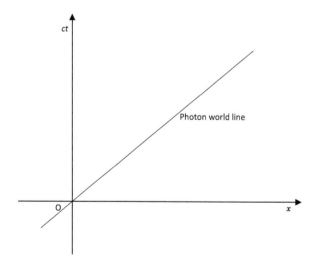

FIGURE 4.1 A simple Minkowski diagram showing the spatial (x) and temporal (ct) axes for a system at rest with respect to an observer placed at the origin. The 45° line is the world line of an object moving through the system at the speed of light.

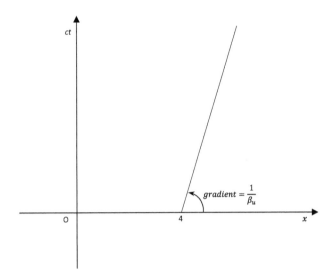

FIGURE 4.2   A Minkowski diagram for an object moving along the *x*-axis with velocity *u* in our system, *S*.

Hence, the *faster* the object moves the *shallower* the world line. In fact, any world line inclined at less than 45° would have to be moving faster than light. Hence, all physical world lines have to be inclined at angles ≥45°.

A world sheet, such as that drawn out by our measuring rod from earlier, would be bounded by two world lines (one at each end of the rod) drawn with equal angles (or the ends of the rod would be moving at different speeds!).

Our next step is to include the axes belonging to another system moving through ours at some constant velocity *v*. To make things easier from now on, we refer to systems as *S*, the stationary system with *x*, *ct*-axes, and *S'* which is moving at speed *v* with *x'*, *ct'*-axes.

The first thing to do is to go back to the Galilean transformations:

$$t' = t$$

$$x' = x - vt$$

and convert them, as we did the raw Lorentz transformations, by setting $\beta_v = v/c$:

$$ct' = ct$$

$$x' = x - \beta_v ct$$

which, incidentally, we can put in matrix form:

$$\begin{pmatrix} ct' \\ x' \end{pmatrix} = \begin{pmatrix} 1 & 0 \\ -\beta_v & 1 \end{pmatrix} \begin{pmatrix} ct \\ x \end{pmatrix}.$$

As is conventional, we agree with the observer embedded into *S'* that their origin point passes through ours at the time $ct = ct' = 0$. Also, as a simplification, we set things up so that *S'* is sliding along the *x*-axis in *S*. So, the *x*- and *x'*-axes will lie on top of each other.

Trap for the Beginner

You might think that the origin point of S' should be sliding along the x-axis as well. However, that would be forgetting that the origin is $x' = 0$ *at* $ct' = ct = 0$ and that event is nailed in place at the centre of the diagram. This is the sort of visualisation jump that you need to make in order to fully understand how Minkowski diagrams work...

With x and x' now in place, we need to turn to the representation of the ct'-axis. You might think that this is quite simple, after all Galilean transformations do not affect time, so that the ct'-axis ought to sit on top of the ct-axis just as the x'-axis sits on top of the x-axis. However, that is not right. The ct'-axis *is the world line of the S' spatial origin*, and so must be represented in the S system as a world line of gradient $= 1/\beta_v$ (Figure 4.3).

Any event in space-time will be rendered as a point, such as E in Figure 4.4, with co-ordinates (x, ct) in the S system.[4] We ascertain these co-ordinates by drawing a line from E parallel to the x-axis in order to find ct (as all points along that line have the same ct value no matter what their x value), and a line parallel to ct in order to find x (as all points along that line have the same x value no matter what their ct value).

Finding the co-ordinates of E in the S' system follows the same pattern. To get the ct' value we need to draw a line parallel to the mutual x-axes passing through E and see where it cuts the ct'-axis. Just by looking at the diagram we can see that this line will cut ct' further from the origin than it cuts the ct-axis $(OB > OA)$. We could mistakenly conclude from this that $ct' > ct$, which would not fit in with the Galilean transformations and the assumption of universal time which forms one of the backbones of Newtonian mechanics. The scale on the ct'-axis must be different to that of the ct-axis, so that the values read from the axis are the same. See how easily we can be misled by not attending to detail.

To get the x' value of E we need to draw a line through E *parallel to the ct'-axis*, not a vertical line as we did in S. A vertical line represents points of the same x no matter what the ct value is in the S system, but in S' it will represent the *same x'* but *changing* values of ct' which cannot be right for an x' co-ordinate at a specific moment in ct'.

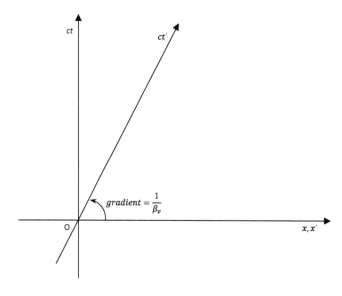

FIGURE 4.3   The ct'-axis is a world line in the S system with a gradient $= 1/\beta_v$.

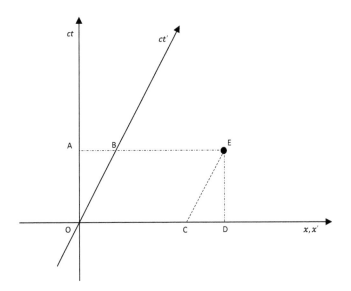

FIGURE 4.4 Finding the co-ordinates of E in the S and S' systems.

In Figure 4.4, the x co-ordinate of E is D and the x' co-ordinate is C. We have already said that the ct'-axis has gradient $1/\beta_v$ and as the line EC is parallel to that axis, it has the same gradient. From the definition of a gradient:

$$\text{gradient of } EC = \frac{ED}{CD} = \frac{1}{\beta_v} = \frac{ct}{CD}.$$

so:

$$CD = \beta_v ct = vt.$$

Hence, the x' value $= OC = OD - vt = x - vt$ in full accord with the Galilean transformations.

4.2.2 Lorentzian Space-time

Now that we have started to build up some familiarity with Minkowski diagrams, it is time to see how they work with the new picture of space-time based on the Lorentz transformations.

The biggest single difference is that, due to the Lorentz transformations, *both* the x'- and ct'-axes are rotated away from the x- and ct-axes, respectively (Figure 4.5). This is similar to the standard rotation transformation from Chapter 3, but in this case although the axes are rotated through the same *angle*, they are not rotated in the same *sense*. This squares with the difference between $\mathfrak{R} = \begin{pmatrix} \cos\theta & \sin\theta \\ -\sin\theta & \cos\theta \end{pmatrix}$

and $L = \begin{pmatrix} \gamma & -\beta\gamma \\ -\beta\gamma & \gamma \end{pmatrix}$: $\mathfrak{R}$ changes signs on the diagonal elements (leading to both axes being rotated in the same sense); L does not change sign on the diagonal elements (leading to rotations in the opposite sense). By the way, this explains the comment at the start of Chapter 3, where I said that Lorentz transformations become 'rotations' in Minkowski space-time.

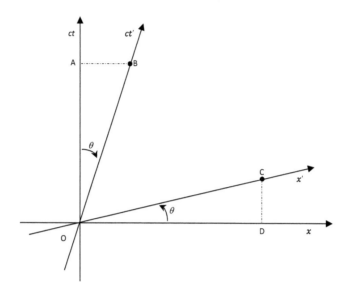

FIGURE 4.5 A Minkowski diagram showing the systems S and S' related to each other by the Lorentz transformations.

Consider the point B on the ct'-axis. Like every point on this axis, it has $x' = 0$ so that its co-ordinates in the S system are:

$$ct = \gamma\left(ct' + \beta_v x'\right) = \gamma ct'$$

$$x = \gamma\left(x' + \beta_v ct'\right) = \gamma\beta_v ct'.$$

The tan of the angle θ between the ct- and ct'-axes is:

$$\tan(\theta) = \frac{AB}{OA} = \frac{\gamma\beta_v ct'}{\gamma ct'} = \beta_v.$$

To do the same for the x- and x'-axes we turn to point C, which has $ct' = 0$ and co-ordinates in S:

$$ct = \gamma\left(ct' + \beta_v x'\right) = \gamma\beta_v x'$$

$$x = \gamma\left(x' + \beta_v ct'\right) = \gamma x'.$$

The tan of the angle between x and x' is then:

$$\tan(\theta) = \frac{CD}{OD} = \frac{\gamma\beta_v x'}{\gamma x'} = \beta_v.$$

The two angles are the same and directly related to the relative velocity between the systems.

Just one more thing about the line representing the ct'-axis: the gradient of that line in S is:

$$\text{gradient of line} = \frac{OA}{AB} = \frac{1}{\tan(\theta)} = \frac{1}{\beta_v}$$

showing that the ct'-axis is again the world line of $x' = 0$.

Minkowski diagrams can be slightly confusing until you get used to them. There is the temptation to read the scales on the boosted axes wrongly. Just to get our bearings, consider Figure 4.6. Here we have the world line of an object moving with $\beta_u = 0.6$ viewed from our stationary system (S) and also from a system (S') moving with $\beta_v = 0.3$. The event A is at co-ordinates $x = 3$, $ct = 1.8$. The x co-ordinate is given by the black dotted vertical line parallel to the ct-axis; the ct co-ordinate is from the black dotted horizontal line parallel to the x-axis.

In the boosted system, we need a line parallel to the ct'-axis to read the x' co-ordinate and a line parallel to the x'-axis to get the ct' co-ordinate. These are the grey dotted lines on the figure. In this case, the boosted values are $x' = 2.58$, $ct' = 0.94$. The scales on those axes look a bit different!

4.3 The Past, the Future and the Elsewhere...

In Section 3.6, we introduced the idea of proper time and with it the Lorentz invariants:

$$s^2 = (ct)^2 - x^2 \quad (\Delta s)^2 = (c\Delta t)^2 - (\Delta x)^2.$$

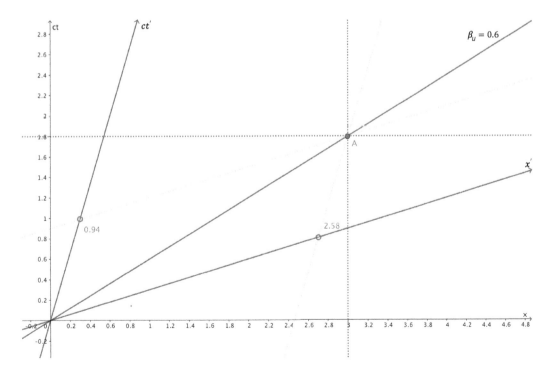

FIGURE 4.6 The co-ordinates of an event read from a Minkowski diagram.

To calculate the second version of the invariant, we pick two events in space–time and get $c\Delta t$ and Δx in a straightforward manner:

$$c\Delta t = ct_2 - ct_1 \quad \Delta x = x_2 - x_1.$$

In principle we can pick any two events – they do not have to be related in any way. It might be that an object of some sort passed from one event to the other, or it might be that there was no matter or energy exchange between them at all. In either case, we can calculate an invariant interval between the events. Given this complete freedom, three possibilities arise:

1. $c\Delta t > \Delta x$ in which case $(\Delta s)^2 > 0$
2. $c\Delta t < \Delta x$ in which case $(\Delta s)^2 < 0$
3. $c\Delta t = \Delta x$ in which case $(\Delta s)^2 = 0$.

In the next few sections, we will look at these different possibilities in detail.

4.3.1 Time-Like Intervals

In the first case, $(\Delta s)^2 > 0$, so *in principle* it is possible for the two events to be linked. After all, $c\Delta t$ is the distance that a photon can travel in the time interval between the two events. If that distance is greater than the spatial separation of the events, then a photon *could* have crossed the space in the time between the events. That does not mean that a photon *did* move from one to the other, just that it would be possible. Also, this does not mean that matter or energy moving more slowly than light would be able to cross the distance in the time. That is an individual circumstance that we would have to calculate in each given situation. This is about making a distinction between events that can in principle influence on one another from those that cannot.

I press a switch on my wall and as a result, within a fraction of a second, the lightbulb on the opposite side of the room turns on. Those two events must have a space–time interval where $c\Delta t > \Delta x$ as they have a *causal influence* on one another. Their spatial separation is less than the distance a photon could travel in the time between the events. In fact, in this case the distance is short enough for the electric field to establish itself once the switch is pressed and before the lightbulb comes on.

Pairs of events where $c\Delta t > \Delta x$ have a space-time invariant interval that is *time-like*. In that situation, it is possible for the two events to be connected by a (fast enough) particle's world line. If there is such a world line, then there is a co-moving system and hence a proper time can be defined. The existence of a proper time is the guarantor of a time-like separation.

If we specify that one of the events is the origin of our co-ordinate system S, then Δx is just the x co-ordinate of the second event and $c\Delta t$ is the time co-ordinate, ct. Consequently, the space-time interval is $s^2 = (ct)^2 - x^2$. As an example, the curve in Figure 4.7 shows all the points sitting at an invariant interval $s = 4$ from the origin.[5]

Clearly the curve cuts the ct-axis at $ct = 4$. This marks the event that has no spatial separation from the origin but happens a time $4/c$ after the origin moment. All the other events on the curve are separated from the origin in both space and time.

A point object departing from the origin and moving with a velocity of $0.3c$ has a world line that crosses the invariant curve at $x \sim 1.3, ct \sim 4.2$ (also on Figure 4.7). In a system co-moving with this object, the elapsed proper time would be $c\Delta\tau = 4$. Using the Lorentz transformations to boost back into our system S gives:

$$c\Delta t = \gamma \left(c\Delta\tau + \beta\Delta x' \right) = \gamma c\Delta\tau = 4\gamma$$

$$\Delta x = \gamma \left(\Delta x' + \beta c\Delta\tau \right) = \gamma\beta c\Delta\tau = 4\beta\gamma$$

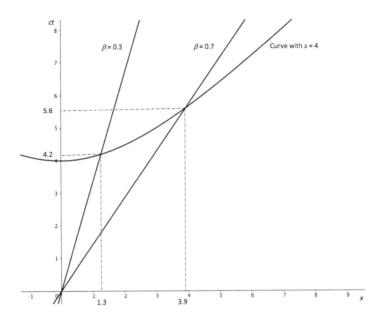

FIGURE 4.7 A Minkowski diagram for a co-ordinate system S showing the curve connecting all events that have a space-time invariant $s = 4$ with respect to the origin. Also shown are world lines for point objects moving at $0.3c$ and $0.7c$ away from the origin.

remembering that for a co-moving system $\Delta x' = 0$.

With $\beta = 0.3$, $\gamma = 1/\sqrt{1 - 0.3^2} \sim 1.05$, hence:

$$c\Delta t = 4\gamma = 4 \times 1.05 = 4.2$$

$$\Delta x = 4\beta\gamma = 4 \times 0.3 \times 1.05 = 1.26 \approx 1.3$$

establishing the co-ordinates that I quoted.

A similar calculation will give the co-ordinates in S where a particle moving at $0.7c$ from the origin achieves a proper time $c\Delta\tau = 4$ (also shown in Figure 4.7).

Both of these events have a time-like invariant separation from the origin, like any event on the curve defined by $s = 4$.

Recall from the previous section that the two lines in Figure 4.7 marked $\beta = 0.3$ and $\beta = 0.7$ will also be the ct'-axes of the co-moving systems embedded with the objects. Hence, we have a way of determining the scale on these axes: in each case, the point $ct' = 4$ will be where the curve cuts the axis. We can draw similar curves for *all time-like separations from the origin*. For example, Figure 4.8 shows curves for $s = 2, 4, 6, \& 8$ and axes for objects moving from the origin with $\beta = 0.3$ (axis ct') and with $\beta = 0.7$ (axis ct_1'). The dots show where the axes cross the curves.

Sitting in our system, S, we watch an object move away from the origin with $\beta = 0.7$. Somehow, we contrive to observe a clock connected with this object tick off time in units of $c\tau$, which is the passage of proper time for that moving object. When that clock ticks around to the value $c\tau = 4$, we note the following:

1. from Figure 4.8, we can see that the time elapsed on our clock is a little more than $ct = 5$ (draw a line horizontally from where the $\beta = 0.7$ line crosses the $s = 4$ curve until it hits the ct-axis);

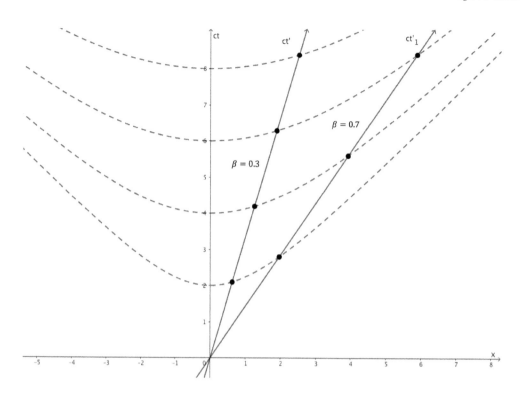

FIGURE 4.8 Curves of constant invariant interval with $s = 2, 4, 6$ and 8. Also shown are the ct'-axes for objects moving through system S with $\beta = 0.3$ and $\beta = 0.7$.

2. by this moment, the object would have moved a little less than $x = 4$ units away from the origin (draw a vertical line down from where $\beta = 0.7$ crosses $s = 4$ until it meets the x-axis);
3. more precise calculation shows that the event $c\tau = 4$ for the moving object corresponds to $ct = 5.601$, $x = 3.921$ in S;
4. from the point of view of the moving object, the two events A (being at the origin) and B (reaching $c\tau = 4$) are separated purely and singly in time, but from our perspective in S the events are separated in time and space;
5. that description would be *equally true in a genuinely Newtonian world*; what makes this distinctively relativistic is that the separation in time observed in S is *different* to the proper time observed on the moving clock;
6. the time between A and B in our system is *always greater* than the proper time recorded in the moving system. We have balanced time against space: as $(\Delta s)^2 = (c\Delta t)^2 - (\Delta x)^2$, the greater time interval from our perspective is counter balanced by the spatial difference that has to be subtracted to form the invariant. Looking at it going the other way, what was space and time in S boosts into just time in S'.

This effect, that the time we record in our system is always greater than the proper time of a moving object, is called *time dilation* and has been experimentally demonstrated (see Section 4.5.1). Time dilation is a third 'tent pole' feature of special relativity.

4.3.2 Absolute Past, Absolute Future

In Figure 4.9, I have extended Figure 4.8 into time periods before the origin moment and added in a couple of photon world lines (the solid lines) – one moving from $-x$ through the origin to $+x$ and another from $+x$ through the origin to $-x$. The dashed lines are the curves of constant interval.

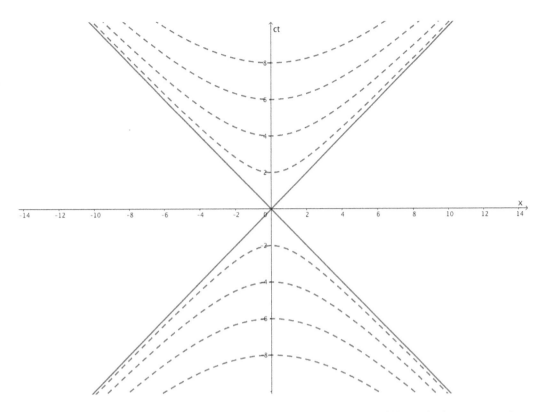

FIGURE 4.9 Curves of constant interval with s = 2, 4, 6 and 8 along with the world lines of a photon moving from −x to +x and another moving from +x to −x.

In this extended diagram, we can see that the curves of constant s are symmetrical about the ct-axis and reflected in the x-axis. This is expected, as $(\Delta s)^2 = (c\Delta t)^2 - (\Delta x)^2$ will give the same answer for positive values of $c\Delta t$ and Δx as for negative values.

Evidently the curves of constant interval are bounded by the world lines of the photons. You can also see how the curves crowd towards each other and get close to the photon lines at higher values of x (and −x). Actually, the photon world lines represent lines of constant s with s = 0. They are the third case we noted at the end of Section 4.3. For a photon, the proper time is always zero and all the events it 'experiences' happen in the same place and all at the same time.

Events in the triangular region bounded by the photon world lines below the origin are in the *absolute past* of the origin point (Figure 4.10). As the interval between any event in this region and the origin is time-like, in principle an event at the origin could be causally influenced by any event in the absolute past. This would certainly be the case if that causality were due to a photon travelling between the events, but in some situations, with the right interval, it could be some other influence mediated by matter or energy.

Equally, the triangular region above the x-axis is the *absolute future* of the origin. An influence setting out from the origin could reach events within this region and have a causal effect on them. The world lines of any object moving through our system from the origin must pass through this region. Figure 4.11 shows the world line of a particle moving at $\beta = 0.3$ and passing through $x = 4$ at $ct = 0$ (co-ordinates (4, 0)). This object will not pass into the absolute future of the origin until it gets to event A on the diagram. If we send a photon from the origin, then it will light up our moving object at event A.

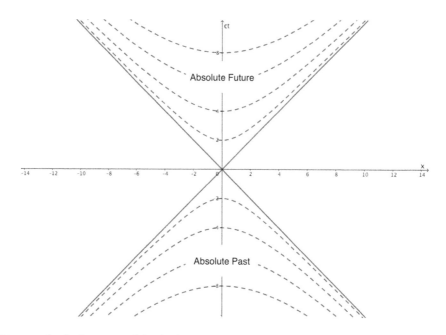

FIGURE 4.10 The absolute past and the absolute future of the origin in *S*.

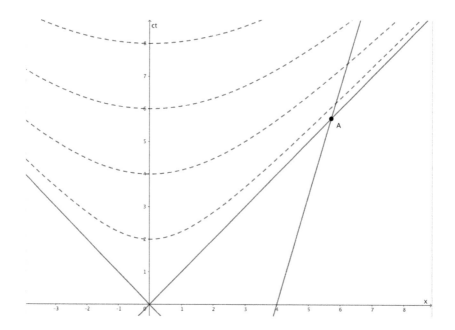

FIGURE 4.11 The world line of a particle passes into the absolute future of the origin when it reaches event A.

Events lying outside of the absolute past and the absolute future can never have any influence on events at the origin, nor can events at the origin have an influence on them. They lie neither in the absolute past, nor the absolute future. That is quite a radical concept. Generally, one thinks of all events as either falling in the past or happening in the future. In relativity, things are not so simple.

4.3.3 Space-Like Intervals

If we pick pairs of events at random, then in many cases $c\Delta t < \Delta x$, in which case the space-time interval is *space-like* (this is option 2 that we listed at the end of 4.3). Such events have a negative value of $(\Delta s)^2$, which means that Δs cannot be evaluated and does not correspond to a proper time. This is hardly surprising as no particle could cross between the events. As a result, the events cannot be causally related.

Evidently, having $(\Delta s)^2 > 0$ for time-like intervals and $(\Delta s)^2 < 0$ for space-like, rather than the other way around, arises from my choice of η_1 rather than η_2 back in Section 3.5. Although the physics is identical with either choice, it seems to me more natural to have $(\Delta s)^2 > 0$ for the physical time-like intervals, allowing us to take the square root and obtain $c\Delta\tau = \Delta s = \sqrt{(\Delta s)^2}$ without having to worry about the square root of negative numbers: that issue is confined to the non-physical space-like intervals.

Looking at the interval in full we have:

$$-\left|(\Delta s)^2\right| = (c\Delta t)^2 - (\Delta x)^2 ,$$

where the $|\ |$ is used to emphasise that the value between these straight brackets is positive. This allows us to switch things around slightly:

$$\left|(\Delta s)^2\right| = (\Delta x)^2 - (c\Delta t)^2 ,$$

which we can use to plot curves of equal interval as before, as s can be evaluated. Given any two events that happen at the same time, so that $c\Delta t = 0$, $\left|(\Delta s)^2\right| = (\Delta x)^2$ and $\Delta s = \Delta x$ which is known as the *proper distance*. If you recall, proper time is the elapsed time in the co-moving system and also the invariant of a time-like interval. Proper distance is the co-ordinate distance between two events evaluated at the same time and is hence the invariant for space-like intervals.

We can plot curves of constant proper distance (Figure 4.12) where we again see that the curves are bounded by the photon world lines and crowd towards them at large $+ct$ and $-ct$.

The triangular regions to the left and right are classed as the *absolute elsewhere*, regions that are out of causal connection with the origin.

Consider event C (Figure 4.13) at co-ordinates $(-5, -4)$. That makes it $4ct$ 'moments' before the origin and 5 distance units away. A photon could not cross that distance in that time. Equally, event B lies at $(5, 2)$ which only gives a photon $2ct$ to reach across a separation of 5 distance units. It's not going to happen.

4.3.4 Light Cones

The boundaries formed by the photon world lines are the edges of a *light cone*. If we allow one of the other spatial co-ordinates back into our system, the Minkowski diagram becomes harder to draw but somewhat more realistic. In this rendering (Figure 4.14), the triangular regions of the 2D diagrams become cones above and below the origin, as in principle the photon world lines could set off from the origin in any direction. The light cones then mark the progress of a ring of light converging on the origin from the absolute past and diverging away into the absolute future. If we could draw this in 3D + 1 (for the time axis), then the 'cones' would track the progress of a shell of light contracting to the origin and expanding away. In any rendering, the absolute future and past of the origin lie within the boundaries of the cones.

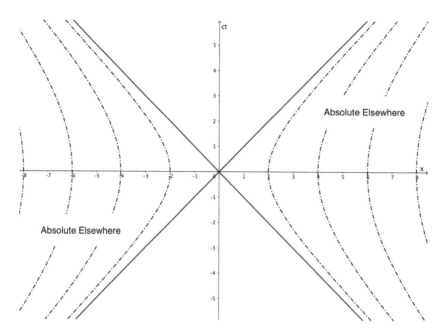

FIGURE 4.12   Curves of equal invariant interval for $s = 2, 4, 6$ and 8, but this time with space-like intervals.

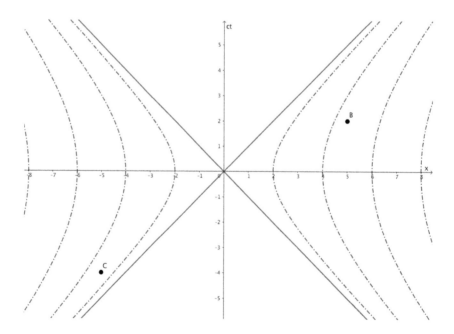

FIGURE 4.13   Two events, B and C, that lie in the absolute elsewhere of the origin.

That leaves all other events in the system embedded in the elsewhere, with the x/y plane being thought of as the present (for the origin).

If we were to step along any world line, say from one event A to another event B on the same world line, then from A's perspective B must lie in the absolute future cone of A and from B's perspective A must lie in the absolute past cone of B. A world line must thread along from one light cone to another.

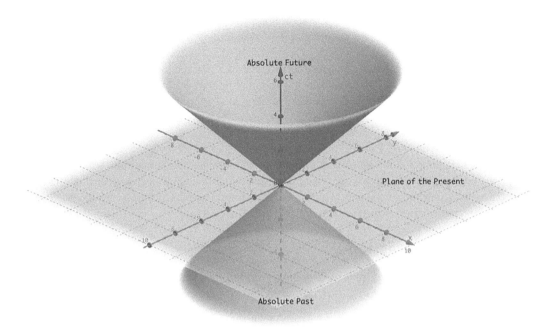

FIGURE 4.14 Including the *y* co-ordinate extends a Minkowski diagram into a 3D rendering where the absolute future lies in a light cone above the origin, the absolute past in the cone below the origin and the *x/y* plane marks the present.

Extending this to three spatial dimensions is impossible in any diagram as the *ct*-axis must be included as well. However, our displacement 4-vector $(c\Delta t, \ \Delta x, \ \Delta y, \ \Delta z)$ when applied to any two points on a world line will 'point' from one to another. In that sense, the world line is a sequence of displacement 4-vectors stepping from event to event.

4.3.5 Time Ordering

Any event that happens in the absolute past of the origin could potentially exert a causal influence on events taking place at the origin. Equally, any origin event can exert causal influences on events in its absolute future. It is therefore crucial that the time ordering of such events is not disturbed by any Lorentz boost. If the sequence of events is $A\,(c\Delta t < 0)$, $O\,(c\Delta t = 0)$, $B\,(c\Delta t > 0)$, then every observer must agree on this time ordering.

The temporal Lorentz transformation is $c\Delta t' = \gamma\,(c\Delta t - \beta\Delta x)$, and given the time-like nature of the intervals between these events, $c\Delta t > \Delta x$, so $c\Delta t'$ must have the same sign (positive or negative) as $c\Delta t$. In the transformation, you are always subtracting a smaller number $(\beta\Delta x)$ from a larger $(c\Delta t)$ (remember that $\beta < 1$), which guarantees the time ordering of the events.

The position co-ordinates of the events can switch, so that the event nearest to the origin and the event furthest from the origin can change places after the boost. The spatial transformation is $\Delta x' = \gamma\,(\Delta x - \beta c\Delta t)$, so $\Delta x' < \Delta x$ if $c\Delta t > 0$.

The same is not true for events in the absolute elsewhere. In this case, $c\Delta t < \Delta x$ so if $c\Delta t < \beta\Delta x$, $c\Delta t'$ can be negative even if $c\Delta t$ is positive. An event that happens after the origin in our system can happen before the origin in another.

Suitably chosen Minkowski diagrams can help us see how this works.

Figure 4.15 shows our standard system *S* and another *S′* moving at a leisurely $\beta_v = 0.1$ through our system. Event A lies in our absolute past at $ct = -3$ and B at $ct = 3$ in our absolute future. Note that the dotted lines mark the grid lines of the *S'* system not *S*, which is why the diagram can look slightly crooked at first glance. Now we move to Figure 4.16, where the *S'* system is moving more briskly at $\beta_v = 0.5$.

Remembering that the dotted lines mark the axis grid of *S'*, we see that A happened a little more recently with $ct' \sim -2.5$, but still in the absolute past. B is also pulled back towards the origin with $ct' = 2$. However, things are very different with C and D. Back in Figure 4.15, event C lies at $ct' \sim -0.5$ and D has $ct' \sim 0.5$. Now if we take a look at the same events in Figure 4.16, $ct' = 1$ for C and $ct' = -1$ for D. Not only has the order for C and D changed, they have switched with regard to the origin.

Even though the fable of absolute time has been dispelled by relativity, it does not follow that all temporal bets are off. The sequence of causality is still stitched into space-time with world lines navigating their paths within light cones from one event to another. However, for events that have no potential causal influence on one another, time ordering has no possible physical consequence and so the Lorentz transformations allow the temporal dice to fall as they may.

4.4 Length Contraction

When we set about measuring the length of an object, we are generally not too punctilious about how it is done: slapping a measuring ruler next to it and taking a reading is normally good enough. However, working with relativity we know that our intuitions are not a good guide and we need to be precise about what we are doing in order to avoid making mistakes. Consequently, we need a defined procedure for measuring length – we record the edges of an object against our *x*-axis scale, being sure to do each edge at the same moment in time. If the object is stationary in our system, you may suspect

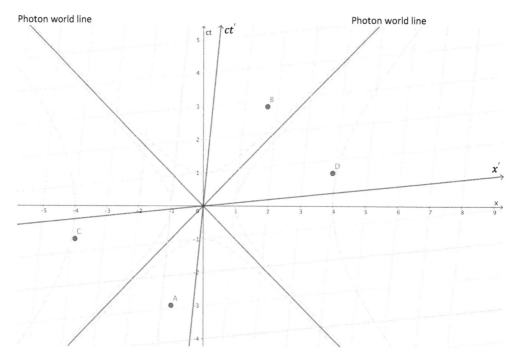

FIGURE 4.15 In this Minkowski diagram, the *ct'*- and *x'*-axes are shown for a system moving at $\beta = 0.3$ with respect to us. Event A lies in the absolute past and B in the absolute future. C happens $ct = -1$ before the origin and D $ct = 1$ after the origin. Both C and D lie in the absolute elsewhere. The dotted lines mark the grid of the *S'* system and the dashed lines are a curves of constant interval for $s = \pm 1$ and $s = \pm 4$.

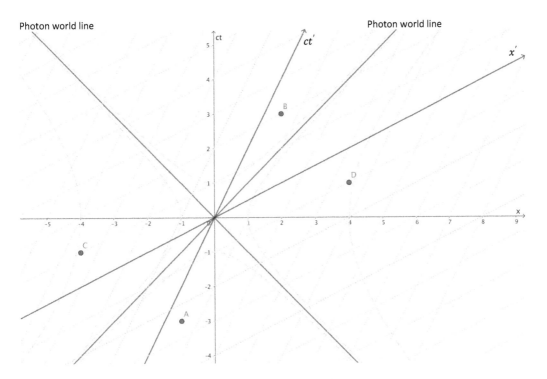

FIGURE 4.16 The same set of events considered from S and a new S' with $\beta = 0.5$.

that we do not have to be so precise about the time. After all, the two ends of the object are not going anywhere...

Well, in one sense they are – they are moving through time, and we know that time is not a universal. Also, the object may well be stationary in our system, but is undoubtedly moving in some other. It would certainly not be right to record the edges against an x'-axis at different times, as they would have moved. If we are going to be sure that we are comparing like with like, we need to stipulate that the edges are recorded at the same time.

We also need to consider the possibility that the object we are trying to measure is so vast that we can't see one edge of it from the other. In that case, we will need a team of trained observers stationed at strategic points along the x-axis, all hoping to catch a glimpse of the object. If it is moving through our system, then it will of course pass every point along the x-axis in its direction of flight. So, our team need synchronised clocks and the understanding that the position of the object will be recorded at an agreed moment in time by any observer that happens to find it in their vicinity. Any other S' will need a similarly trained set of acolyte observers with synchronised clocks. This is precisely the sort of system that we first set up in Section 2.3.

The length of an object recorded in any co-moving system is known as the *proper length*,[6] l_0, where $l_0 = \Delta x$ when $\Delta t = 0$. Boosting into a moving system we find:

$$\Delta x' = \gamma \left(\Delta x - \beta c \Delta t \right) = \gamma \Delta x = \gamma l_0$$

and

$$c \Delta t' = \gamma \left(c \Delta t - \beta \Delta x \right) = -\beta \gamma \Delta x.$$

However, $\Delta x'$ is not the length recorded in the moving system – it can't be as the positions on the x'-axis have not been recorded at the same time in that system: $\Delta t' \neq 0$.

What we have calculated here is the S' system's view of *our* measurement. They think that we have got it wrong as we have not recorded the edges at the same moment, as seen from their system.

We can make this somewhat clearer in a diagram, such as Figure 4.17. Here we find an object stationary in S with the world lines of its edges being marked by the ct-axis and the vertical line through B and C on the diagram. The object is measured as having a length of 2.5 units in this system ($\Delta x = 2.5$ as measured at $t = 0$ and consequently $\Delta t = 0$).

The same object is observed from the vantage point of the S' system moving with respect to S with $\beta_v = 0.6$. In this system, our attempt to record the positions of the edges of the object, events A and B in S, seems to have gone hopelessly wrong, as event B happens at a time $c\Delta t' = -\beta\gamma\Delta x$, i.e. some $1.9ct$ units *before* event A. In fact, the co-ordinates of B as recorded in S' are the points d and e on the diagram (drawn as open circles to show that they are single co-ordinates, not events). Hence, the length Ae is $\Delta x'$ and Ad is $c\Delta t'$. In S' the correct measurement of length derives from events A and C. In that system, the observer at the origin must tell his acolytes to ignore event B and wait until the right-hand edge of the object has reached event C, which will be at the same ct' as A in their system.

Note that waiting in this way will have the effect of moving e nearer to C along the x' line (remember that the system $S'$is moving away from the origin of S). Given that the acolyte observers will have to wait a time $ct' = \beta\gamma\Delta x$ in order to allow B to reach C, we would expect e to have moved along a distance $vt' = \beta ct' = \beta \times \beta\gamma\Delta x = \beta^2\gamma\Delta x$. This is the distance we need to *subtract* from our original $\Delta x'$ in order to obtain the length, l', recorded in S':

$$l' = \Delta x' - \beta^2\gamma\Delta x = \gamma\Delta x - \beta^2\gamma\Delta x = \gamma\Delta x\left(1-\beta^2\right) = \frac{\Delta x}{\gamma} = \frac{l_0}{\gamma},$$

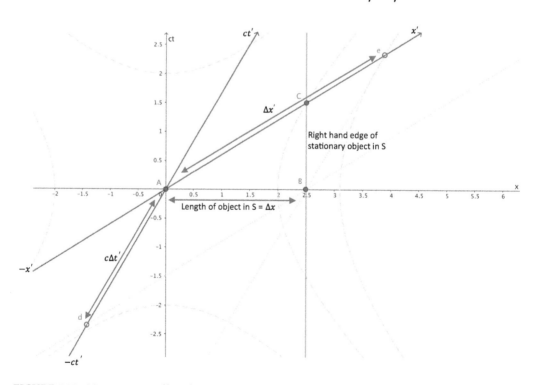

FIGURE 4.17 Measurements of length as seen in two systems. In S the object is stationary and the positions of its edges recorded at the same time – events A and B. The world lines of its stationary edges are the ct-axis and the vertical line through events B and C, respectively. As the S' system is designed to have the same origin, observers in that system agree on event A. However, for them our event B took place some $1.9ct$ units before (with $\beta_v = 0.6$) at co-ordinate d. Their measurement of the object derives from events A and C.

which is the famous length contraction formula.

With all this talk about edges and lengths, it is easy to lose site of the underlying reality: *space is being compressed along the axis of motion*, at least as viewed by another observer.

The symmetry of spatial distortion is made clearer by Figure 4.18. Here we have our stationary object in S with the edges of its world lines being the ct-axis and the vertical line through D at $x = 2.5$. As before, the length of this object will be AC in the S′ system, which as we can see from the curve of constant interval for $s = 2$, is a length of 2 units.

On the same diagram, we also have the world lines of the edges of an object moving at $\beta_u = 0.6$ through S. The line BC is the right-hand world line, and the ct'-axis is the other as we are assuming that S′ is co-moving with this second object. In this system, A and C mark the measurements of the edges at the same time, giving the object a proper length of 2. However, in S the correct measurements are A and B, so we measure the length as 1.6. Space moving relative to us appears compressed. From the co-moving system space appears fine, and our space appears compressed.

4.5 Time Dilation

We have already seen how time runs more slowly for objects observed to be in motion. In fact, in Section 3.71 we showed that the relationship between the proper time $\Delta \tau$, to the 'moving time' Δt, is:

$$\Delta \tau = \frac{\Delta t}{\gamma_u} \text{ or } \Delta t = \gamma_u \Delta \tau.$$

As with length contraction, the effect is symmetrical between observers in relative motion. Another Minkowski diagram will help to illustrate this: Figure 4.19. Sitting stationary in system S, we allow

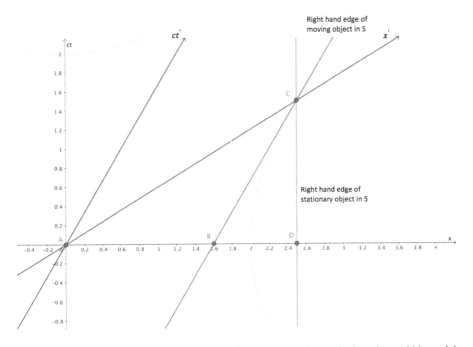

FIGURE 4.18 Length contraction – passing through the points B and C – which is the world line of the right-hand end of a rod moving at $\beta = 0.6$ through system S; the world line of the left-hand end of the rod is the ct'-axis. The line passing through points D and C is the world line of the right-hand end of a rod stationary in S, the ct-axis being the world line of the left-hand end of the same rod.

our proper time to tick over until $ct = 2$. In the S' system, we can see that the curve of invariant proper time $s = 2$ crosses the ct'-axis at e. The co-ordinate of our $ct = 2$ is actually at c where $ct' = 2.5$. From the perspective of S' we are the moving system and our clocks are running slower: 2.5 units of their proper time corresponds to 2 units of ours. On the other hand, if you look at $ct' = 2$ in their system (e), we see that as an elapsed time of $ct = 2.5$. We see their time running slow by exactly the same factor as they see ours.

The symmetry between these two perspectives has been exploited in a famous argument known as *the twin paradox*. According to this contrived situation, a pair of twins are separated. One remains in our stationary system, while the other hitches a lift on the moving system. According to each twin, the other suffers from dilated time and so ages more slowly compared to them. After a suitable period, the moving twin (from our perspective) reverses course and returns to meet with his sibling. On the way back, the same argument applies, each believes that the other is ageing more slowly. However, once we have them standing next to each other, we ought to be able to determine which one has in fact grown older. In reality, no paradox ensues as the situation is not truly symmetrical. One of the twins had to reverse course and to do so required an acceleration. If one twin were on Earth and the other in a spacecraft, then that ship would have to use some form of propulsion to reverse its travel and return to Earth so that the twins could be compared. Although the twin on the ship sees the Earth move away and then approach again, the Earth did not accelerate during the manoeuvre. Only one of the twins suffers an acceleration and as we develop our ideas of curved time, we will see that acceleration produces its own time dilation. The returning twin has been in a non-inertial system (see Section 9.3 for further discussion of this).

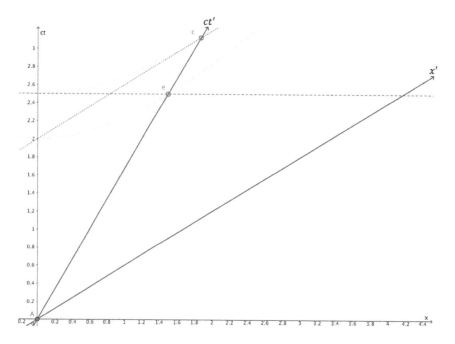

FIGURE 4.19   Relative time dilation for observers in motion.

4.5.1 Experimental Confirmation

The effects of time dilation have been confirmed in various ways over the years. Experiments using lithium ions accelerated to one third of light speed have shown that energy level transitions by electrons within the ions take place at a slower rate compared with stationary ions. Atomic clocks flown around the equator have been compared with stationary clocks[7] and the Global Positioning System (GPS) satellites need to have time dilation effects factored into their clocks (Section 8.2.1). However, the classic experimental confirmation involves the decay of particles known as *muons*.

These subatomic particles are often created by cosmic rays hitting particles in the upper atmosphere (about 15 km above the surface). Muons are unstable and will decay with an average lifetime of 2.2 μs. Even moving at 0.995c, muons would take ~ 50 μs to reach the ground, giving ample time for most of them to decay. Yet they can be detected on the surface quite easily.

From the point of view of a co-moving system alongside the muons, the particles are stationary and an average 2.2 μs of proper time can happily tick by before they decay. The Earth is rushing towards the particles, but the distance to the ground is length contracted and many muons can reach the surface over this contracted distance in the time they have.

From the ground's point of view, we observe time for the moving muon tick past more slowly than it does for us (time dilation), so that if we wait 2.2 μs of our time, somewhat less than that time appears to have passed for the moving muons. Hence, it is not surprising that some of them arrive at the ground. Notice how symmetrical the explanation is: from the co-moving system distance is contracted; from the ground-based system, time is dilated.

One of the best experiments of this type was conducted in 1963 by D H Frisch and J H Smith.[8] By detecting muons on Mount Washington and 1907m lower in Cambridge Massachusetts, they were able to determine that the muons' mean initial velocity was ~0.995c dropping to ~0.988c by the time they had reached Cambridge. At such speeds, the muons should travel the distance between detection points in 6.4 μs, which is nearly three of their lifetimes. Consequently, even though 563 muons per hour were picked up on Mount Wilson, only ~27 muons/h should have been detected at Cambridge. In fact, the team had a regular flux of 412 muons/h at Cambridge: an apparent time dilation factor (γ) of 8.8 ± 0.8. Given the range of speeds experienced by the muons, the calculated time dilation comes to 8.4 ± 2, so there is good confirmation.

This striking effect is emphasised in Figure 4.20, where the number of muons that should be detected, given a starting flux of 563 per hour, against time of flight (lower curve) is contrasted with the line showing the detectable flux with time dilation factored in. The results clearly provide strong confirmation for the relativistic effect.

4.6 Note for the Beginner

Length contraction and time dilation can sometimes get a little confusing. I find that the problem is keeping straight which is the time/length in the stationary system and which is the time/length in the moving one. We have called the time elapsed in a co-moving system the proper time and the length measured in the co-moving system the proper length. The formulas are then:

$$\text{time dilation}: \Delta t = \gamma \Delta \tau \text{ time elapsed in moving system} = \gamma \times \text{proper time}$$

$$\text{length contraction}: \Delta x' = \Delta x / \gamma \text{ length in moving system} = \text{proper length}/\gamma.$$

Always keep in mind that $\gamma > 1$ and you will see that the moving time is *more* than the proper time and the moving length is *less* than the proper length.

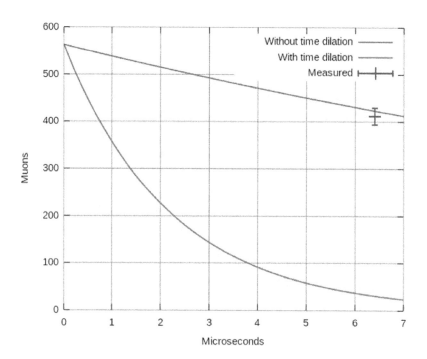

FIGURE 4.20 The Frisch and Smith experimental results showing the number of muons that should be detected against flight time, with and without time dilation. The experimental results show strong confirmation of the time dilation effect. (D.H. Own work – Created with Gnuplot. Based on data given in Frisch/Smith, *American Journal of Physics*, 31(5): 342–355 (1963); licenced under CC BY-SA 3.0: https://creativecommons.org/licenses/by-sa/3.0/.)

4.7 Metrics and the Geometry of Space-time

For most of us the word *geometry* conjures up pictures of lines, circles, triangles and various mathematical theorems to do with their properties. However, mathematicians have long since pushed past the basic properties of shapes to consider more fundamental aspects. One of the background assumptions of geometry is that the shapes are embedded in some space which, in turn, has geometrical properties limiting what is possible within that space.

Any triangle that you draw on a flat sheet of paper must have an angle sum ~180° (the only thing stopping it being an exact sum is our inability to draw and measure to ultimate precision). However, if you were to use a marker pen to draw a triangle on the surface of a ball, then you would not only find that triangles have a different angle sum depending on their size, but also that all triangles have an angle sum > 180°. For the most extreme example, draw two dots on the equator of a ball separated by 90° of latitude and a third at one of the poles. Connect the three dots together with lines and you have just drawn a triangle that contains three right angles. This is still 2D geometry as we are playing on the surface of the ball and it is not dark magic.

The 2D space in which the triangle is embedded has its own geometrical properties, in this case the 2D space curves around in three dimensions and closes on itself. A flat piece of paper does not do that. Its geometry is fundamentally different and somewhat simpler.

These ideas can be extended further into three dimensions, but in ways that are much harder to visualise. The 3D analogue of a piece of paper is a 'flat' space with a simple geometry where the length of a line[9] is determined from the co-ordinates of its end points via:

$$\left(\Delta l\right)^2 = \left(\Delta x\right)^2 + \left(\Delta y\right)^2 + \left(\Delta z\right)^2.$$

However, and this is where imagination fails us, other 3D spaces are possible. In fact, we can have a 3D space that curves around on itself and closes up into a 4-sphere. In this case the 3D space would be the surface of the 4-sphere just as the 2D space was the surface of the ball we were using earlier.

Several points need to be spelled out at this juncture to avoid potential confusions going forward:

- The surface of a sphere is not a 2D space in reality. Any line that we draw on such a surface has to have some thickness and hence stands up from the surface into the third dimension. Genuine 2D spaces are mathematical constructs.
- Creatures living in a 3D space that folds back onto itself would not necessarily be directly aware of the fact. Just because their space curves in a fourth dimension does not give them access to that fourth dimension – they are not able to walk off the surface. Equally, a genuinely 2D creature sliding around the 2D surface of a ball would not be able to step off that surface into the ball or into the room in which it was resting; they lack the dimensionality to do so.
- It would be possible for creatures in a 3D curved space to figure out the geometry of their space by studying the properties of the objects within it. Our 2D creatures grow up thinking that triangles can have angle sums $> 180°$, but a sufficiently bright 2D mathematician could come up with an abstract theory of geometry which imagined a flat 2D surface where all triangles had $180°$ angle sums. By extending that theory they could deduce the fundamental geometry of the space they were living in.
- Don't get ahead of yourself: the embedding fourth dimension of my curved 3D space is not time!

One possible marker we can use to categorise the fundamental geometry of a space is the calculation that determines the length of a line element drawn within that space. In our flat 3D space, we have $(\Delta l)^2 = (\Delta x)^2 + (\Delta y)^2 + (\Delta z)^2$.

Using (non-Lorentz) vectors we can write this formula as $(\Delta l)^2 = \begin{pmatrix} \Delta x & \Delta y & \Delta z \end{pmatrix} M_F \begin{pmatrix} \Delta x \\ \Delta y \\ \Delta z \end{pmatrix}$

where $M_F = \begin{pmatrix} 1 & 0 & 0 \\ 0 & 1 & 0 \\ 0 & 0 & 1 \end{pmatrix}$.

Adding the matrix M_F into things does not seem to achieve very much, however M_F takes this form[10] precisely because of the simple geometry of our flat 3D space. In other spaces, M would look very different. For example, for a 2D space on the surface of a 3D sphere, the length of a line element is given by:

$$(\Delta l)^2 = (R\Delta\theta)^2 + (R\sin(\theta)\Delta\varphi)^2$$

using *spherical-polar co-ordinates* as defined in Figure 4.21.

In matrix form this would be $(\Delta l)^2 = \begin{pmatrix} \Delta\theta & \Delta\varphi \end{pmatrix} M_S \begin{pmatrix} \Delta\theta \\ \Delta\varphi \end{pmatrix}$ with $M_S = \begin{pmatrix} R^2 & 0 \\ 0 & R^2\sin^2(\theta) \end{pmatrix}$.

In space-time, the equivalent of a line element's length is the invariant interval, Δs, which we have already constructed from $(\Delta s)^2 = \begin{pmatrix} c\Delta t & \Delta x \end{pmatrix} \eta \begin{pmatrix} c\Delta t \\ \Delta x \end{pmatrix}$ with $\eta = \begin{pmatrix} 1 & 0 \\ 0 & -1 \end{pmatrix}$ or in its full 4-vector

form $\eta = \begin{pmatrix} 1 & 1 & 0 & 0 \\ 0 & -1 & 0 & 0 \\ 0 & 0 & -1 & 0 \\ 0 & 0 & 0 & -1 \end{pmatrix}$.

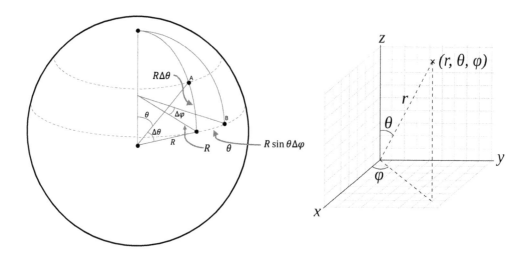

FIGURE 4.21 Right: the spherical-polar co-ordinate system which is an alternative to Cartesian co-ordinates. Left: the length of a line element on the 2D surface of a 3-sphere. If $\Delta\theta$ and $\Delta\varphi$ are very small, then the length of the line connecting points A and B, drawn over the surface of the sphere is $(\Delta l)^2 = (R\Delta\theta)^2 + (R\sin(\theta)\Delta\varphi)^2$.

All of the matrices that we have mentioned in this section are examples of *metrics,* a term that we introduced back in Section 3.5 where we set up the Minkowskian metric, η. They can be a little confusing until you get used to them, as their exact form depends on the co-ordinate system that is chosen. For example, η looks like:

$$\eta = \begin{pmatrix} 1 & 0 & 0 & 0 \\ 0 & -1 & 0 & 0 \\ 0 & 0 & -1 & 0 \\ 0 & 0 & 0 & -1 \end{pmatrix}$$

if we use so-called *Cartesian co-ordinates* x, y, z supplemented with ct. However, in spherical-polar co-ordinates (r, θ, φ) (Figure 4.21) and ct, the metric looks like:

$$\eta_{sp} = \begin{pmatrix} 1 & 0 & 0 & 0 \\ 0 & -1 & 0 & 0 \\ 0 & 0 & -r^2 & 0 \\ 0 & 0 & 0 & -r^2 \sin^2(\theta) \end{pmatrix}$$

giving:

$$(\Delta s)^2 = (c\Delta t)^2 - (\Delta r)^2 - r^2 (\Delta\theta)^2 - r^2 \sin^2(\theta)(\Delta\varphi)^2.$$

Note that the co-ordinate elements $(c\Delta t, \ \Delta r, \ \Delta\theta, \ \Delta\varphi)$ are not included in the matrix representing the metric. We have to keep a close eye on the co-ordinate systems being used in order to evaluate what the metric is telling us. The complexity of this metric, for example, is nothing to do with the structure of the space-time, which is still the 'flat' Minkowskian one. On the other hand, the *Schwarzschild metric,* is a genuine space-time metric, curved by the presence of mass and hence showing effects we used to attribute to the force of gravity:

$$g_{\text{Sch}} = \begin{pmatrix} \left(1 - \dfrac{2GM}{c^2 r}\right) & 0 & 0 & 0 \\ 0 & -\left(1 - \dfrac{2GM}{c^2 r}\right)^{-1} & 0 & 0 \\ 0 & 0 & -r^2 & 0 \\ 0 & 0 & 0 & -r^2 \sin^2(\theta) \end{pmatrix}.$$

This metric gives rise to the interval:

$$(\Delta s)^2 = \left(1 - \frac{2GM}{c^2 r}\right)(c\Delta t)^2 - \frac{(\Delta r)^2}{\left(1 - \dfrac{2GM}{c^2 r}\right)} - r^2(\Delta\theta)^2 - r^2 \sin^2(\theta)(\Delta\varphi)^2.$$

One of the key aspects of this metric is the dependence on co-ordinates within the terms containing $\left(1 - 2GM/c^2 r\right)$, which is telling us real physics about the space-time and the changes in the 'curvature' between different parts of that space-time, not just related to the choice of co-ordinate system.

Right through this section, we have been constructing intervals using the general matrix formula:

$$(\Delta s)^2 = X^T g X,$$

where g is the metric dictated by the co-ordinate system chosen and the geometry of the space-time concerned. This idea is worth committing to memory, as it will be embedded in much that we do going forward.

Notes

1. Minkowski, H. (1909). "Raum und Zeit" [Space and Time]. In *Jahresberichte der Deutschen Mathematiker-Vereinigung*. B.G. Teubner (ed), pp. 1–14.
2. Not truly 'in space' of course due to the suppression of the y and z axes.
3. Note that although the object is moving along the x-axis, the world line is not drawn along that axis in the Minkowski diagram. This is due to the time axis.
4. Do not get co-ordinates such as (x, ct) mixed up with vectors such as $(ct\ x)$ as that could lead to a lot of confusion. A set of co-ordinates includes a "," to separate the variables whereas a vector contains no comma and the variables are more spread out. By convention co-ordinates list the x co-ordinate first, and in our vectors we put ct first as that is more convenient for the invariance…
5. The equation for this curve is simply $4^2 = (ct)^2 - x^2$.
6. Do not get proper length and proper distance confused. The proper length of an object is evaluated in a co-moving system. The proper distance is a space-like invariant distance between two events evaluated in the system where they happen at the same time.
7. In this case time dilation and general relativistic effects need to be taken into account to get the precise result.
8. Frisch, D.H and Smith, J.H., 1963. Measurement of the relativistic time dilation using μ-Mesons. *American Journal of Physics*, **31**, pp. 342–355.
9. We call this 3D space 'flat' as the length of a line takes the same form as it does for a 2D line drawn on a flat piece of paper, where $(\Delta l)^2 = (\Delta x)^2 + (\Delta y)^2$.
10. Yes, I know: it is the identity matrix…

5

Mass, Energy and Dust

5.1 4-Velocity and Newtonian Velocity

In Chapter 3, we introduced the 4-velocity:

$$U = \frac{\Delta X}{\Delta \tau} = \begin{pmatrix} \gamma_u c \\ \gamma_u u_x \\ \gamma_u u_y \\ \gamma_u u_z \end{pmatrix} = \begin{pmatrix} U_0 \\ U_1 \\ U_2 \\ U_3 \end{pmatrix}$$

which, when we supress the y and z components (U_2, U_3), looks like this:

$$U = \begin{pmatrix} \gamma_u c \\ \gamma_u u_x \end{pmatrix} = \begin{pmatrix} U_0 \\ U_1 \end{pmatrix}.$$

This vector was constructed specifically to have a comparable form to Newtonian velocity, yet with similar transformation properties to the 4-displacement, X, under the Lorentz transformations.

While this is all very well and good for mathematical elegance, hard-nosed physicists are going to want to see if it pays off in other regards: does it predict new physics or does it explain physics that is not otherwise catered for?

It is certainly the case that the spatial component[1] of our vector is consistent with the Newtonian view of velocity, when the values involved are small compared to the speed of light. To see this, we deploy a useful approximation:

$$(1+x)^n \approx 1 + nx + \frac{n(n-1)x^2}{2}$$

which works nicely if x is small.[2] If x is *very* small, you can sometimes get away with an even more brutal approximation $(1+x)^n \approx 1 + nx$.

Just to convince you:

$$(1+0.001)^3 = 1.003003001 \text{ when you work it out exactly}$$

$$(1+0.001)^3 \approx 1 + 3 \times 0.001 = 1.003 \text{ when you use the brutal approximation}$$

$$(1+0.001)^3 \approx 1 + 3 \times 0.001 + \frac{\left(3 \times 2 \times 0.001^2\right)}{2} = 1.003003 \text{ if you want to be a little more precise.}$$

In this example, n is a whole number, but the approximation works for fractional and even negative fractional powers.

As the spatial component of our 4-velocity is:

$$U_1 = \gamma_u u = \frac{u}{\sqrt{1 - \beta_u^2}} = u \times \left(1 - \beta_u^2\right)^{-\frac{1}{2}}$$

we can take $x = -\beta_u^2$ and $n = -1/2$ and use the brutal approximation to get:

$$\left(1 - \beta_u^2\right)^{-\frac{1}{2}} = 1 + \frac{\beta_u^2}{2}.$$

So:

$$U_1 = \gamma_u u = \frac{u}{\sqrt{1 - \beta_u^2}} = u \times \left(1 - \beta_u^2\right)^{-\frac{1}{2}} \approx u \times \left(1 + \frac{\beta_u^2}{2}\right) = u + \frac{u\beta_u^2}{2} = u + \frac{u^3}{2c^2} \approx u$$

when the velocities involved are small compared with c.

All of which simply confirms the impossibility of distinguishing between the Newtonian definition of velocity and our relativistic one, until we experiment with objects moving at good fractions of the speed of light. In truth, velocity was *never* the Newtonian form: it was just a good enough approximation to fool us at domestic speeds.

We can, of course, apply the same approximation to the temporal component, $\gamma_u c$. If we do that, the result is:

$$U_0 = \gamma_u c \approx \left[1 + \frac{u^2}{2c^2}\right] c \approx c.$$

Hence, the full velocity 4-vector will shrink down into an approximate form, when the velocities are non-relativistic, and look like $U = \begin{pmatrix} \gamma_u c \\ \gamma_u u \end{pmatrix} \rightarrow \begin{pmatrix} c \\ u \end{pmatrix}$.

Where our new 4-vector starts to pay off with new physics is revealed when we expand our discussion to include *momentum* and *energy*.

5.1.1 Momentum

In Section 2.2.5, I introduced Newtonian momentum:

$$\text{momentum of object} = \text{mass of object} \times \text{velocity of object} \quad \text{or} \quad p = mu$$

and how useful it is in some situations due to its being *conserved* (the total momentum in the universe is always the same).

If we are to move from an old-fashioned Newtonian view of velocity to a shiny new 4-vector one, then we need to port momentum over as well. With that in mind, we define 4-momentum as:

$$P = m \begin{pmatrix} \gamma_u c \\ \gamma_u u_x \end{pmatrix} = \begin{pmatrix} P_0 \\ P_1 \end{pmatrix}$$

which has a spatial component:

$$P_1 = m\gamma_u u_x = \frac{mu_x}{\sqrt{1 - \beta_u^2}} = \frac{mu_x}{\sqrt{1 - u_x^2/c^2}}. \tag{5.1}$$

I have completely unpicked this expression as the final form, $P_1 = mu_x / \sqrt{1 - u_x^2/c^2}$, is very commonly quoted in other developments of relativity.

As before, one of the first things that we need to do with a 4-vector is to see what it gives us as an invariant. We already know that the invariant of the 4-velocity is c^2, so it is simple to see that:

$$\left(\begin{array}{cc} m\gamma_u c & m\gamma_u u_x \end{array} \right) \eta \left(\begin{array}{c} m\gamma_u c \\ m\gamma_u u_x \end{array} \right) = m^2 c^2.$$

Equally, given that the 4-velocity shrinks down to:

$$U = \left(\begin{array}{c} \gamma_u c \\ \gamma_u u \end{array} \right) \xrightarrow{u \ll c} \left(\begin{array}{c} c \\ u \end{array} \right),$$

it follows that the 4-momentum must become:

$$P = m \left(\begin{array}{c} \gamma_u c \\ \gamma_u u \end{array} \right) \xrightarrow{u \ll c} \left(\begin{array}{c} mc \\ mu \end{array} \right).$$

The spatial component has reduced to the expected Newtonian form for momentum, but the temporal component sticks out. On the one hand, we have said that nothing can travel at the speed of light other than light[3] and yet the temporal component seems to be a 'momentum' as if the object were moving at the speed of light. We need to identify the physical meaning of this component.

5.1.2 Enter Energy, Stage Left, Looking Tired...

It will pay us to look in more detail at how the temporal component behaves as we reduce the speed to non-relativistic values. Starting with $P_0 = m\gamma_u c$, which looks completely unfamiliar when compared to any Newtonian expression sitting on the shelf, we apply our approximation:

$$P_0 = mc\left(1 - \beta_u^2\right)^{-\frac{1}{2}} \approx mc\left(1 + \frac{\beta_u^2}{2}\right) \approx mc + \frac{1}{2}mc\frac{u^2}{c^2} = mc + \frac{1}{2}m\frac{u^2}{c},$$

so that:

$$P_0 \approx \frac{mc^2 + \frac{1}{2}mu^2}{c}$$

or slightly more tidily:

$$cP_0 \approx mc^2 + \frac{1}{2}mu^2. \tag{5.2}$$

Allowing that second term to come through, points us in the right direction when it comes to understanding the meaning of the temporal component. The $\frac{1}{2}mu^2$ part should be familiar from school physics as it is the kinetic energy of an object of mass m moving at speed u:

$$\text{kinetic energy} = \frac{1}{2}mu^2.$$

If that second term is an energy, then the first term, mc^2, must be a form of energy as well, as you can't add an energy to something that is not an energy.

Of course, I am almost certainly not surprising anyone here. If you have any passing familiarity with physics (and it is odd to be reading this book if you don't...although welcome aboard!), then you will have seen or heard of the famous *Einstein equation* $E = mc^2$, which carries the dubious distinction of being the most well know and least well understood equation in physics.

Parking the deep meaning of this for the moment, we identify the combination:

$$E \approx mc^2 + \frac{1}{2}mu^2$$

as the approximate value of the *relativistic energy*, E, which takes its full form from:

$$cP_0 = E = c\left(m\gamma_u c\right) = m\gamma_u c^2,$$

so that

$$P_0 = E/c = m\gamma_u c.$$

Our momentum 4-vector is then conveniently written in the form:

$$P = \begin{pmatrix} E/c \\ m\gamma_u u \end{pmatrix} = \begin{pmatrix} E/c \\ p \end{pmatrix}$$

with an invariant:

$$\begin{pmatrix} \dfrac{E}{c} & p \end{pmatrix} \eta \begin{pmatrix} E/c \\ p \end{pmatrix} = \left(E/c\right)^2 - \left(p\right)^2.$$

To find the value of this invariant, we evaluate it in the co-moving system which has $p = 0$ and $E/c = m\gamma_u c = mc$ at rest. Hence, the full invariant is

$$\left(E/c\right)^2 - \left(p\right)^2 = m^2 c^2$$

or, as many people prefer, as it is a little tidier:

$$E^2 - p^2 c^2 = m^2 c^4. \tag{5.3}$$

This equation opens up a relationship between energy, momentum and mass that is not found in Newtonian physics and which is rich with features that we will explore over the next few sections. First, however, we need to take a short diversion into the transformation of wave properties.

5.2 A 4-Vector for Waves

A wave is characterised by its *amplitude*, *frequency*, *wavelength* and *speed*. For a classical wave, the amplitude and frequency determine its energy content, while the frequency relates, for example, in the case of sound, to the *pitch*. The frequency is defined as the number of oscillations completed by the wave medium each second. The wavelength is the shortest distance between equivalent points on the wave and the two are tied together by:

$$\text{wave speed} = \text{frequency} \times \text{wavelength} \quad \text{or} \quad v = f\lambda.$$

Now, in 1900 Max Planck discovered a link between energy and frequency for light which Einstein then extended and used to explain a physical effect that had defied classical explanation on the basis of light being a wave.[4] From this point of view, the particle aspect of light (or any electromagnetic wave) is related to its wave aspect as follows:

$$\text{energy of photon} = h \times \text{frequency of the wave} \quad \text{or} \quad E = hf$$

$$\text{momentum of photon} = h/\text{wavelength of the wave} \quad \text{or} \quad p = h/\lambda$$

where $h = 6.626 \times 10^{-34}$ kg m^2/s is the fundamental constant known as *Planck's constant*. Accepting these relationships without discussion for the moment, we use them as a clue to a possible 4-vector for wave properties. After all, for a particle we have defined the 4-momentum to be $P = \begin{pmatrix} \gamma mc \\ \gamma mv \end{pmatrix} = \begin{pmatrix} E/c \\ p \end{pmatrix}$. For a photon (particle of light) this would be:

$$P_{\text{photon}} = \begin{pmatrix} E/c \\ p \end{pmatrix} = \begin{pmatrix} hf/c \\ h/\lambda \end{pmatrix} = h \begin{pmatrix} f/c \\ 1/\lambda \end{pmatrix}.$$

To tidy this 4-vector up somewhat, it is conventional to re-dress frequency and wavelength using two new variables, ω (the angular frequency) and k (the wave number) defined by:

$$\omega = 2\pi f \quad k = 2\pi/\lambda \quad c = f\lambda = \frac{\omega}{2\pi} \times \frac{2\pi}{k} = \frac{\omega}{k}.$$

With these variables, the Planck–Einstein relationships become:

$$E = hf = \frac{h\omega}{2\pi} \quad p = \frac{h}{\lambda} = \frac{hk}{2\pi}.$$

In turn, this gives us:

$$P_{\text{photon}} = h \begin{pmatrix} f/c \\ 1/\lambda \end{pmatrix} = h \begin{pmatrix} \omega/2\pi c \\ k/2\pi \end{pmatrix} = \frac{h}{2\pi} \begin{pmatrix} \omega/c \\ k \end{pmatrix}.$$

Suggesting that the correct *frequency 4-vector* for the wave aspect of light and by extension for any wave is $F = \begin{pmatrix} \omega/c \\ k \end{pmatrix}$. A Lorentz boost of this 4-vector would be $F' = \begin{pmatrix} \omega'/c \\ k' \end{pmatrix} = L \begin{pmatrix} \omega/c \\ k \end{pmatrix}$

or in longhand:

$$\omega'/c = \gamma \left(\omega/c - \beta k \right)$$

$$k' = \gamma \left(k - \beta \omega/c \right).$$

Flicking back to the particle view of light for a moment, the energy transformation of a photon then becomes:

$$E' = \frac{h\omega'}{2\pi} = \frac{hc}{2\pi} \left(\omega'/c \right) = \frac{hc}{2\pi} \gamma \left(\omega/c - \beta k \right) = \frac{hc}{2\pi} \gamma \left(\frac{\omega}{c} - \beta \frac{\omega}{c} \right) = \frac{h\omega}{2\pi} \gamma \left(1 - \beta \right)$$

or finally:

$$E' = \gamma E \left(1 - \beta \right) \quad \text{or} \quad f' = \gamma f \left(1 - \beta \right)$$

which looks somewhat different from our normal transformation, due to the extra connection the wave properties force between k and ω. This result will come in useful shortly.

5.2.1 Redshift

While we are on the subject, and because it will be useful later (e.g. Chapter 8), we can take a brief look at the *relativistic Doppler effect*.

The transformation $k' = \gamma\left(k - \beta\omega/c\right)$ is most neatly framed in this form, but can also be written in terms of wavelength, using $k = 2\pi/\lambda$:

$$\frac{1}{\lambda'} = \gamma\left(\frac{1}{\lambda} - \beta\frac{f}{c}\right) = \gamma\left(\frac{1}{\lambda} - \beta\frac{1}{\lambda}\right) = \frac{\gamma}{\lambda}(1-\beta)$$

or, more elegantly:

$$\lambda' = \frac{\lambda}{\gamma(1-\beta)} = \lambda\sqrt{\frac{1+\beta}{1-\beta}},$$

where the last step has used $\gamma = 1/\sqrt{1-\beta^2}$. This expression tells us how the wavelength of a photon is viewed from two systems in relative motion at $\beta = v/c$ with respect to each other, assuming that the stationary system is co-moving with the light source. In school we learn how classical waves suffer a shift in frequency and wavelength due to the relative motion of the transmitting and receiving objects; this is called the *Doppler effect*. Here we have the relativistic version of this, applied to light.

We can explore how our relativistic relationship works for $v \ll c$ by applying our $(1+x)^n \approx 1 + nx$ approximation in a couple of steps:

$$\sqrt{1+\beta} = (1+\beta)^{\frac{1}{2}} = 1 + \frac{\beta}{2} \qquad \frac{1}{\sqrt{1-\beta}} = (1-\beta)^{-\frac{1}{2}} = 1 + \frac{\beta}{2},$$

so that:

$$\lambda' = \lambda\sqrt{\frac{1+\beta}{1-\beta}} \approx \lambda\left(1+\frac{\beta}{2}\right)\left(1+\frac{\beta}{2}\right) = \lambda\left(1+\beta+\frac{\beta^2}{4}\right) \approx \lambda(1+\beta).$$

Note that the wavelength observed in the moving system is *longer* (shifted to the red) than that in the stationary one (relatively speaking).

Traditionally, we express the shift in wavelength in terms of *redshift*, which is defined using the formula:

$$z = \frac{\Delta\lambda}{\lambda} = \frac{\lambda'-\lambda}{\lambda} \approx \frac{\lambda(1+\beta)-\lambda}{\lambda} = \beta = \frac{v}{c},$$

which is the classical expression for the Doppler shift of light.

5.3 Relativistic Energy

Relativity exposes a deep relationship between mass and energy that was not apparent until we started manipulating particles that move about at good fractions of the speed of light and interact with each other to produce new forms of matter. This is another 'tent pole' feature of relativity and perhaps the hardest to conceptualise clearly.

Our temporal component of the 4-momentum vector, in its approximate form, has two terms:

$$cP_0 \approx mc^2 + \frac{1}{2}mu^2, \tag{5.4}$$

the second of which is evidently the Newtonian kinetic energy. Hence, we argued that mc^2 must also be an energy, as you can't add an energy to something that is not an energy. Whatever this energy is, it clearly does not disappear at rest, as in the co-moving system $cP_0 \approx mc^2$. There are no other physical variables involved here (no charge, no current; nothing like that) so we have no real clues leading us to the nature of this energy.

There is a temptation to say that this is a form of energy due to the object's mass, but we can't easily sustain this as the mass also comes into the second term.

Equally, it is not true to say that energy and mass are 'interchangeable' – that mass can be converted into energy and vice versa. Mass and energy are two related but distinct physical quantities.

The correct interpretation is that mass is a *property of energy*. Hence, the mass of an object depends directly on its energy content. However, before we develop this idea any further it is useful to confirm it from another direction, via an elegant argument put forward by Einstein in a follow-up paper published in 1905.

5.3.1 Energy and Mass

Imagine that we have a stationary particle in our co-ordinate system with energy E_i (initial energy). When observed from another inertial system, this same particle has energy E_i'.

The particle sends out two photons in opposite directions and along the same axis as the relative motion between the systems.[5] Each photon carries energy $L/2$ and as a result, the emitting particle remains stationary in S, but now has energy E_f (final energy). Simple conservation of energy gives us:

$$E_i = E_f + \frac{L}{2} + \frac{L}{2} = E_f + L.$$

Viewed from S' and given that the principle of relativity dictates that the energy in this system must be conserved as well, we have:

$$E_i' = E_f' + \gamma \frac{L}{2}(1 - \beta_v) + \gamma \frac{L}{2}(1 + \beta_v) = E_f' + \gamma L,$$

where I have used the photon energy transformation derived earlier, noting that as the photons are travelling in opposite directions we need $\pm\beta_v$ values. Subtracting gives us:

$$(E_i' - E_i) = (E_f' - E_f) + (\gamma - 1)L.$$

The difference $(E_i' - E_i)$ can only be due to the relative motion of the two systems and so represents the kinetic energy of the object in the moving system. To be on the safe side, we will even add in a constant, as energy differences are all we can really measure. So,

$$(E_i' - E_i) = KE_1 + C.$$

The same argument applies to the difference $(E_f' - E_f)$, which tells us that:

$$(E_f' - E_f) = KE_2 + C.$$

As a consequence:

$$(KE_1 + C) - (KE_2 + C) = (\gamma - 1)L.$$

Evidently, in the moving system the object has lost kinetic energy equal to:

$$KE_1 - KE_2 = (\gamma - 1)L.$$

Using our approximation for small speeds compared to light:

$$\frac{1}{\sqrt{1-\beta_v^2}} = \left(1-\beta_v^2\right)^{-1/2} = 1 + \frac{\beta_v^2}{2}$$

gives us:

$$KE_1 - KE_2 \approx \left(1 + \frac{\beta_v^2}{2} - 1\right)L = \frac{1}{2}\left(\frac{L}{c^2}\right)v^2.$$

As the Newtonian formula for $KE = \frac{1}{2}mv^2$, this strongly suggests that the difference in the kinetic energy of the object has come about from a change in *mass*. In fact, *emitting photons of total energy L has reduced the object's mass by an amount L/c^2*.

As Einstein puts it,[6] '*If a body gives off the energy L in the form of radiation, its mass diminishes by L/c^2*. The fact that the energy withdrawn from the body becomes energy of radiation evidently makes no difference, so that we are led to the more general conclusion that the mass of a body is a measure of its energy-content; if the energy changes by L, the mass changes in the same sense by $[L/c^2]$'.

5.3.2 Internal Energy

We have been conditioned to think of mass in terms of the 'quantity of matter in an object', so that mass has a tangible feel related to 'stuff'. It is hard to think of energy in terms of 'stuff', so even harder to think of energy as having mass.

In truth, mass is correctly related to an object's inertia (its 'reluctance' to change its state of motion). The definition we teach to pupils in their earlier years ('mass is the quantity of matter in an object[7]') is really a desperate attempt to say something, rather than a sustainable line of thought. In any case, with our modern view of matter in terms of electron clouds, tiny nuclei and subatomic particles, matter has lost a good deal of its 'stuff-ness'. It now looks a lot less tangible than the little marble view of particles that was in vogue in the early 1900s.

It is hard to believe that adding energy to something increases its mass. Yet, it is true: a cup containing hot coffee (baring evaporation) will have more mass than the same cup with (the same amount of) cold coffee. However, the additional mass is so small we could not hope to measure the increase. Remember that $E = mc^2$, so that the change in mass is $\Delta m = \Delta E/c^2$. With $c^2 \sim 9 \times 10^{16}\,\text{m}^2/\text{s}^2$ it takes a very large energy change, ΔE, to add any significant mass. Our only hope for detecting a mass increase is to start with something that has a tiny mass in the first place and to add to it by applying an appropriate amount of energy. subatomic particles, such as electrons, are ideal. However, it is tricky to directly measure the mass of an object in flight. What we can do is to allow a charged particle to pass through a magnetic field and measure the extent to which its path is curved by the magnetic force. That gives us a measure of the particle's momentum, and if we know the velocity then the mass can be inferred. If we do that, we find that the mass of the electron increases with energy.

The first demonstration of this effect, with sufficient precision to rule in favour of Einstein compared to other models that were around at the time, was carried out in 1940. Rogers et al.[8] used electrons emitted from a radium source and their results are reproduced in Figure 5.1. Since then this dependence is routinely confirmed in particle physics experiments, and factored into the design of particle accelerators such as the LHC at CERN.[9]

Having confirmed that the energy of an object dictates its mass, we are lead to ponder the nature of the energy content of an object at rest, separated from any sources of potential energy, such as other masses (gravitational potential energy) or charges (electrostatic potential energy). In this situation, some form of energy must be 'inside' the object. I'm going to call this *internal energy*.

There are two cases to consider: objects which have some structure inside them (composite particles) and fundamental particles which have no internal structure.

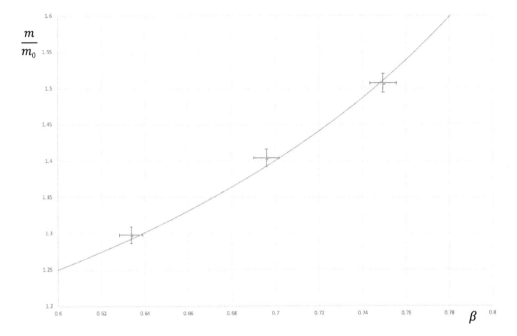

FIGURE 5.1 The variation of mass with velocity for electrons at three velocity values. The *y*-axis is showing the ratio of the mass compared to the mass at rest, and the *x*-axis the ratio of the velocity to that of light. The solid line is the theoretical prediction from the equation $m/m_0 = 1/\sqrt{1-\beta^2}$. Data are taken from Rogers et al.

Since the mid-1960s, we have been aware of fundamental particles called *quarks* which bind together to form composite particles, like protons and indeed neutrons. The internal energy of a proton is a complex mix, involving the kinetic energy of the moving quarks flying around inside the volume of the proton, the interaction energies between the quarks and finally the internal energy within the quarks themselves. Great progress has been made in recent years using computer calculations and we now have an accurate description of how these various energies form the mass of the proton. Of all the factors involved, the internal energy of the quarks turns out to be the least significant.

Moving up in scale, the energy content of an atom is that of the protons and neutrons in the nucleus, modified by the fields involved and the orbiting electrons with their field energy and kinetic energy. All of these factors combine to determine the mass of the atom. Of course, atoms bind together in various ways to form matter and the atoms have kinetic energy either due to vibrational motion or translation through space. Such energy would also contribute mass to the total. Heating matter up (my coffee again) serves to raise the average energy of the atoms and molecules involved, which is how the (tiny) mass change comes about.

Throughout all this detail, we are still left with a fundamental problem: identifying the internal energy of particles that are not composite: quarks for one and electrons for another.

Relativity tells us that such energy exists,[10] but says nothing about its nature. Einstein was of course unhappy about this state of affairs and worked hard to try and explain the internal energy of particles in the context of the general theory of relativity, especially by incorporating Mach's principle. Those efforts have been somewhat circumvented by more modern developments in the field of particle physics, which we will come to shortly. We will also talk more about Mach's principle in Chapter 6.

5.3.3 Mass and Rest Mass

Whatever this internal energy is, it is parcelled up inside the particle and gives it the mass we measure when it is at rest: the *rest mass, m_0*.

As we add energy to the particle, by accelerating it for example, its mass will increase. We can see this from our momentum equation if nothing else:

$$P_1 = \gamma m_0 u = \frac{m_0 u}{\sqrt{1 - u^2/c^2}} = m_u u,$$

where $m_u = m_0 / \sqrt{1 - u^2/c^2}$ is the mass at a speed u, sometimes called the *relativistic mass*.

While this is a perfectly valid approach to the issue of mass, it is not the only possible view. Our colleagues in the particle physics community refer to rest mass as *the* mass of the particle, as the value is a characteristic property of the species concerned. For a particle in motion, particle physicists tend to work with the energy and momentum rather than worrying about the mass in such circumstances. We will follow their pattern.[11]

Interestingly, Einstein was not keen on the idea of relativistic mass as the quantity cannot be directly measured:

It is not good to introduce the concept of the mass $M = m / \sqrt{1 - v^2/c^2}$ of a moving body for which no clear definition can be given. It is better to introduce no other mass concept than the 'rest mass' m. Instead of introducing M it is better to mention the expression for the momentum and energy of a body in motion.[12]

In a comparatively modern text on the subject, Taylor and Wheeler[13] take the view:

The concept of "relativistic mass" is subject to misunderstanding. That's why we don't use it. First, it applies the name mass – belonging to the magnitude of a 4-vector – to a very different concept, the time component of a 4-vector. Second, it makes increase of energy of an object with velocity or momentum appear to be connected with some change in internal structure of the object. In reality, the increase of energy with velocity originates not in the object but in the geometric properties of space-time itself.

Their argument starts from the energy-momentum 4-vector, $P = \begin{pmatrix} E/c \\ p \end{pmatrix}$ with the invariant:

$$\begin{pmatrix} E/c & p \end{pmatrix} \eta \begin{pmatrix} E/c \\ p \end{pmatrix} = \left(\frac{E}{c}\right)^2 - p^2$$

which, as we saw when we evaluated it the co-moving system, takes the value $m^2 c^2$. In their view, this is the correct place to define the mass of a particle. Looked at in this manner, the mass becomes a scalar quantity. Moving from one system to another has no impact on the energy parcelled up *within* the particle. It *can* impact on the *kinetic energy* of the particle, as velocity transforms between systems. If the 'total' mass of a moving particle were to be determined, that would be different between systems, but the component of that mass due to the internal energy would not have changed. All in all, it is simpler to:

- acknowledge that the energy content of an object does alter that mass if the object is composite, but
- otherwise treat the mass as a scalar and
- do not worry about the mass of a moving object.

That is why in the equation for momentum, $p = \dfrac{m_0 u}{\sqrt{1 - u^2/c^2}}$, I preferred to bind the Lorentz factor, γ, to the definition of the 4-velocity rather than use it to define a mass that transforms, e.g. $p = m_0 (\gamma_u u)$ rather than $p = (\gamma_u m_0) u$. In fact, from now on I will drop the subscript, 0, on the mass, so that m_0 is just written as m.

So, we have chased the issue of internal energy down to fundamental particles and identified that for them to have mass, there must be some internal energy involved which is not related to dynamic factors. This is where the *Higgs field* comes in.

5.3.4 Higgs Fields and Higgs Particles

For various reasons associated with the theory of fundamental particles, the issue of particle mass came to the fore in the 1970s. At the time, various extremely elegant theories were being developed to explain the interaction of particles. All of them, however, suffered from a serious issue: their full mathematical elegance, based on what appeared to be sound principles, fell apart if the fundamental particles had mass. An ingenious solution was devised independently by three groups: Robert Brout and François Englert; Peter Higgs; and by Gerald Guralnik, C. R. Hagen and Tom Kibble.

The *Higgs mechanism*, as it is now known, relies on the existence of a Higgs field that permeates the whole of space-time. In that restricted sense, it is reminiscent of the aether proposed to explain the wave nature of light. In this theory, all fundamental particles are inherently massless, but some species acquire mass by interacting with the Higgs field and 'carving out' an interaction energy that rides with them.

The real secret of the Higgs mechanism is the way it shades the underlying mathematical structure of the theories rather than destroying it. The interaction with the Higgs field gives rise to terms in the equations which look just like 'proper' mass terms but do so by forcing the equations to fall into one of several equivalent possibilities, in the process hiding the underlying principles at work. One, very limited, analogy is to think of a pencil balanced on its end. If all is set up perfectly well, the mathematics of the situation allows the pencil to remain vertical indefinitely. However, in practice some gentle fluctuation will make the pencil fall over. The underlying principles are still there, in that the pencil can fall over in any direction around its point, but the final result masks the symmetry that was there: the pencil is on its side in some orientation. The Higgs terms force the equations to 'fall-over' from a position of perfect symmetry into one particular instance from the set of possibilities under the symmetry.

Now, as this Higgs field is fundamentally a quantum field, it displays the same sort of granularity that other quantum fields do: it has wave/particle aspects. Consequently, given enough energy and the right experimental context, the Higgs field can be prompted to display its particle aspect. It can be exited into ripples in its otherwise 'smooth' surface, which show up as particles. Finding the *Higgs Boson*, the particle associated with the Higgs field, was crucial in demonstrating that the Higgs field existed and so that the Higgs mechanism was more than a nifty mathematical dodge. After decades of work and technological development, the discovery of the Higgs Boson was announced in Switzerland in 2012. As one of the key scientific discoveries of the 21st century, this led later that year to Higgs and Englert being awarded the Nobel Prize.[14]

Our picture then is of a universe containing fundamental particles that, in the absence of a Higgs field, would have no internal energy and hence no mass. However, the Higgs field does exist and for most fundamental particles their interaction with this field carves out an energy that is bound to them and gives rise to their mass, even in the absence of other forms of energy. However, some fundamental particles do not interact with the Higgs field and so are genuinely massless. The photon being the archetypal example.

5.3.5 The Photon Loop-Hole

As an object is accelerated by applying a force, energy is transferred to the object (work is done). In Newtonian physics, that energy goes to increase the kinetic energy of the object. This is also true in relativity. However, in relativity the dependence on velocity is not as simple as $KE = \frac{1}{2}mv^2$. Briefly, the faster the object is moving the harder it is to make it move any faster.

We can see this in Figure 5.2, where I have calculated the relativistic energy of an electron, $E = \gamma m_e c^2$, for different values of $\beta = v/c$.

The energy clearly rises rapidly for $\beta \to 1$. In fact, the amount of energy needed to increase the speed is also rising. Consider going from $\beta = 0.1$ to $\beta = 0.2$, which requires an energy increase of 4.4×10^4 J. Going from $\beta = 0.91$ to $\beta = 0.92$ is an energy rise of 3.9×10^5 J and from $\beta = 0.991$ to $\beta = 0.992$ is a rise of 1.3×10^6 J. You can see how incremental changes in speed are getting further and further out of reach.

This is why it is impossible to accelerate an object up to the speed of light.

As a contrast to this, Figure 5.2 also shows the variation of Newtonian kinetic energy with speed (as a fraction of the speed of light). The calculation has been cut off at $v = c$, even though there is no speed limit in Newtonian physics. Note how the incremental increases in velocity do not produce a rapidly increasing energy requirement.

Given the increasing energy gains required to advance the velocity in small steps, there is an obvious question: how do photons manage to move at the speed of light?

The first thing to say is that photons do not *accelerate*, in that their velocity does not change with time.[15] They are quantum objects – the granularity of the electromagnetic field, if you like. Photons are emitted from atoms, for example, when orbital electrons change their energy state. In that process, some energy is transformed from one form to another and dumped into the electromagnetic field. That energy excites the field and a photon is formed. At the moment of formation, *the photon is created moving at the speed of light*. There is no period of acceleration. Indeed, there is no period during which the photon is 'created'.

However, this does not address the parallel issue which is how a photon can travel at light speed without an unlimited supply of energy. The most obvious problem comes from the energy equation itself:

$$E = \gamma mc^2 = \frac{mc^2}{\sqrt{1 - v^2/c^2}}.$$

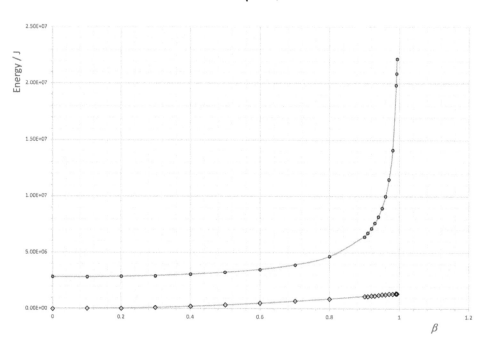

FIGURE 5.2 The variation of energy with speed. The horizontal axis shows the speed of an object in fractions of the speed of light (v/c), and the vertical axis shows the relativistic energy of the object at that speed. This has been calculated for an electron. Hence, the energy tends to $m_e c^2$ as the speed approaches zero. Also shown on this graph is the same calculation for the Newtonian kinetic energy (lower line with diamond shaped markers).

Should $v = c$, we end up with $mc^2/0$ and the equation 'blows up' or, in more mathematically rigorous terminology $E \to \infty$ as $v \to c$.

To see our way through this, it helps to examine how this energy equation comes about, then we can see its range of applicability.

If we go back to the 4-momentum invariant:

$$E^2 - p^2 c^2 = m^2 c^4,$$

an obvious re-arrangement is:

$$E^2 = m^2 c^4 + p^2 c^2.$$

Now we insert into this equation the formula for momentum, $p = \gamma m v$, yielding:

$$E^2 = m^2 c^4 + \left(\gamma m v\right)^2 c^2 = m^2 c^2 \left(c^2 + \gamma^2 v^2\right) = m^2 c^2 \left(\frac{\left(1 - v^2/c^2\right) c^2 + v^2}{1 - v^2/c^2}\right) = \gamma^2 m^2 c^4$$

or

$$E = \gamma m c^2.$$

Clearly the substitution of the momentum expression is at the crux of this. So, what if there was another expression for momentum different to $p = \gamma m v$, but consistent with it? Such a possibility exists. If we go back to Planck's formulas for the energy and momentum of a photon:

$$E = hf \quad p = \frac{h}{\lambda}$$

they can be combined to give $E = pf\lambda = pc$, or $p = E/c$ which is completely consistent with the invariant $E^2 = m^2 c^4 + p^2 c^2$, if $m = 0$.

We now have this interesting idea: the photon can have energy and momentum and can travel at the speed of light *provided it has no mass*. The only form of energy that a photon has is kinetic energy. Without that, it vanishes. As it does not have a mass, it can never be at rest. Consequently, it does not accelerate to light speed and so does not need an indefinite amount of energy to get there.

In one sense, the photon exploits a loop-hole, for if the momentum of a particle is:

$$p = \gamma m v = \frac{m v}{\sqrt{1 - v^2/c^2}}$$

and a photon has $m = 0$, we have:

$$p = \frac{0}{\sqrt{1 - v^2/c^2}},$$

which does not 'blow up' as $v \to c$. A better way of looking at it, though, would be to say that this formula for momentum only works for $m \neq 0$ otherwise you have to use the version $p = E/c$.

5.4 Mass, Energy and Gravitation

Now that we are getting used to the relationship between energy has mass, we need to address some related questions, such as does potential energy have mass (yes, with caveats) and if so does that alter the mass of a particle (no, with caveats)?

At school we learn about gravitational potential energy, which is the energy stored in a gravitational field when an object is raised to a certain height. In the general theory of relativity, the issue of gravitational potential energy is not straight forward as we have replaced the idea of a gravitational field with curvature in space-time. This is something that we need to come back to.

The electromagnetic field also contains a form of potential energy, but we treat that as separate to the objects embedded into the field. The only exception to this comes from composite particles where the field energy inside the boundary of the composite is counted towards that composite's mass. As mentioned earlier, the mass of a proton gets a substantial contribution from the fields at work inside the proton.

The distinction is important. After all, if forms of energy have mass, we expect them to be sources for a gravitational field!

Here we are starting to tiptoe into the general theory of relativity, so we need to step cautiously. Newtonian theory regards mass as the source of a gravitational field, while also being subject to the effects of a gravitational field. Newton himself wrestled with understanding how two objects could exert forces on each other even when separated by large distances (the Earth and the Moon for example). Forces, in his experience, required contact between the objects concerned. The rather spooky 'action at a distance' central to gravitation was an issue that he never resolved to his own satisfaction. Indeed, Newton wrote[16]:

> It is inconceivable that inanimate brute matter should, without the mediation of something else which is not material, operate upon and affect other matter without mutual contact... That gravity should be innate, inherent, and essential to matter, so that one body may act upon another at a distance through a vacuum, without the mediation of anything else, by and through which their action and force may be conveyed from one to another, is to me so great an absurdity that I believe no man who has in philosophical matters a competent faculty of thinking can ever fall into it.

Later developments allowed a rather neat solution. One mass (say the Earth) acts as the source of a gravitational field extending out into space (in principle without limit). Another mass (the Moon) placed into that field interacts with the field *at its location* and a force is created pointing towards the source mass. There is no action at a distance, the fields and masses interact *locally*.

Mathematically, we express this in the following way. Newton's law of gravity is:

$$F = -\frac{GM_1M_2}{r^2}.$$

If we regard the mass M_1 as the source of the gravitational field, we can write:

$$g = -\frac{GM_1}{r^2},$$

where we are using g to represent the strength of the gravitational field a distance r from the mass.

The mass M_2 is sitting in this field and so a force arises:

$$F = M_2g = -\frac{GM_1M_2}{r^2}.$$

This would just be a mathematical dodge, if the field did not have some demonstrable independent reality. Of course, when it comes to the general theory this whole idea is replaced by space-time curvature, but a similar argument can be followed for the electrical field surrounding a charge and that leads to the idea of electromagnetic waves, which are direct evidence for the existence of fields and so justify this approach.

Anyway, if the general theory of relativity is to be consistent with Newtonian gravity (and we expect that it must for small speeds and small, on some scale, masses) then the source of space-time curvature must be energy. If all forms of energy have mass, then we expect that space-time will be curved by the presence of any energy, even if that energy is contained in an electromagnetic field, for example. Indeed, we expect that gravitational energy itself will be a source of gravitation...

5.4.1 Poisson's Equation

Newton's law of gravity, as quoted in the last section, applies to point masses, which we seldom come across in nature. In order to apply this law to celestial bodies, Newton had to prove that a spherical mass has the same gravitational effect as a point mass of equal magnitude, placed at the centre of the sphere. You can also treat an extended object of some random shape as a collection of point objects and do some intricate mathematics to calculate their overall effect. However, the law of gravity expressed in this form is fundamentally limited and if we are going to explore how Newton's theory morphs into general relativity, we need a somewhat different mathematical expression to work with.

Back in Chapter 2, we introduced Maxwell's equations and in particular the first equation for electric fields:

$$\frac{\Delta E_x}{\Delta x} + \frac{\Delta E_y}{\Delta y} + \frac{\Delta E_z}{\Delta z} = \frac{\rho}{\varepsilon_0}.$$

At the time, I explained that this equation related how the (x, y, z) components of the field changed over small distances $(\Delta x, \Delta y, \Delta z)$, to the density of charge, ρ, in that region. The similar equation for magnetic fields is:

$$\frac{\Delta B_x}{\Delta x} + \frac{\Delta B_y}{\Delta y} + \frac{\Delta B_z}{\Delta z} = 0$$

as there are no magnetic charges to act as sources or sinks of the field.

Now, such equations have a more general application than the single context of electromagnetism. In fact, for any field,[17] we can write:

rate of change of field with distance = density of field sources

and applying the idea to the gravitational field gives us:

$$\frac{\Delta g_x}{\Delta x} + \frac{\Delta g_y}{\Delta y} + \frac{\Delta g_z}{\Delta z} = -4\pi G\rho,$$

where g_x, g_y, g_z are the components of the gravitational field, ρ is the mass density in the region and G the universal gravitational constant. The 4π enters by convention and is related to the field being spherically symmetrical.[18] The minus sign forces the gravitational field to be attractive for a positive mass (the only kind of course...), so that field lines run towards the mass. In the version for electrical fields, the field lines run away from a positive charge; hence, there is no minus sign in that equation.

It is also possible to write the gravitational force in terms of the potential energy associated with it:

gravitational force = − the rate at which the potential energy changes with distance

or

$$F = -\frac{\Delta GPE}{\Delta r}$$

the minus telling us that the force points from regions of high GPE towards those with lower GPE, which is exactly the direction you would expect an object to move.

Here I am using r to refer to a distance in some random direction.

If you are not familiar with this idea, then consider lifting an object in a gravitational field. To do that, you have to apply a force to overcome gravity. Applying that force and moving the object a vertical distance is doing work on the object, which transfers energy. Where does that energy go? Normally when you do work, a change in kinetic energy is the result. However, in this case while the object might have a bit of kinetic energy while you are lifting it, it is not enough to account for the work done. The energy you transfer to the object via the force you apply is drained away from the object by the force of gravity. That energy is then stored in the gravitational field (Figure 5.3).[19]

On being released, the object falls back towards the ground. In the process, the stored energy in the gravitational field gets transferred back, increasing the kinetic energy of the object. So, the object will naturally fall from a region of high potential energy towards a region where the potential energy is lower. Also, a large change in kinetic energy in a short distance must result from a big change in the potential energy over that distance and can only be mediated by a larger force. Hence the gradient of the potential energy with distance is equal to the magnitude of the force.

Not so long ago we extracted the idea of the gravitational *field* from the gravitational *force* by writing $g = -\dfrac{GM_1}{r^2}$ and $F = M_2 g$. We can use the same trick to extract the gravitational *potential*, ϕ, from the potential energy, U:

$$\phi = -\frac{GM_1}{r} \quad \text{and} \quad U = M_2\phi.$$

The gravitational potential being, in effect, the potential energy per unit mass, as the field can be thought of as the force per unit mass, which if you think about it is the same as the acceleration due to gravity. Putting this together we come up with[20]:

$$g = -\frac{\Delta\phi}{\Delta r} \quad \text{or} \quad g_x = -\frac{\Delta\phi}{\Delta x} \quad g_y = -\frac{\Delta\phi}{\Delta y} \quad g_z = -\frac{\Delta\phi}{\Delta z}.$$

Armed with these expressions, we can go back to our equation relating the field change to the mass density and write:

$$\frac{\Delta}{\Delta x}\left(-\frac{\Delta\phi}{\Delta x}\right) + \frac{\Delta}{\Delta y}\left(-\frac{\Delta\phi}{\Delta y}\right) + \frac{\Delta}{\Delta z}\left(-\frac{\Delta\phi}{\Delta z}\right) = -4\pi G\rho$$

or:

$$\frac{\Delta}{\Delta x}\left(\frac{\Delta\phi}{\Delta x}\right) + \frac{\Delta}{\Delta y}\left(\frac{\Delta\phi}{\Delta y}\right) + \frac{\Delta}{\Delta z}\left(\frac{\Delta\phi}{\Delta z}\right) = 4\pi G\rho. \tag{5.5}$$

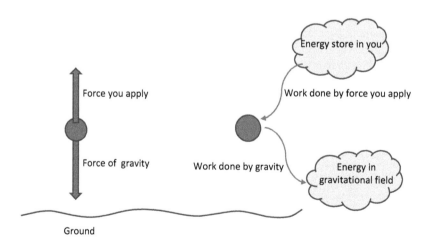

FIGURE 5.3 The force that you apply to an object in order to lift it transfers energy from you. The force of gravity transfers that energy away from the object to be stored as gravitational potential energy.

The expressions $\Delta/\Delta x(\Delta\phi/\Delta x)$, etc. tell us how much the change in ϕ with x changes with x.

Think of it like this: if speed is the rate of change of distance with time, $\Delta x/\Delta t$, and acceleration[21] is the rate of change of speed with time $\Delta u/\Delta t$, then we could write acceleration as $\Delta/\Delta t(\Delta x/\Delta t)$ which is the rate at which the change of distance with time is changing with time. I've just done a similar thing for rates of change of distance in Equation 5.5.

Mathematicians like using the most compact notation possible, so they have invented a symbol for

$$\frac{\Delta}{\Delta x}\left(\frac{\Delta\phi}{\Delta x}\right)+\frac{\Delta}{\Delta y}\left(\frac{\Delta\phi}{\Delta y}\right)+\frac{\Delta}{\Delta z}\left(\frac{\Delta\phi}{\Delta z}\right)$$ and would write this as $\nabla^2\phi$, which makes our gravitational equation:

$$\nabla^2\phi = 4\pi G\rho,$$

which is also known as *Poisson's equation*[22] and is the most useful form of the law of gravity when it comes to crossing into general relativity. We will see how the left-hand side of this equation makes the transition in a later chapter. For the moment, we focus on the right-hand side and look for a suitable replacement for the mass density, now that we know that mass comes from energy.

5.5 The Physics of Dust

The general theory of relativity, in all its glory, can be used to explain the behaviour of exotic objects like black holes and active galaxies. It can be applied to the universe as a whole and produce workable models of the evolution and eventual fate of reality. So, when we are looking for an appropriate structure to act as the source of gravitation, we need to look at extended objects containing many particles, rather than a single point-particle source. This is why we developed Poisson's equation.

The simplest extended structure is a collection of non-interacting point masses, each of which has the same mass and all of which are stationary. Colloquially, such a structure is known in the community as *dust*.[23] The properties of such a simple structure amount to the masses of the individual particles, m, the number of particles, N, and hence the total mass of the structure, $M = Nm$, as well as the mass density, ρ, which would be crucial in determining the local curvature of space-time.

Tantamount to knowing the density would be to know the *number density* of the particles: the number of particles in a unit volume surrounding a point of interest:

$$n = \frac{N_{xyz}}{\Delta x\Delta y\Delta z},$$

where N_{xyz} is the number of particles in an imaginary rectangular box $\Delta x\Delta y\Delta z$ surrounding the point (x,y,z). In a slightly more complicated situation, it is very likely that the number density would vary from place to place inside the distribution of dust, making this a function of position and potentially time, $n(t,x,y,z)$.

The number density is invariant under Galilean transformations and so in Newtonian physics was considered to be a fixed quantity independent of vantage point. This is not true in relativity: space is contracted along the boost direction so that $\Delta x' = \Delta x/\gamma$ making $n' = \gamma n$. The total number of particles in the dust is, of course, the same for all observers.

The number density is not the only aspect of the dust which will be different for an observer moving with respect to a system that is stationary within the dust. The *flux* of particles, which is the number of particles passing through a certain area in a certain time, will be different as well. If the particles are moving in the x direction, then the flux will be the number of particles passing through the side $\Delta y\Delta z$ of our imaginary rectangular box, per second.

Viewed from the system S' in which the particles are moving (to the *right,* so S' is moving to *left* from the point of view of the dusty system), the flux is simply the number of particles within a volume *ut'* by $\Delta y' \Delta z'$ (Figure 5.4). Any particle that starts off further away than *ut'* from the area of interest will not have reached that area in the time *t'* and so will not be counted as part of the flux.

Consequently:

$$\text{number of particles crossing area in time } t' = n'ut'\Delta y'\Delta z',$$

so that:

$$\text{flux of particles} = \frac{\text{number of particles crossing area in time } t'}{t'\Delta y'\Delta z'} = n'u$$

and, as $n' = \gamma_u n$

$$\text{flux of particles} = \gamma_u nu.$$

We are uncovering here something rather elegant and illuminating about the way that relativity expands our view in physics. If we take our familiar velocity 4-vector and multiply by *n*, the number density in the system where all the dust particles are at rest, we get:

$$\mathbb{N} = nU = n \begin{pmatrix} \gamma_u c \\ \gamma_u u \end{pmatrix} = \begin{pmatrix} \gamma_u nc \\ \gamma_u nu \end{pmatrix}.$$

The spatial component is the flux and the temporal component certainly looks something like the number density, but for the factor *c*.

We are used to the idea that flux would be vary from viewpoint (system) to viewpoint as velocity does alter under the Galilean transformations. The notion that number density depends on system is a novelty introduced by relativity. However, if we rotate our thinking for a moment, we can view the number density *as the flux of particles through time*.

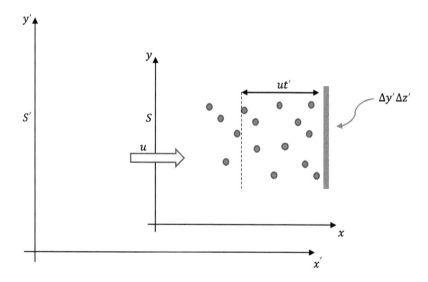

FIGURE 5.4 The flux of particles. Viewed from the *S'* system, *S* is sliding to the right along with its collection of dust particles (which are stationary in *S*). If we place an imaginary area $\Delta y' \Delta z'$ in their path, then only those dust particles closer than *ut'* to the area will have passed through in the time *t'*.

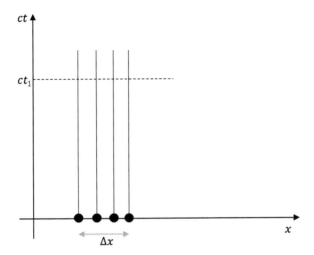

FIGURE 5.5 The number density in the rest system is the flux of stationary dust particles through time.

If we follow the same argument as we did for the spatial flux, then we can see that only particles within a volume $ct_1\Delta x\Delta y$ will 'pass through' the area $\Delta x\Delta y$ after time t_1 (Figure 5.5).

So, arguably:

$$\text{temporal 'flux' of particles} = \frac{nct_1\Delta x\Delta y}{\Delta x\Delta yt_1} = nc$$

in the rest system of the dust. Plausible though this argument is, we should not get too hung up on the thought that 'particles are moving through time at speed c'. That would not be a helpful, physically justifiable or philosophically coherent extension of our thinking.

Our new 4-vector carries the slightly clumsy title 4-number-flux and, just for completeness, the invariant obtained from it is:

$$\left(\begin{array}{cc} \gamma_u nc & \gamma_u nu \end{array} \right)\eta\left(\begin{array}{c} \gamma_u nc \\ \gamma_u nu \end{array} \right) = \left(\gamma_u nc\right)^2 - \left(\gamma_u nu\right)^2 = \gamma_u^2 n^2 c^2\left(1 - u^2/c^2\right) = n^2 c^2.$$

5.5.1 The Energy of Density of Dust

Sitting in the dust's rest system, the energy density due to the stationary particles is $\mathbb{T} = nmc^2$. Viewed in a system moving with the respect to the dust, this energy density would have to be supplemented by the kinetic energy of the particles, determined from their momentum in that system. Their relativistic energy-momentum is contained in the momentum 4-vector.

Arguably, what we need as a representation of the total energy density of the dust is a product of the number density and the 4-momentum of the particles. Perhaps something like this:

$$\mathbb{T} = n\left(\begin{array}{c} \gamma_u c \\ \gamma_u u \end{array} \right)m\left(\begin{array}{c} \gamma_u c \\ \gamma_u u \end{array} \right).$$

What would we make of such a construction?

Firstly, this is not the sort of 'product' of two 4-vectors that we have been using up to now. That product has the metric, η, sandwiched in the middle and has the express purpose of combining two vectors together to get an invariant quantity. So far, we have only done this when the two vectors have been the same, but it works to produce an invariant for any pair of vectors.

Whatever this new object is, it would presumably need two Lorentz transformations to boost it into a new system; one for each 4-vector. That is the clue we need, as we have seen an object that needs two Lorentz matrices before: it is a *tensor*.

Whichever way we build a tensor from this product (and I am coming to that in a second...) we can guess that one component will be $n\gamma_u m\gamma_u c^2 = n'E'$, which is just what we would expect in an energy density: the particle energy times the number density of particles.

When we worked with tensors before, we represented them as a matrix and we can specify each component of a matrix using two subscripts. If a 4-vector takes the form:

$$A = \begin{pmatrix} A_0 \\ A_1 \\ A_2 \\ A_3 \end{pmatrix} \quad \text{or} \quad B = \begin{pmatrix} B_0 \\ B_1 \\ B_2 \\ B_3 \end{pmatrix}$$

then a matrix is:

$$C = \begin{pmatrix} C_{00} & C_{01} & C_{02} & C_{03} \\ C_{10} & C_{11} & C_{12} & C_{13} \\ C_{20} & C_{21} & C_{22} & C_{33} \\ C_{30} & C_{31} & C_{32} & C_{33} \end{pmatrix}.$$

This would suggest that our tensor-making-product (or *tensor product* for short) builds component ij of the matrix by multiplying component i of one vector by j of the other:

$$C_{ij} = A_i B_j$$

$$C = \begin{pmatrix} A_0 B_0 & A_0 B_1 & A_0 B_2 & A_0 B_3 \\ A_1 B_0 & A_1 B_1 & A_1 B_2 & A_1 B_3 \\ A_2 B_0 & A_2 B_1 & A_2 B_2 & A_2 B_3 \\ A_3 B_0 & A_3 B_1 & A_3 B_2 & A_3 B_3 \end{pmatrix}.$$

Following this pattern, our energy density tensor would be:

$$\mathbb{T} = n\begin{pmatrix} \gamma_u c \\ \gamma_u u_x \\ \gamma_u u_y \\ \gamma_u u_z \end{pmatrix} m \begin{pmatrix} \gamma_u c \\ \gamma_u u_x \\ \gamma_u u_y \\ \gamma_u u_z \end{pmatrix} = nm \begin{pmatrix} \gamma_u c \gamma_u c & \gamma_u c \gamma_u u_x & \gamma_u c \gamma_u u_y & \gamma_u c \gamma_u u_z \\ \gamma_u u_x \gamma_u c & \gamma_u u_x \gamma_u u_x & \gamma_u u_x \gamma_u u_y & \gamma_u u_x \gamma_u u_z \\ \gamma_u u_y \gamma_u c & \gamma_u u_y \gamma_u u_x & \gamma_u u_y \gamma_u u_y & \gamma_u u_y \gamma_u u_z \\ \gamma_u u_z \gamma_u c & \gamma_u u_z \gamma_u u_x & \gamma_u u_z \gamma_u u_y & \gamma_u u_z \gamma_u u_z \end{pmatrix}$$

or, tidying things up a bit:

$$\mathbb{T} = nm \begin{pmatrix} \left(\gamma_u\right)^2 c^2 & \left(\gamma_u\right)^2 cu_x & \left(\gamma_u\right)^2 cu_y & \left(\gamma_u\right)^2 cu_z \\ \left(\gamma_u\right)^2 cu_x & \left(\gamma_u\right)^2 u_x^2 & \left(\gamma_u\right)^2 u_x u_y & \left(\gamma_u\right)^2 u_x u_z \\ \left(\gamma_u\right)^2 cu_y & \left(\gamma_u\right)^2 u_y u_x & \left(\gamma_u\right)^2 u_y^2 & \left(\gamma_u\right)^2 u_y u_z \\ \left(\gamma_u\right)^2 cu_z & \left(\gamma_u\right)^2 u_z u_x & \left(\gamma_u\right)^2 u_z u_y & \left(\gamma_u\right)^2 u_z^2 \end{pmatrix}.$$

Just so we can see the wood for the trees, let's cut things down by assuming that the only velocity component is in the x direction, so that $u_y = u_z = 0$. This creates a bit of mayhem in the tensor's components, reducing it to something a little more tractable (while we get our bearings):

$$\mathbb{T} = nm \begin{pmatrix} \left(\gamma_u\right)^2 c^2 & \left(\gamma_u\right)^2 cu_x & 0 & 0 \\ \left(\gamma_u\right)^2 cu_x & \left(\gamma_u\right)^2 u_x^2 & 0 & 0 \\ 0 & 0 & 0 & 0 \\ 0 & 0 & 0 & 0 \end{pmatrix}. \tag{5.6}$$

Now it becomes illuminating to circle back on the problem from another direction (in an attempt to surround it).

Suppose we started by assuming that our energy density had to be represented in tensor form. There are good grounds for suspecting this, as the metric is a tensor, so if gravity is to be a curvature of spacetime represented by some metric and the source of that curvature is to be an energy density, then that should probably be in tensor form as well.

In the rest system of the dust, we could reasonably assume that the tensor would look like this:

$$\mathbb{T} = \begin{pmatrix} nmc^2 & 0 & 0 & 0 \\ 0 & 0 & 0 & 0 \\ 0 & 0 & 0 & 0 \\ 0 & 0 & 0 & 0 \end{pmatrix}.$$

The transformation of this tensor (assuming it to be contravariant) would be $\mathbb{T}' = L\mathbb{T}L$, however there is a fiddly aspect to this that we need to address. Throughout the previous argument we had S' moving *to the left* compared to the system embedded with the dust, whereas a standard boost has S' moving to the *right*. So, to get an answer we can compare, we need to switch $\beta \to -\beta$. This makes the transformation:

$$\mathbb{T}' = L(-\beta_u)\mathbb{T}L(-\beta_u) = \begin{pmatrix} \gamma_u & \beta_u\gamma_u & 0 & 0 \\ \beta_u\gamma_u & \gamma_u & 0 & 0 \\ 0 & 0 & 1 & 0 \\ 0 & 0 & 0 & 1 \end{pmatrix} \begin{pmatrix} nmc^2 & 0 & 0 & 0 \\ 0 & 0 & 0 & 0 \\ 0 & 0 & 0 & 0 \\ 0 & 0 & 0 & 0 \end{pmatrix} \begin{pmatrix} \gamma_u & \beta_u\gamma_u & 0 & 0 \\ \beta_u\gamma_u & \gamma_u & 0 & 0 \\ 0 & 0 & 1 & 0 \\ 0 & 0 & 0 & 1 \end{pmatrix}$$

$$= \begin{pmatrix} \left(\gamma_u\right)^2 nmc^2 & \beta_u\left(\gamma_u\right)^2 nmc^2 & 0 & 0 \\ \beta_u\left(\gamma_u\right)^2 nmc^2 & \beta_u^2\left(\gamma_u\right)^2 nmc^2 & 0 & 0 \\ 0 & 0 & 0 & 0 \\ 0 & 0 & 0 & 0 \end{pmatrix}$$

reducing to:

$$\mathbb{T}' = \begin{pmatrix} \left(\gamma_u\right)^2 nmc^2 & \left(\gamma_u\right)^2 nmcu & 0 & 0 \\ \left(\gamma_u\right)^2 nmcu & \left(\gamma_u\right)^2 nmu^2 & 0 & 0 \\ 0 & 0 & 0 & 0 \\ 0 & 0 & 0 & 0 \end{pmatrix} = nm \begin{pmatrix} \left(\gamma_u\right)^2 c^2 & \left(\gamma_u\right)^2 cu & 0 & 0 \\ \left(\gamma_u\right)^2 cu & \left(\gamma_u\right)^2 u^2 & 0 & 0 \\ 0 & 0 & 0 & 0 \\ 0 & 0 & 0 & 0 \end{pmatrix}$$

which is exactly what we obtained before (Equation 5.6).

So, guessing an energy density tensor in the rest system and boosting it into a moving system gets the same answer as building an energy density tensor from the tensor product of the 4-number-flux and the 4-momentum. It seems that we are on the right lines, at least for describing dust.[24]

Any practical structure is going to involve many particles moving at different velocities and interacting with each other. We can build the energy density tensor for such a structure by adding together distinct parts:

$$\mathbb{T}_{\text{complicated}} = n_1 \begin{pmatrix} \gamma_u c \\ \gamma_{u_1} u_1 \end{pmatrix} m_1 \begin{pmatrix} \gamma_u c \\ \gamma_{u_1} u_1 \end{pmatrix} + n_2 \begin{pmatrix} \gamma_u c \\ \gamma_{u_2} u_2 \end{pmatrix} m_2 \begin{pmatrix} \gamma_u c \\ \gamma_{u_2} u_2 \end{pmatrix} + n_3 \begin{pmatrix} \gamma_u c \\ \gamma_{u_3} u_3 \end{pmatrix} m_3 \begin{pmatrix} \gamma_u c \\ \gamma_{u_3} u_3 \end{pmatrix} + \ldots$$

with n_1 being the number of particles of mass m_1 moving at velocity u_1, etc. The interactions would be dealt with via the fields present. I don't want to dive into justifying the energy density tensor for various fields as that is a complication beyond what we can achieve in a book like this. However, for completeness I can tell you that the energy density tensor of the electromagnetic field looks like this:

$$\mathbb{T}_{\text{electromagnetic field}} = \begin{pmatrix} \frac{1}{2}\left(\varepsilon_0 E^2 + \frac{1}{\mu_0}B^2\right) & \frac{1}{c\mu_0}\left(E_x + v_y B_z - v_z B_y\right) \\[2ex] \frac{1}{c\mu_0}\left(E_x + v_y B_z - v_z B_y\right) & \frac{1}{2}\left(\varepsilon_0 E^2 + \frac{1}{\mu_0}B^2\right) - \left(\varepsilon_0 E_x^2 + \frac{1}{\mu_0}B_x^2\right) \\[2ex] \frac{1}{c\mu_0}\left(E_y + v_z B_x - v_x B_z\right) & -\left(\varepsilon_0 E_x E_y + \frac{1}{\mu_0}B_x B_y\right) \\[2ex] \frac{1}{c\mu_0}\left(E_z + v_x B_y - v_y B_x\right) & -\left(\varepsilon_0 E_x E_z + \frac{1}{\mu_0}B_x B_z\right) \end{pmatrix}$$

$$\begin{pmatrix} \frac{1}{c\mu_0}\left(E_y + v_z B_x - v_x B_z\right) & \frac{1}{c\mu_0}\left(E_z + v_x B_y - v_y B_x\right) \\[2ex] -\left(\varepsilon_0 E_x E_y + \frac{1}{\mu_0}B_x B_y\right) & -\left(\varepsilon_0 E_x E_z + \frac{1}{\mu_0}B_x B_z\right) \\[2ex] \frac{1}{2}\left(\varepsilon_0 E^2 + \frac{1}{\mu_0}B^2\right) - \left(\varepsilon_0 E_y^2 + \frac{1}{\mu_0}B_y^2\right) & -\left(\varepsilon_0 E_y E_z + \frac{1}{\mu_0}B_y B_z\right) \\[2ex] -\left(\varepsilon_0 E_y E_z + \frac{1}{\mu_0}B_y B_z\right) & \frac{1}{2}\left(\varepsilon_0 E^2 + \frac{1}{\mu_0}B^2\right) - \left(\varepsilon_0 E_z^2 + \frac{1}{\mu_0}B_z^2\right) \end{pmatrix}$$

which will teach you for asking...

We now have a robust proposal for the energy density tensor to represent the source of gravitation in general relativity. However, no matter how plausible these arguments may be, we always need to remember that the final arbiter is Nature. We can construct all the elegant equations that we wish, but if they do not produce predictions that correspond with Nature, then we are merely entertaining ourselves, not doing science.

5.5.2 The Physical Meaning of the Tensor's Terms

The mass energy density component of this tensor we expect to be a traditional source of gravitational effects as we now understand that mass derives from energy. However, that is just one component in the tensor – there are 15 others. What are they all doing? Here is a situation where our intuitions do not help us and we need to develop new ones. There are a few points that can help foster this:

1. Space-time curvature is encoded into the metric. In the case of 'flat' space-time, we can use Cartesian co-ordinates, giving a metric of the form $\eta = \begin{pmatrix} 1 & 0 & 0 & 0 \\ 0 & -1 & 0 & 0 \\ 0 & 0 & -1 & 0 \\ 0 & 0 & 0 & -1 \end{pmatrix}$.

   The η_{00} component of this metric sets the coefficient of the $c\Delta t$ part of the space-time interval, and the other components factor into the interval in turn:

   $$(\Delta s)^2 = (c\Delta t)^2 - (\Delta x)^2 - (\Delta y)^2 - (\Delta z)^2 = \eta_{00}(c\Delta t)^2 + \eta_{11}(\Delta x)^2 + \eta_{22}(\Delta y)^2 + \eta_{33}(\Delta z)^2.$$

   In the case of moderately large masses (planet or star size) moving at speeds which are not comparable to that of light, the temporal part of the interval is going to dominate, because of the factor of c^2. Under these conditions, the gravity field is 'weak' (still hurts tho...) and can be regarded as a small change in the flat metric, mostly via a change to η_{00}. Formally we would write:

   $$g_{00} = \eta_{00} + h_{00}$$

   with g being the appropriate metric for gravitation in that vicinity of space-time and h the small part we add on to the flat metric to get the curvature.

   In this situation, the energy density tensor reduces to the mass energy density and everything basically becomes Newton's law of gravity.

   In short, our standard daily and astronomical experience of gravity boils down to curvature in the *time component of the metric* fed by mass energy density.

   As we get to more exotic situations, such as with giant mass black holes and fast-moving objects, the spatial components of curvature play more of a part, as those coefficients in the metric have grown to a similar size as the time component. These situations lie (far) outside our daily experience, so it is not surprising that our intuitions of how gravity works (based on our experience of the effects and causes of temporal curvature) are not very helpful.

   As the other components of the energy density tensor have no impact on gravitational effects in the weak field case, we have not grown up to think of them as 'sources' of a 'gravitational field'. Widening our experience horizons must bring with it an acceptance that the other terms are equally valid sources of gravitation.

2. As we have seen, the energy density tensor transforms when we move from one system to another. In our very simple case, in the co-moving system, the tensor had only one non-zero component, which we expected as a gravitational source. However, once we transform to a new system, other components rotate into view out of the original non-zero component. *We have to accept that these are then equally valid sources of gravitational effects.* Once again, the laws of physics should not depend on the system from which the universe is viewed.

Structurally, the energy density tensor breaks into several regions with slightly different physical interpretations.

$$\mathbb{T} = nm \begin{pmatrix} \gamma_u c \gamma_u c & \gamma_u c \gamma_u u_x & \gamma_u c \gamma_u u_y & \gamma_u c \gamma_u u_z \\ \gamma_u u_x \gamma_u c & \gamma_u u_x \gamma_u u_x & \gamma_u u_x \gamma_u u_y & \gamma_u u_x \gamma_u u_z \\ \gamma_u u_y \gamma_u c & \gamma_u u_y \gamma_u u_x & \gamma_u u_y \gamma_u u_y & \gamma_u u_y \gamma_u u_z \\ \gamma_u u_z \gamma_u c & \gamma_u u_z \gamma_u u_x & \gamma_u u_z \gamma_u u_y & \gamma_u u_z \gamma_u u_z \end{pmatrix}.$$

Let us take a look at them a term at a time, starting with the top left component:

$$\mathbb{T}_{00} = nm\gamma_u c\gamma_u c = (n\gamma_u)(\gamma_u mc^2) = \text{energy density}.$$

Other components along the top row take the form:

$$\mathbb{T}_{0i} = nm\gamma_u c\gamma_u u_i = (n\gamma_u)(\gamma_u mu_i)c = \text{`time flux' of the } i \text{ momentum density}.$$

Components on the left-hand vertical column are:

$$\mathbb{T}_{i0} = nm\gamma_u u_i\gamma_u c = (n\gamma_u)(\gamma_u mu_i)c = \text{`time flux' of the } i \text{ momentum density}.$$

Along the left to right centre diagonal we have:

$$\mathbb{T}_{ii} = nm\gamma_u u_i\gamma_u u_i = (n\gamma_u)(\gamma_u mu_i)u_i = \text{`spatial flux' of the } i \text{ momentum density along direction } i.$$

All other terms are:

$$\mathbb{T}_{ij} = nm\gamma_u u_i\gamma_u u_j = (n\gamma_u)(\gamma_u mu_i)u_j = \text{`spatial flux' of the } i \text{ momentum density along direction } j.$$

The last combination is slightly counter-intuitive, as you would not think that a velocity in one direction would carry momentum in another, but Figure 5.6 should hopefully make this clearer.

The different regions of the energy density tensor are shown in Figure 5.7.

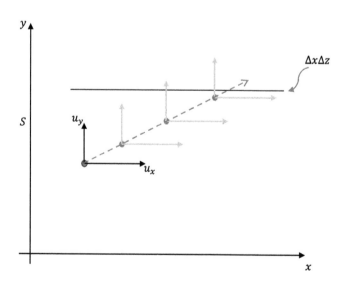

FIGURE 5.6   How a particle's motion in the y direction brings about a flux of *x* momentum through an area $\Delta x \Delta z$.

$$
\begin{pmatrix}
\boxed{(n\gamma_u)(m\gamma_u c^2)} & (n\gamma_u)(\gamma_u m u_x)c & (n\gamma_u)(\gamma_u m u_y)c & (n\gamma_u)(\gamma_u m u_z)c \\
\boxed{(n\gamma_u)(\gamma_u m u_x)c} & \boxed{(n\gamma_u)(\gamma_u m u_x)u_x} & (n\gamma_u)(\gamma_u m u_x)u_y & (n\gamma_u)(\gamma_u m u_x)u_z \\
\boxed{(n\gamma_u)(\gamma_u m u_y)c} & (n\gamma_u)(\gamma_u m u_y)u_x & \boxed{(n\gamma_u)(\gamma_u m u_y)u_y} & (n\gamma_u)(\gamma_u m u_y)u_z \\
\boxed{(n\gamma_u)(\gamma_u m u_z)c} & (n\gamma_u)(\gamma_u m u_z)u_x & (n\gamma_u)(\gamma_u m u_z)u_y & \boxed{(n\gamma_u)(\gamma_u m u_z)u_z}
\end{pmatrix}
$$

FIGURE 5.7   The regions of the energy density tensor. Very light grey is momentum temporal flux, dark grey is momentum density flux in the direction of momentum, light grey is momentum density flux in a different direction to the momentum.

5.5.3 Deriving General Relativity

Special relativity explores the consequences of the Lorentz transformations, which you can derive from the assumption that the speed of light is the same for all observers. General relativity pivots around Einstein's field equations, which cannot be derived in the same fashion. As we will see in the next chapter, Einstein was motivated by certain philosophical beliefs and tried various educated and enlightened guesses until he honed in on the correct formulation. One fixed guidepost was the need for the new theory to reduce to Newtonian gravity as a limiting case, the same way that special relativity eases back into Newtonian mechanics for speeds very much less than light.

Newton's theory of gravity is enshrined in Poisson's equation:

$$
\frac{\Delta}{\Delta x}\left(\frac{\Delta\phi}{\Delta x}\right) + \frac{\Delta}{\Delta y}\left(\frac{\Delta\phi}{\Delta y}\right) + \frac{\Delta}{\Delta z}\left(\frac{\Delta\phi}{\Delta z}\right) = 4\pi G\rho
$$

but as the general theory is more encompassing, we can only use this as a guide for what the new equations might look like. Given that gravitational force is replaced by curved space-time, we might expect something like:

$$
\text{Something clever to do with metrics} = \kappa\mathbb{T}
$$

where κ is a constant that will need to be determined, which we will do by insisting that the field equations reduce to Poisson's equation in the weak field case.

Unfortunately, to fill in the left-hand side, we need to develop the mathematics of geometry and curvature. That will be a major part of Chapter 7, in the meantime we are going to take a break from calculations and spend Chapter 6 discussing the principles Einstein set down as the foundations for the general theory of relativity.

Notes

1. I introduced this phraseology in Chapter 3.
2. I'm just using x here to be a variable, this is NOT supposed to be a spatial co-ordinate...
3. Or any other massless particle.
4. The *photoelectric effect* whereby ultraviolet light shone on certain metals could release electrons from the surface of the metal. The energy of these electrons did not change with the brightness of the UV (as you would expect with a wave) but did change with the frequency (which you would not expect with a wave). Einstein's description in terms of packets of energy (photons) striking the metal explained all the observed effects.
5. Einstein's full argument is developed for photons emerging at any angle, but we do not need that complication here.

6. Does the inertia of a body depend upon its energy-content? 1905, A Einstein English translation of the paper first published as Ist die Trägheit eines Körpers von seinem En- ergiegehalt abhängig?, *Annalen der Physik*, **18**, p. 639, https://pdfs.semanticscholar.org/44a9/b84ebb88b4c09261b5bc-67c5b294fa5bb106.pdf.

7. I cringe every time I use this definition and back-peddle as much as I can...

8. McReynolds, A.W., Rogers, F.T., 1940. A determination of the masses and velocities of three radium b beta-particles: The relativistic mass of the electron, *Physical Review*, **57**(5), pp. 379–383.

9. CERN is the main European centre for particle physics, just outside Geneva. The LHC, or Large Hadron Collider is currently (2019) the most powerful particle physics accelerator on the planet.

10. Relativity would, of course, work perfectly well if all objects were (rest) massless. To include particles that have no sub-structure, but yet mass, relativity dictates that there has to be a form of internal energy for the particle. It does constrain or dictate what that energy is.

11. I trained as a particle physicist and like to keep to the club rules...

12. Albert Einstein in letter to Lincoln Barnett, 19 June 1948 as quoted in Okun, L.B., 1989, The concept of mass. *Physics Today*, **42**(6), pp. 31–36, p. 42.

13. Taylor, E.F., Wheeler, J.A., 1992. *Spacetime Physics* (second edn.), New York: W.H. Freeman and Company, pp. 248–249, ISBN 0-7167-2327-1.

14. Unfortunately, Brout died in 2011 and the Nobel Prize is not awarded posthumously.

15. Their momentum can change, if they scatter off electrons for example.

16. Third letter to Bentley, 25 February 1693. Quoted in *The Works of Richard Bentley, D.D.* (1838), Vol. 3, pp. 212–213.

17. Subject to some technical restrictions that should not worry us.

18. It does not appear in Maxwell's equations again by convention, but will pop up in calculations done based on those equations. You can't avoid it...

19. Clearly, this argument is completely couched in Newtonian terms...

20. If you are sharp eyed you might spot that the minuses do not quite square up here. In truth they do, but only if you follow the complete derivation using calculus. Think of the minus we end up with as showing the direction in which the force acts.

21. Strictly, acceleration is the rate of change of velocity with time, as you probably know...

22. Siméon Denis Poisson (1781–1840).

23. It is not clear why this sort of system was awarded the name *dust*. It is possible that the physicist who first coined the term was having a bad shelf day...

24. Strictly, to make the argument water-tight, we should examine what happens when $u_y, u_z \neq 0$.

6

Generalising Relativity

Grossmann, you must help me, or else I'll go crazy!

A Einstein[1]

Now it came to me: ... the independence of the gravitational acceleration from the nature of the falling substance, may be expressed as follows: In a gravitational field (of small spatial extension) things behave as they do in a space free of gravitation. ... This happened in 1908. Why were another seven years required for the construction of the general theory of relativity? The main reason lies in the fact that it is not so easy to free oneself from the idea that coordinates must have an immediate metrical meaning.

A Einstein[2]

Every boy in the streets of Göttingen understands more about four-dimensional geometry than Einstein. Yet, ... Einstein did the work and not the mathematicians.

D Hilbert[3]

6.1 The Problem with History...

As a young student, I vividly remember buying a book of articles on particle physics from a respected science journal. I ended up being terribly frustrated by the book, as I could not follow a thread of argument through the articles. It stayed on my shelf all through my time at school and as an undergraduate. When it came time to do my PhD, I took the book down again, looking for a reference. Suddenly it dawned on me why I had found the book so confusing. The articles spanned a period of about 20 years during the development of the subject. The consistent line of argument and understanding that I had been looking for was not there as ideas, experimental data and approaches had changed radically during that time. It was an archive of a remarkable period of rapid development, not a compendium of correct thinking.

It seems to me that this is an issue that all students sometimes face. The keen and enthusiastic relativist learning their way in the subject will go back to the source material. If you want to know about general relativity, who better to consult than Einstein?

The problem is Einstein's ideas changed considerably while he was developing the general theory. Between 1907 and 1915 (when the final theory appeared), he regularly published position papers outlining his latest views and beyond 1915, developmental and review articles continued. Consequently, something published one year was not necessarily consistent with what was in print from the years before. Einstein was well aware of this, and on one occasions wryly remarked[4]:

it's convenient with that fellow Einstein, every year he retracts what he wrote the year before.

In this chapter, we are going to take a break from the mathematical development of relativity and examine the foundational principles of the general theory, both from Einstein's perspective, that of his contemporaries and the 20:20 position of modern-day hindsight.

6.2 The Quest for a General Theory of Relativity

The years of searching in the dark for a truth that one feels but cannot express, the intense desire and the alternations of confidence and misgiving until on breaks through to clarity and understanding are known only to him who has himself experienced them.

A Einstein[5]

In 1905, Einstein published two papers on the special theory of relativity: his original paper *On the electrodynamics of moving bodies* and a second developing the relationship between energy and mass, although a specific statement of $E = mc^2$ did not appear in print until 1907. At the time, he was working at the Patent Office in Bern, a post he was to hold until 1909. Through 1906 and 1907, his output focussed on issues in the physics of heat (thermodynamics and statistical thermodynamics), photons and further developments of the special theory in various applications. However, in October 1907 the first paper on his quest to draw gravity into the new picture of space-time and to extend the relativity principle to accelerating systems[6] was published. In this work, Einstein presents the principle of equivalence and uses it to calculate how much a beam of light will be deflected by passing close to a large mass, (eventually leading to the 1919 Eclipse expedition) and demonstrates that clocks run more slowly the closer they are to a gravitating mass.

By 1909, Einstein's ambition to move from his work in the Patent Office to a full-time academic post had been fulfilled and he had been appointed to the University of Zurich. In September 1909, he attended a conference in Salzburg and finally met various luminary physicists, including Max Planck, that he had been corresponding with.

Although Einstein continued to publish in his favourite areas, it is not until 1911 that he returns to gravity.[7] By this stage, he has realised that Newton's theory of gravity and the special theory of relativity both need to be replaced by an encompassing theory that reduced to them in limiting cases. A simple extension of Newton's ideas would not work.[8]

1911 saw him move to a more prestigious full professorship at the University of Prague, having accepted Austrian citizenship in order to qualify for the post.

Two more papers follow in 1912 where Einstein points out that the Lorentz transformations have to be generalised and that a new theory of gravitation has to be non-linear, as the energy in a gravitational field must itself act as a source of gravitation. At this stage Einstein is working with the curvature of time but keeping space metrically flat. Significantly, his 1912 return to Zurich to take a post at the university where he had studied (the Swiss Federal Institute of Technology, where he remained until 1914) allowed him to rekindle a friendship with his old classmate, Marcel Grossmann, who in 1907 had been appointed full Professor of Descriptive Geometry at the Federal Polytechnic School in Zurich. As he was an expert in geometry and tensors, Einstein was able to turn to his friend for assistance with the mathematical side of his developing ideas.

The breakthrough paper emerged in 1913 and was co-authored with Grossmann. In *Outline of a Generalized Theory of Relativity and of a Theory of Gravitation*, Einstein and Grossmann replace the traditional gravitational field with the components of a symmetric 4x4 metric tensor along with ten *field equations* that allow these components to be calculated.[9] However, the field equations are not yet in their ultimately correct form.

In July 1913, Einstein was elected to membership of the Prussian Academy of Sciences and offered a professorship at the University of Berlin. This new post would be without any teaching or administrative duties, which was a great attraction to him and as a result he moved to Germany in 1914.

Given his struggles with the theory, Einstein was interested in any experimental confirmation that could convince himself and others that he was on the right track. The 1914 eclipse expedition to the Crimea, to try and measure the deflection of light, so intrigued him that he offered to contribute to the costs. As it happened, other private donations made it possible. However, the outbreak of war had an unfortunate influence, with the German team being captured by the Russian army and their equipment confiscated.

Einstein's calculations at the time showed a deflection that was the same as that of Newton's theory (as he was not including spatial curvature). So, if a successful measurement of the deflection had been taken in 1914, it would have found a value twice that size, undermining Einstein's work. Fortunately for history, cloudy conditions on the day meant that the American team in the area did not get any useable pictures.

A year later, Einstein was to make the completely correct calculation that was the focus of the 1919 eclipse and its confirmation of general relativity.

Other developmental papers appeared through 1913 and 1914, significantly one entitled *Formal Foundations of the General Theory of Relativity*[10] in which the geodesic equation for the motion of point particles through metrical space-time is derived (Section 7.4) and the deflection of light re-calculated. However, Einstein is still using the incorrect field equations from 1913.

From 1913 until the Autumn of 1915, Einstein worked primarily to develop the first draft of the general theory of relativity and apply it in a variety of cases. However, he became increasingly convinced that the theory was fundamentally flawed. Part of this conviction arose from discovering that his draft theory could not account for an historically important quirk in Mercury's orbit (which could not be explained by Newtonian theory either: see Section 9.5). Also, his earlier work had focussed on a special set of co-ordinates, something that Einstein now sought to replace by adapting the mathematics so that the theory would be truly general and work in all co-ordinate systems.

November 1915 became a feverishly brilliant period, during which four lectures were delivered to the Prussian Academy of Sciences (one per week on a Thursday and published in the *Proceedings of the Prussian Academy of Sciences*) with updates on Einstein's battle to save the space-time theory of gravitation. The third of these[11] (18 November 1915) explained how a revised theory *could* explain the anomalous motion of Mercury's orbit. It also includes the completely correct calculation of the deflection of light passing close to the Sun (incorporating both curvature of space and time). The final field equations of the general theory of relativity appeared in print for the first time on 25 November 1915.[12]

Einstein wrote many more papers and review articles on the theory after 1915. Two in particular stand out: *Approximative Integration of the Field Equations of Gravitation*[13] contains the first prediction of gravitational waves and *Cosmological Considerations in the General Theory of Relativity*[14] launched the modern science of cosmology by demonstrating how, with certain simplifying assumptions, the general theory of relativity can be used to describe the large-scale structure of the universe.

Although the First World War meant that the work of scientists in the Central Powers[15] was only easily available to other scientists in the same countries, some of Einstein's work did reach the United Kingdom and the USA via the efforts of de Sitter, Lorentz and Ehrenfest.[16] As we have seen, Eddington was a recipient of this information and became a major champion of Einstein's work, largely due to his unique combination of mathematical skill to understand the theory, talent in astronomy to contribute to the 1919 eclipse expedition and style of communication that made him an easy populariser.

Now that the dust has settled, historians have rightly questioned why it took Einstein so long to complete his quest.[17] The simple answer, that creating a new theory of gravity is not child's play, will not do as throughout the development years and especially with the 1913 paper authored with Grossmann, the correct field equations were so nearly in view. Indeed, in 1915 David Hilbert,[18] a hugely significant figure in the history of mathematics, became interested in the problem (largely due to a series of lectures delivered by Einstein at the University of Göttingen where Hilbert worked) and had the technical skill to convert the 1913 Einstein and Grossmann draft into a final theory. He could have beaten Einstein to the final prize. Some have speculated that this was a significant motivation behind Einstein's frantic activity that November.

It appears that Einstein was lead in various directions by confusions relating to the principles that he believed formed the guideposts to the correct theory. Indeed, long after the final form of the general theory was discovered, debates between Einstein and a range of critics forced him to periodically rethink the principles he claimed were its cornerstones.

Even now, more than 100 years after the general theory was first published, there is no universally agreed assessment of the correct set of foundational principles. Equally, a coherent modern representation of Einstein's own view is hard to come by, as different commentators stress different phases of the evolution that Einstein went through. There are variations in both the content of the foundational principles chosen and the emphasis placed on them. Some express views broadly consistent with Einstein's development, even if those sands shifted somewhat over the years. Others are critical of one or more aspects of Einstein's thinking, even while they express admiration for the beauty and elegance of the theory itself.

6.3 Einstein's Principles of General Relativity

At various times, Einstein referred to the following as fundamental to the development of general relativity:

1. The *principle of equivalence*, which asserts an equivalence between an inertial system in the presence of a uniform gravitational field and another system in uniform acceleration.
2. The *general principle of relativity/general covariance* holds that the laws of nature can be expressed in a form that preserves their mathematical structure when transforming from one set of co-ordinates to another, without any restriction on the co-ordinates chosen, and that this removes the spectre of absolute motion.
3. *Mach's principle* that the inertia of a body derives from its interaction with the broad distribution of matter in the universe.

Over the next few sections, we will take a look at each of these in turn and the criticism that they have received, before boiling things down to a residue that will serve us moving forward.

6.3.1 The Principle of Equivalence

According to Einstein's own later account, the principle of equivalence occurred to him while working on a review article about relativity at the Patent Office in Bern. For whatever reason, he was moved to think about someone jumping off a roof. While falling, Einstein realised, that person would experience *no evident effects of gravity*. Any object that they released during the fall would float in front of them, as if gravity had disappeared, an experience guaranteed by Galileo's declaration that all objects fall with the same acceleration in a gravitational field, irrespective of their mass.[19] Needless to say, this illusion would not last the whole flight, the impact with the ground being a particular wake-up call.

This thought had a profound impact on Einstein. As he wrote in an article that never made it to publication,[20]

> Then there occurred to me the 'glücklichste Gedanke meines Lebens,' the happiest[21] thought of my life, in the following form. The gravitational field has only a relative existence in a way similar to the electric field generated by magnetoelectric induction. *Because for an observer falling freely from the roof of a house there exists – at least in his immediate surroundings – no gravitational field.* Indeed, if the observer drops some bodies then these remain relative to him in a state of rest or of uniform motion, independent of their particular chemical or physical nature (in this consideration the air resistance is, of course, ignored). The observer therefore has the right to interpret his state as 'at rest'.
>
> **(Einstein's emphasis)**

Removing the effects of gravitation by transforming to a different co-ordinate system seemed to Einstein reminiscent of his earlier observation regarding electrical and magnetic fields: that you could switch from one to another via a change in co-ordinate system. He hoped that this equivalence would dissolve the absolute nature of acceleration: as well as being able to transform away the effects of a gravitational field, you can also mount a fair impersonation of one by switching to a system that is accelerating.

In such a system, an observer could release an object and watch it 'fall' in the opposite direction to the system's acceleration. Indeed, all objects released would fall at the same rate.

Another observer, in inertial motion, would attribute these observations to the effect of the system itself being accelerated. As soon as an object is released, it is no longer under the influence of the force responsible for accelerating the system, so it falls behind as the system gains speed. It is then hardly surprising that all released objects should fall behind at the same rate, as this reflects the rate at which the system itself is accelerating.

An observer within the accelerating system, deprived of other clues (such as seeing the engine strapped to their system), could very well decide that the simplest explanation for the falling objects would be the presence of a gravitational field.

The principle of equivalence then has two aspects:

- Transforming to a co-ordinate system that is in free fall in a gravitational field removes all evidence of that gravitational field. A set of observers embedded in such a system would not be able to distinguish between their specific system and another that was in inertial motion in free space away from any source of gravitation.[22]
- Equally, a set of observers (denied other clues) in inertial motion in the presence of a uniform gravitational field would not be able to distinguish their situation from the parallel one of a system in deep space away from gravitational fields but in uniform acceleration.

There is a relativism here that Einstein hoped to exploit.

In his 1911 paper, he included the following statement of the principle of equivalence, although it was not formally named as a principle until 1912[23]:

> In a homogeneous gravitational field (acceleration of gravity g) let there be a stationary system of co-ordinates K, oriented so that the lines of force of the gravitational field run in the negative direction of the axis of z. In a space free of gravitational fields let there be a second system of co-ordinates K', moving with uniform acceleration (g) in the positive direction of the z axis...
>
> ... we arrive at a very satisfactory interpretation of this law of experience; if we assume that the system K and K' are physically exactly equivalent, that is, if we assume that we may just as well regard the system K as being a space free from gravitational fields, if we then regard K as uniformly accelerated. *This assumption of exact physical equivalence makes it impossible for us to speak of the absolute acceleration of the system of reference*, just as the usual theory of relativity forbids us to talk of the absolute velocity of a system; *and it makes the equal falling of all bodies in a gravitational field a matter of course.*

(My emphasis)

Stated in this fashion, the equivalence principle is open to some criticism.

First, aside from various hints that would be hard to suppress (engine noise, accompanying vibrations, etc.) the lack of an evident mass or energy acting as a source would surely convince observers that they were accelerating, rather than being in a gravitational field. Conjuring a gravitational field without an evident physical source is no different to deploying fictitious or inertial forces to explain accelerative effects. Einstein responded to this criticism by referring to Mach's principle, which we will come to in due course.

Second, when we drop objects on Earth we tend to think of them as falling vertically towards the ground, whereas they all fall towards the centre of the planet. In an extreme case, a cricket ball dropped in England[24] and a cricket ball dropped in Australia will, from a distant perspective, move in almost opposite directions. In a less extreme case, two balls held a distance apart at the same height and then released will hit the floor slightly closer together as their lines of flight are not parallel. Performing this simple experiment would tell an observer that they were in the presence of a planetary gravitational field, rather than in some spacecraft in accelerated motion, for in the latter case, the flight of the balls would be exactly parallel (given the normal provisos regarding air resistance, etc.). So, in general, gravitation and acceleration are not fully equivalent, but they can be *locally* i.e. when the separation of the balls is not far enough to measure a difference when they hit the floor. This is why Einstein specifies that the gravitational field is *homogeneous* (although we will see that such a field does not exist in practice, see Section 9.2.2).

In modern parlance, a *local equivalence principle* is often used to justify the *local flatness* of the geometry inherent in any region of space-time. After all, if the effects of gravitation are truly down to the curvature of space-time and if gravitation can be removed by transforming to a system in free fall, then working in that system must show a flat geometry, at least in a small vicinity around our current location.

An analogy is commonly used to explain this. Over a small enough region of the Earth, it is possible to view the surface as broadly flat (baring the obvious hills and valleys that show up from time to time) in that the surface curvature is not a factor when walking around. However it does become a factor when flying to another country: the surface can appear locally flat, while being globally curved.

A problem with this analogy is that it encourages us to become fixated on the *spatial* aspects of flatness. A local equivalence principle suggests that the local geometry of space-time can be 'flattened' by transforming to a system in free fall – i.e. one that is *moving*. We are talking about *space-time* being flat, so temporal curvature is just as important. The co-ordinate system you transform to has to flatten out both the spatial and temporal curvatures.

For my taste, the best example of the equivalence principle re-tasked for a modern audience appears in Steven Weinberg's[25] magisterial development, in which he writes[26]:

> At every space-time point in an arbitrary gravitational field it is possible to choose a 'locally inertial co-ordinate system' such that, within a sufficiently small region of the point in question, the laws of nature take the same form as in an unaccelerated Cartesian co-ordinate system in the absence of gravitation.

Weinberg's Cartesian co-ordinates are the standard co-ordinates used in the special theory. Indeed, Weinberg amplifies this point by clarifying his assertion about the laws of nature, by which he means: *the form given to the laws of nature by special relativity.*[27]

There are two advantages to this version of the equivalence principle which are, to my mind, decisive. First, it stresses the *physics* of the situation by focussing on the form of the laws of nature, rather than tying the equivalence principle to geometry and co-ordinate systems. We don't need a local equivalence principle to justify the local flatness of space-time.[28] Second, it promises that we do not have to throw out all our work on the special theory. Indeed, understanding how to express the laws of nature within that context is vital. The laws of nature can be discovered by working in inertial systems far from gravitational effects (or in free fall). If we then write them in co-ordinate independent fashion by using all the apparatus of vectors, tensors and metrics, the same laws of nature will work in any system and any context, as gravitation is folded into the mix via the metric. In turn, the metric is determined by the one law of nature that falls outside this programme: Einstein's field equations relating the metric's components to the energy density tensor.

Arguably, Einstein never saw the equivalence principle as a means to justify any form of geometry for space-time, local or otherwise. Aside from being a guide that helped him on the path to the general theory, he always held it as *a relativity principle*. That much should be clear from his statement of the principle quoted before. However, to re-enforce the point, I will use one further remark from his writings:

> The assumption of the complete physical equivalence of the systems of co-ordinates, K and K', we call the 'principle of equivalence'; ... [it] *signifies an extension to the principle of relativity to*

co-ordinate systems which are in non-uniform motion relative to each other. In fact, through this conception we arrive at the unity of nature of inertia and gravitation.[29]

(My emphasis)

Although his optimism regarding unifying inertia (i.e. the mass of an object and the inertial forces in an accelerated system) with gravitation via Mach's principle is contested, and he abandoned the idea eventually, the import he ascribes to the principle of equivalence is clear. Modern relativists do not see it the same way.

6.3.2 The General Principle of Relativity

The special theory fulfils its promise to deliver a restricted (special) relativity. As it is not possible to distinguish between an inertial system at rest and one in motion, the very notions 'at rest' and 'in motion' lose any *absolute* meaning and can only be specified in a *relative* sense. Bluntly, a statement such as 'system S is at rest' is devoid of any physical meaning and must be replaced with 'system S is at rest *relative to system S'*.

Special relativity guarantees this by bringing the electromagnetic field into the purview of the Lorentz transformations and, once Minkowski's space-time approach is assimilated, the 4-vectors and tensors of the theory allow us to write the laws of nature in a way that preserves their mathematical form under Lorentz transformations. The laws of nature are hence *covariant* under the Lorentz transformation (although by this they mean something slightly different to the covariance of a 4-vector as previously discussed.) Specifically, the argument in Chapter 3 confirming:

$$\frac{1}{c}\left(\frac{\Delta E_x}{c\Delta t}\right) = \frac{\Delta B_z}{\Delta y} - \frac{\Delta B_y}{\Delta z} \xrightarrow{\text{Lorentz Transformation}} \frac{1}{c}\left(\frac{\Delta E'_x}{c\Delta t'}\right) = \frac{\Delta B'_z}{\Delta y'} - \frac{\Delta B'_y}{\Delta z'}$$

demonstrated covariance for that equation from Maxwell's arsenal.

So, in special relativity the covariance of the correctly formed laws of nature guarantees that there are no observable consequences when you switch between systems in relative inertial motion. This, in turn, prevents their state of motion from being determined in any absolute sense.

The special principle of relativity is synonymous with Lorentz covariance.

Unfortunately, a system in *accelerated motion* is distinguishable from another which is in inertial motion. Looking from outside, physical effects (falling balls for example) are easily attributed to the acceleration of the system itself. Within the system, inertial forces are used to explain these effects. Either way, it is possible to say, with physical meaning, 'that system is accelerating' without reference to any 'relative' system.

Einstein hoped to expand the relativity principle to include accelerated systems. He felt that the principle of equivalence had taken him part way by blurring the distinction between inertial systems in deep space and those in free fall within a gravitational field. The principle also prevents us from separating out inertial systems in the presence of a uniform gravitational field from those in uniform acceleration. To express the principle of equivalence mathematically, the covariance of the laws of nature had to be expanded from simply Lorentz covariance to a broader basis that included transforming from one system to another in this context.

If you take that step, you may as well expand covariance to encompass transforming from *any* arbitrary system into another, whatever the state of motion, whatever the arrangement of co-ordinates.

Physically, though, that last move is huge.

Within the special theory, covariance and the relativity principle are one and the same, as the covariance is restricted to the physical context of the Lorentz transformations. To say that *general covariance* also guarantees a *general relativity principle,* in that all states of motion are relative, is far from clear. If we follow the analogy of the special theory and Lorentz covariance, we would expect that a generalised principle of relativity would certainly require covariance under transformations between systems in arbitrary states of motion (not simply between inertial systems). General covariance extends this even further, however, by requiring covariance under *all* transformations *including those that are not at all related to different states of motion*. A good example of this sort of thing would be transforming between Cartesian and spherical polar co-ordinates in flat space-time, something that has no physical relevance whatsoever.

It is possible that Einstein felt obliged to go the extra step, as a natural way of restricting covariance to just transformations between differing states of motion eluded him.

In his 1916 review of general relativity, Einstein uses various examples (and we will discuss one shortly) to conclude that within an accelerating system geometry is no longer flat, which causes complications when it comes to the choice of co-ordinate system used. Consequently, he sees no choice but to extend covariance to a fully general form, accounting for any possible set of co-ordinates[30]:

> The method hitherto employed for laying co-ordinates into the space-time continuum in a definitive manner thus breaks down, and there seems to be no other way which would allow us to adapt systems of co-ordinates to the four-dimensional universe so that we might expect from their application a particularly simple formulation of the laws of nature. So there is nothing for it but to regard all imaginable systems of co-ordinates, on principle, as equally suitable for the description of nature. This comes to requiring that:-
>
> *The general laws of nature are to be expressed by equations which hold good for all systems of co-ordinates, that is, are co-variant with respect to any substitutions whatever (generally covariant).*
>
> It is clear that a physical theory which satisfies this postulate will also be suitable for the general postulate of relativity. For the sum of all substitutions in any case includes those which correspond to all relative motions of three-dimensional systems of co-ordinates.

(Einstein's emphasis)

However cast, the connection between general *covariance* and a general *relativity* is open to criticism, ultimately decisively so as Einstein himself later conceded. Some very significant commentators have expressed negative views about general relativity/covariance: for example, Sir Hermann Bondi[31]:

> It is rather late to change the name of Einstein's theory of gravitation, but general relativity is a physically meaningless phrase that can only be viewed as a historical memento of a curious philosophical observation.
>
> …one may surely admire and embrace Einstein's theory of gravitation while rejecting his route to it, however heuristically useful he himself found it.

As early as 1917 Kretschmann[32] pointed out that it only needed mathematical ingenuity to produce a generally covariant theory, so the principle was not adding anything to physics. In Chapter 2, we constructed co-ordinate systems with a clock at every point in space. This allowed us to characterise an event as a collection of co-ordinates obtained locally by finding an object at a particular spatial location at a time determined on the clock at that position. Transferring to any other system of co-ordinates must retain that coincidence of an object being at that place at that time. So, if the data on which a theory depends and the predictions that it makes are nothing more than a collection (or list it you like) of events at particular times and places, and if those events can be effectively specified in any co-ordinate system, the fact that we can shift from one system to another is adding nothing to the theory. So, while a general *relativity* might be desirable if you wanted to abolish absolute motion, general *covariance* has no guarantee of establishing that.

In Einstein's 1918 reply to Kretschmann, he effectively concedes the point, while retaining a more than mathematical significance to general covariance[33]:

I believe Herr Kretschmann's argument to be correct, but the innovation proposed by him not to be commendable. That is, if it is correct that one can bring any empirical law into generally covariant form, the principle (a) still possesses a significant heuristic force.

Einstein strongly believed that natural law was inherently simple and beautiful: something that can be revealed by the natural and elegant manner in which the theory presents itself in a generally covariant form. So, in Einstein's view, the joint search for *simplicity* and *natural general covariance* can be significant guides (*'significant heuristic force'*) when looking for new theories. So, while abandoning general *covariance* as a means of guaranteeing general *relativity*, Einstein holds that the principle still has great value as a guide to finding new physics.

Weinberg presents a variation on the principle of covariance that is equally instructive as his principle of equivalence. He suggests[34]:

that a physical equation holds in a general gravitational field, if two conditions are met:

1. The equation holds in the absence of gravitation; that is, it agrees with the laws of special relativity when the metric field g_{ab} equals the Minkowski tensor M_{ab} and when the affine connection Γ^a_{bc} vanishes.
2. The equation is generally covariant; that is, it preserves its form under a general co-ordinate transformation $x \to x'$.

We will meet the *affine connection* in the next chapter, although we will refer to it by a different name. Weinberg points out the linkage between this principle of general covariance and his version of the principle of equivalence and goes on to write[35]:

It should be stressed that general covariance by itself is empty of physical content. Any equation can be made generally covariant by writing it in any one co-ordinate system, and then working out what it looks like in other arbitrary co-ordinate systems...The significance of the Principle of General Covariance lies in its statement about the effects of gravitation, that a physical equation by virtue of its general covariance will be true in a gravitational field if it is true in the absence of gravitation.

In Weinberg's development, this general covariance clips together with his principle of equivalence to give us a technique for developing physics in inertial systems which will then work in all contexts.

6.3.3 Mach's Principle

Einstein's thinking never strayed far from the specific aim of removing all notion of the absolute from physics. Both the principle of equivalence and the principle of general covariance were separately and together tasked with removing an observer's ability to tell with absolute conviction that their system is accelerating.

This debate goes back to Newton's time and is frequently illustrated by an experiment Newton made reference to (and said that he had performed for himself) involving a bucket.

Newton envisaged a bucket suspended from a string and part filled with water. With the bucket stationary, the surface of the water would be flat. Now imagine using the rope to spin up the bucket about an axis through the rope. At first, the water surface remains flat, as friction between the water and the bucket has not as yet transferred any element of the motion to the water. Gradually, the water starts to spin along with the bucket, and as it does so the surface starts to curve with the level of water higher at the edges and lower in the middle. Once the water is spun up fully, the curved surface remains, even though the water and the bucket are no longer in relative motion (Figure 6.1).

There are two ways of explaining this.

From one perspective, an element of the water does not have sufficient *inward* acting force on it (*centripetal* force), so it drifts to the edge of the bucket. As more water arrives at the edge, the level

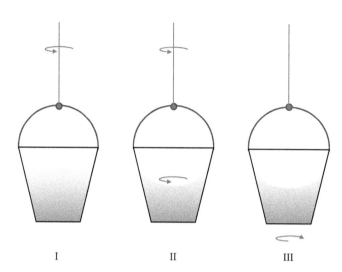

FIGURE 6.1 Newton's bucket experiment. (I) The bucket is rotating, but the water has not yet spun up. (II) The water is now rotating with the bucket. (III) From the point of view of an observer in/on the bucket, both it and the water are stationary and the surroundings are rotating.

climbs and with it the pressure in the water at the base of the growing pile. Water pressure acts equally in all directions, so specifically acts horizontally on elements of water, providing the necessary centripetal force. So, the level of water at a given distance from the centre of the bucket adjusts itself to provide water pressure acting centripetally, allowing the water to stabilise in circular motion with the bucket.

The other explanation, pertinent to an observer sitting (damply) within the bucket is that outwards facing (centrifugal) forces spring into existence that push the water horizontally away from the centre of the bucket, explaining the climb in water level.

In Newton's view, the centrifugal forces (which are examples of the fictitious or inertial forces mentioned in Chapter 2) arise due to the *absolute rotation* of the water with respect to an assumed universal system, *absolute space*.

Einstein's objection to this notion was part physics, part philosophy and is probably best summarised by the following quotation[36]:

> …it is contrary to the mode of thinking in science to conceive of a thing (the space-time continuum) which acts itself, but which cannot be acted upon.

In other words, if absolute space did exist then its action on an accelerating system would be one sided, or asymmetrical – the absolute space acting on the system, but the system having no reciprocal action on the absolute space.

Einstein's *relativity of inertia* developed from the principle of equivalence, suggested that the inertial effects in one system were the gravitational effects in another. Some modern proponents[37] interpret Einstein's ideas by suggesting an inertio-gravitational field, which incorporates the gravitational field of one system and an inertial field in another, fulfilling Einstein's analogy of electrical and magnetic fields being subsumed under an electromagnetic field. Such a construction is possible as, viewed from within an accelerating system, all observed masses appear to have an additional inertial force acting on them, proportional to their mass. The inertial field would explain this uniformity, just as a gravitational field does from another perspective. In general relativity, both fields boil down to the space-time curvature encoded into an appropriate metric.

However, as we suggested earlier, this relativity would demand that both gravitational field and the inertial field be determined by some mass/energy source, which is not immediately evident in the case of an accelerating system. Einstein's solution to this drew from his reading of Mach's[38] *The Science of Mechanics*, in which we find this, slightly cryptic but suggestive, remark[39]:

> Newton's experiment with the rotating vessel of water simply informs us, that the relative rotation of the water with respect to the sides of the vessel produces *no* noticeable centrifugal forces, but that such forces *are* produced by its relative motion with respect to the mass of the earth and the other celestial bodies... The one experiment only lies before us, and our business is, to bring it into accord with the other facts known to use, and not with the arbitrary fictions of our imagination.

(Mach's emphasis)

In Mach's view, it is not valid to conclude the existence of absolute space from the bucket experiment. The best one can do is to propose that rotation is relative to the 'celestial bodies', or in this age we might say 'the general distribution of matter in the universe'. From the point of view of the celestial bodies the bucket and water are rotating in their system. From the point of view of the water, when spun up, the celestial bodies are rotating about it. The situation certainly appears relative, with no need to call upon absolute space.

Newton's absolute space is the background for existence: it would still be present even if the universe were devoid of matter.[40] Mach is suggesting the background matter in the universe is a more tangible and more directly observable reference for acceleration to be viewed against.

By 1918, Einstein had incorporated this idea into his defining principles for the general theory, although he was clearly thinking about it earlier, as various letters and calculations indicate.

Indeed, in September 1915 he was able to show that his draft theory did not reproduce some Machian results that he was expecting, which was probably the final straw: shortly after he rejected the draft field equations and went in search of fully covariant ones, culminating in his November 1915 outpourings.

Einstein's 1918 version of his foundational principles reads as follows[41]:

a. Principle of relativity. The laws of nature are only assertions of timespace coincidences; therefore they find their unique, natural expression in generally covariant equations.

b. Principle of equivalence. Inertia and weight are identical in essence. From this and from the results of the special theory of relativity, it follows necessarily that the symmetric 'fundamental tensor' $g_{\mu\nu}$ determines the metric properties of space, the inertial relations of bodies in it, as well as gravitational effects. We will call the condition of space, described by the fundamental tensor, the 'G-field'.

c. Mach's principle. The G-field is determined without residue by the masses of bodies. Since mass and energy are equivalent according to the results of the special theory of relativity and since energy is described formally by the symmetric energy tensor $(T_{\mu\nu})$ this means that the G-field is conditioned and determined by the energy tensor.

Einstein also adds in a footnote:

> Up to now I have not distinguished principles (a) and (c) and that caused confusion. I have chosen the name 'Mach's principle' since this principle is a generalization of Mach's requirement that inertia be reducible to an interaction of bodies.

So, in the absence of an obvious gravitational source, the metric applicable to an accelerating system is determined by the general distribution of matter in the universe: Mach's 'other celestial bodies'. Acceleration is not with respect to absolute space, but rather the general distribution of matter in the universe. Equally, inertial systems would not be those at rest with respect to absolute space, or in uniform motion with respect to it, but rather those in inertial motion relative to the background distribution of matter in the universe.

Consequently, Einstein believed that a single test particle in an otherwise empty universe would exhibit no inertial properties and furthermore have no definable state of motion. Crucial in this would be the field equations of general relativity which, if you recall, determine the components of the metric. It was Einstein's hope that his field equations would not have any solutions in the absence of matter, hence his point above about the *G* field being determined 'without residue' by mass/energy.

However, in this as well, Einstein was to be disappointed, although he did not give up the struggle without a fight.

Detailed solutions of Einstein's field equations (when we eventually get to them) lie well outside the ambitions of this book, but we need to consider one important aspect of how one sets about solving such equations: the notion of *boundary conditions*. The field equations allow us to calculate the metric in a region of space-time from the energy density tensor applicable in that region. In any real situation, however, the space-time within the region has to fold into the metric applicable outside of that region. Things have to match up at the edges, so to speak. The values of the metric far away (potentially at infinity) from the region we are interested in are called the *boundary conditions* for the specific problem.

When Einstein started to make calculations with his draft theory, to see if it obeyed Mach's principle, he considered a rotating spherical shell of matter and tried to see if the metric it produced in its interior would explain the curvature of water in a bucket at the centre. He assumed that the boundary conditions, in this case far outside of the shell, were such that the metric became Minkowskian at infinity. Unfortunately, this did not give him the desired answer. Not only was the resulting metric inadequate to explain the curvature of the water's surface, only a small part of it was due to the shell: most of the metric came from the boundary condition that it folds into Minkowski space at infinity. Consequently, if we think instead of the inertia of any particles within the shell, we find that this is mostly determined by the boundary conditions, not the shell of matter. This is not a very Machian result. The boundary conditions smell uncomfortably like absolute space.

Armed with his new field equations, Einstein set out to tackle the problem of boundary conditions and came up with a remarkably novel solution: he *removed the boundaries*. In his 1917 paper, which today is hailed as the founding paper for modern cosmology, he proposed a model universe that was spatially closed on itself. In such a universe, there would be no spatial infinity. If you set off walking in one direction, you would eventually end up back where you started. There would be no boundary and the condition on the metric at infinity would be replaced by one that assured the metric was equal to itself once you had gone around the loop of the universe.

Unfortunately, such a universe was not a static solution to the field equations as they stood, so Einstein modified them by introducing a new term, which has become known as the *cosmological constant*. Einstein justified this new term via the gravitational repulsion it produced, which would keep the model universe in balance against the gravitational attraction of the matter within.[42] As this was before Hubble's crucial discovery of the expanding universe,[43] the consensus astronomical view was that the universe was, broadly, in a static and unchanging state. Hence Einstein justified the cosmological constant as a necessary addition to provide a static universe, his real agenda, however, was to tackle the boundary condition problem.

Alas, this manoeuvring did not work. Eddington[44] was able to show that, even with the cosmological constant at work, Einstein's closed universe was unstable and prone to either expansion or collapse, depending on small fluctuations of the matter within it.

In a letter to De Sitter[45] in 1917, Einstein[46] expressed his support for Machian ideas once again:

> It would be unsatisfactory, in my opinion, if a world without matter were possible. Rather, it should be the case that the $g_{\mu\nu}$ field *is fully determined by matter and cannot exist without the latter*. This is the core of what I mean by the requirement of the relativity of inertia.

(Einstein's emphasis)

Unfortunately, De Sitter was able to find a solution to the same field equations for a universe that did not contain any matter. In his 1918 publications, Einstein tried to argue that the De Sitter universe did in fact contain matter, pushed out to the edges as it were. Various physicists were able to show that Einstein was wrong, and that (solutions to the field equations without matter) were possible. Gradually, Einstein's resistance wore away and, in the end, he was forced to give up the idea that his general theory of relativity embodied Mach's principle and removed all notions of absolute motion.

While we now understand general relativity in far greater detail than Einstein himself and his contemporaries (work has been going on for the last 100 years…), the ultimate question of the extent to which general relativity is indeed Machian remains somewhat in dispute. A celebrated entrance and exit pole amongst relativists attending a conference in 1993 produced the following results[47]:

Question	Answer	Entrance Poll	Exit Poll[48]
Was Mach advocating a mere re-description (MRD) of Newtonian mechanics without a change of physical content or was he advocating a genuinely new theory (GNT)?	MRD	12	10
	GNT	26	16
Is general relativity perfectly Machian?	Yes	2	3
	No	30	21
Is general relativity with appropriate boundary conditions of closure of some kind perfectly Machian?	Yes	14	9
	No	18	14
Is general relativity with appropriate boundary conditions of closure of some kind very Machian?	Yes	19	14
	No	18	7

Make of that, what you will…

In the end Einstein was reasonably content. While he accepted that general relativity probably did not fully express Mach's principle and that it retained a notion of absolute space, via the boundary conditions, he was satisfied that his fundamental objection had at least been mitigated. Newtonian absolute space, if you recall, was supposed to influence matter via inertial forces, but not have any reciprocal effect acting on it. Einsteinian space-time definitely influences matter: the metric determines the shape of the space-time, and hence the path of the matter. However, the distribution of matter curves the space-time, setting up the metric (with boundary conditions…). This is a reciprocal action, of a form.

6.3.4 Modern Perspective

In truth there is a perfectly reasonable modern view to all of this. The foundations of the general theory, or rather the relativistic theory of gravitation, do not lie in the equivalence principle, general covariance or Mach's principle. We can get away with two fundamental notions:

- Space-time takes a particular geometrical form (technically called semi-Riemannian) with metric tensors expressing its curvature and gravitational effects are due to that curvature,
- The metric is determined by Einstein's field equations.

Inertial systems are then, simply, local systems that happen to be in free fall.

6.4 Einstein's Disc

Moving from the notion of gravitation being mediated by a field of force to it being a distortion of space-time is such a remarkable jump it seems amazing that anyone should come up with such an idea. While we can't be sure of the process that went on in Einstein's head, we can search through his written output looking for clues, most fruitfully in his 1916 review paper[49] where he makes reference to an elegant thought experiment.

Einstein has us imagine a flat disc rotating at some constant angular speed. One observer is standing on the disc, and so rotating with it. Another is stationary next to the disc and not experiencing any rotation. From this observer's perspective, the person standing on the disc is clearly in rotation and so subject to centrifugal forces. The person on the disc will believe that they are stationary in a gravitational field (from the principle of equivalence) and that the observer outside the disc is in free fall.

A sequence of standard measuring rods laid out end to end, and not overlapping, along a radius of the disc will measure that radius with a value of, say, R. These rods will not be contracted along their length, so the measurement of the radius should agree with any measurement taken with the disc at rest. However, at the circumference things are rather different. Here the measuring rods are contracted, so that their length, as seen by someone stationary next to the disc, is l_0/γ. Consequently, more of these contracted rods would fit around the circumference. If the rods are $1/\gamma$ shorter, then the circumference that they measure must be γ longer. In other words, the circumference of the disc will be $\gamma 2\pi R$, so that the ratio of circumference to diameter of the disc is no longer π. This is a change in the geometry. By the principle of equivalence, what is true for an accelerating system is also true for a stationary one in a gravitational field. Hence, gravitational fields (in old money, as it were) relate to a distortion of geometry (the new money view).

As we have become more conversant with the technicalities of the general theory, so the analysis of the rotating disc has become more detailed. It becomes an excellent example of how things can be rather trickier in general relativity than they appear at first glance.

For example, we can ask an apparently straightforward question such as: are the measuring rods contracted from the point of view of an observer stationary on the disc? The answer depends on the exact definition of what is meant in the question. If the observer on the disc uses the same time co-ordinates as the stationary observer next to the disc, then when observing a rod, they *will* see a length contraction. However, if they take as their time co-ordinates those of a system instantaneously co-moving with one of the rods, then there will be no observed length contraction.

As you step along the radius, you are actually moving from system to system, so any measure of the radius is formed from stitching together proper lengths measured separately in each system. The same is true of the circumference measure. As there is no one system applicable across the whole disc, it is not possible to construct a general spatial geometry for the disc (such as a slice through 3-space taken at equal times across the whole disc).

There are also tricky problems to deal with in getting the disc spun up in the first place. We have to remember that the disc is no more than a collection of atoms bound to each other by interatomic forces. As the disc is slowly accelerated from rest, we have to work out how these atoms respond across the surface, accounting for the space-time distortions of relativity along with the stresses induced in the disc and the time delays imposed by the limitation of the speed of light (so not all parts of the disc commence moving at the same moment). In relativity, there is no such thing as a rigid object.[50]

There is an extensive body of literature associated with the rotating disc in general relativity, not to mention some 'debates' on the internet about it. For our purposes, the significance is simply the suggestion that it put into Einstein's head: that gravitation is geometry.

As I have already mentioned, Einstein's early tinkering with gravitation focussed on the geometry of space-time being expressed solely in a curvature of the time co-ordinate. This lead him to various insights, most prominently that clocks run more slowly closer to a mass/energy source.[51]

Clocks laid out along the circumference of the rotating disc are moving much more rapidly than a partner clock at the centre (which is basically rotating about an axis through its middle). The circumferential clocks, due to special relativistic time dilation, will run more slowly than that at the centre (Figure 6.2).

The gravitational effects experienced by the observer on the disc suggest that masses are arranged somewhere off-disc and pulling objects towards them. Hence, the clocks at the circumference are nearer to the source of gravity and as a result are running more slowly. In this instance, Einstein's form of the principle of equivalence has borne fruit. This time dilation effect has been amply confirmed by experiment, as have other effects we will discuss when we come to the weak field case in Chapter 8.

All that, however, needs us to have a much firmer grasp of the mathematical side of the general theory, which is the subject of the next chapter.

Disk from perspective of
stationary observer

Disk from perspective of an
observer on the disk

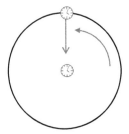
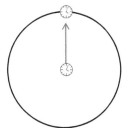

Centripetal acceleration acting
due to the rotation

Gravitational field acting

FIGURE 6.2   Two perspectives on a rotating disc. In both cases, a clock situated at the edge of the disc will run more slowly than one at the centre. In the right hand diagram, this effect is gravitational as the disc is not moving in this system.

Notes

1. Einstein to Grossmann regarding the mathematics of curved space-time, As cited in Abraham Pais, *Subtle is the Lord* (1982, 2005), p. 212.
2. From Paul Arthur Schilpp, 1949. 'Autobiographical Notes', *Albert Einstein: Philosopher-Scientist*, pp. 65–67.
3. *Hilbert,* Constance Reid, Springer Springer-Verlag; ISBN-10: 0387049991, 1983
4. A Einstein to P Ehrenfest, 26 December 1915 Collected Papers of Albert Einstein, 8, Doc. 173, as quoted in Janssen, Michel (2004) *Einstein's First Systematic Exposition of General Relativity* http://philsci-archive.pitt.edu/id/eprint/2123.
5. Letter to M Besso, 10 December 1915 as quoted in *Subtle is the Lord*, A Pais, Oxford University Press, (2005) ISBN 0-19-280672-6.
6. Einstein, A. 1907., On the relativity principle and the conclusions drawn from it. *Jahrbuch der Radioaktivität*, **4**, pp. 411–462.
7. Einstein, A., 1911. On the influence of gravitation on the propagation of light. *Annalen der Physik* (ser. 4), **35**, pp. 898–908.
8. That is, incorporating the idea of a field that mediated gravitational forces as per the electrical field and which responded to changes in matter distribution at a rate determined by the speed of light.
9. Exactly what counts as the 'gravitational field' in modern general relativity will be discussed in more detail in the next chapter.
10. Einstein, A., 1914. *Formal Foundations of the General Theory of Relativity*: Preussische Akademie der Wissenschaften, Sitzungsberichte, (part 2), pp. 1030–1085.
11. Einstein, A., 1915. *Explanation of the Perihelion Motion of Mercury from the General Theory of Relativity*: Preussische Akademie der Wissenschaften, Sitzungsberichte, (part 2), pp. 831–839.
12. Einstein, A., 1915. *The Field Equations of Gravitation*. Preussische Akademie der Wissenschaften, Sitzungsberichte, (part 2), pp. 844–847.
13. Preussische Akademie der Wissenschaften, Sitzungsberichte, 1916 (part 1), 688–696.
14. Preussische Akademie der Wissenschaften, Sitzungsberichte, 1917 (part 1), 142–152.
15. This grouping was formed from Germany, Austria, Hungary, Bulgaria and the Ottoman Empire.
16. Hendrik Lorentz (1853–1928) of Lorentz transformation fame, Willem de Sitter (1872–1934) mathematician, physicist and astronomer and Paul Ehrenfest (1880–1933), physicist.

17. I am indebted to the following authors and their articles for their influence on my thinking in this chapter: John D Norton: *General Covariance and the Foundations of General Relativity: Eight Decades of Dispute*, Rep Prog Phys 56 (1993) 791458; Michel Janssen: *No Success Like Failure; Einstein's Quest for General Relativity, 1907–1920* (2008) http://philsci-archive.pitt.edu/id/eprint/4377; *Einstein's first systematic exposition of General Relativity* (2004) http://philsci-archive.pitt.edu/id/eprint/2123.

18. D Hilbert (1862–1943), one of the most influential and significant mathematicians in the 19th and early 20th centuries.

19. And, of course, provided that the air resistance on them is negligible.

20. As quoted in Pais, A., 1982, *Subtle is the Lord*, Oxford, ISBN-13: 978-0192806727 p. 178.

21. In this translation from the original German, 'happiest' has the context of 'most fortunate'.

22. There is a small caveat to this, which I will come to shortly.

23. As per 7.

24. At the time of writing, it would appear that in England, cricket balls are often dropped when they should be caught.

25. Steven Weinberg (1933–), 1979 Nobel Prize winner in Physics for his work on the interactions of fundamental particles.

26. Weinberg, S., 28 July 1972. *Gravitation and Cosmology: Principles and Applications of the General Theory of Relativity* (1st edn). John Wiley & Sons, p. 68. ISBN-13: 978-0471925675.

27. As per 26.

28. Note for the technically well-informed. Here I am simply saying that once you have accepted that the geometry of space-time takes the form of a Reimannian manifold, the flatness follows. It is possible to justify that choice on grounds of simplicity rather than the equivalence principle.

29. Einstein, A., *The Meaning of Relatively* (5th edn). Princeton University Press: Princeton, NI. P57-8. My emphasis.

30. Einstein, A., 1916. The foundation of the general theory of relativity. *Annalen der Physik (ser. 4)*, **49**, 769–822.

31. Bondi, H., 1919. *Is "General Relativity" Necessary for Einstein's Theory of Gravitation?* In: Relativity Quanta, and Cosmology in the Development of the Scientific Thought of Albert Einstein (Vol. 1). M. Pantaleo and F.D. Finis (eds). New York: Johnson Reprint Corporation, pp. 179–186.

32. Erich Justus Kretschmann (1887–1973) physicist.

33. Einstein, A., 1918. On the Foundations of the General Theory of Relativity. *Annalen der Physik (ser. 4)*, **55**, pp. 241–244.

34. As per 26, p. 92.

35. As per 26, p. 92.

36. Einstein, A., 1922. *The Meaning of Relatively* (5th edn). Princeton University Press: Princeton, NI, p. 55.

37. Notably, John Stachel (physicist and philosopher of science, head of Boston University Center for Einstein Studies) in the 1990s, since when it has been picked up by others including Michel Janssen (Professor, History of Science, University of Minnesota).

38. Ernst Waldfried Josef Wenzel Mach (1838–1916) Physicists and philosopher. The ratio of the speed of a moving object to the speed of sound (Mach number) is named after him.

39. Mach, E., 1919. *The Science of Mechanics*. The Open Court Publishing Co.

40. As it turns out, universes devoid of matter are possible in general relativity, but this is not quite as significant as it might sound as the space-time is curved by the energy in the curved space time. As we have noted before, gravitational potential energy is capable of exerting a gravitational effect.

41. Einstein, A., 1918. On the foundations of the general theory of relativity. *Annals of Physics*, **55**, pp. 240–244.

42. This is couched in terms of gravitational forces, for familiarity, really we should be talking about positive and negative curvatures.

43. Edwin Hubble discovered the recession of distant galaxies in 1929. We will spend a lot more time on this topic in Chapter 12.

44. Eddington, A.S, (1930). On the instability of Einstein's spherical world. *Monthly Notices of the Royal Astronomical Society*, **90**, pp. 668–678.

45. Willem de Sitter (1872–1934), mathematician and physicist.

46. As per 4 Doc 317.

47. *Mach's Principle: From Newton's Bucket to Quantum Gravity* (11 August 1995). J.B. Barbour, H Pfister (eds). Springer Science & Business Media. p. 106.

48. In case anyone is worried about the non-conservation of physicists, some had to leave before the exit poll…

49. As per 30, although the disk was also mentioned in 1912 and Einstein was clearly thinking about it as far back as 1909.

50. Max Born (1882–1970) did a lot of early work on this and a convenient view, based on his thinking, is that objects can *move rigidly*, in the sense that they maintain their shape (subject to appropriate constraints) even though they are not rigid in the familiar sense.

51. Sometimes people say that clocks run more slowly in a gravitational field. I don't find this a helpful phrase and try and avoid it and others like it. My problem with this expression is that it suggests that the gravitational field is something separate from the curvature of time. In fact in a weak field, and that around the Earth is classed as weak, the curvature of time is causing the effects that we normally ascribe to a gravitational field.

7

The Theory of General Relativity

7.1 A Toolkit for the General Theory

The mathematical apparatus of the general theory is far more complicated than that for the special theory. Consequently, we can't dive straight in and describe the Einstein field equations without spending some time developing the terminology and the geometrical meaning behind some of the mathematical objects in use. Also, as the field equations are generally covariant, we will be working in *generalised co-ordinate systems*, which is a good place to start our development.

7.2 Generalised Co-ordinates

In Figure 7.1, a familiar system of co-ordinates (x, y, z) is being used to specify the location of a point (x_1, y_1, z_1) lying within a 2D surface. If we have an equation, $f(x, y, z) = $ constant, for that surface, we can fix one of these co-ordinates, say y_1, while allowing the other two to vary, subject to that equation. This will draw a line across the surface. We could do this for any pair of the co-ordinates while fixing the third and end up with a set of lines, one for each choice of fixed co-ordinate and value chosen for the fixing. These lines form a grid, which can also be used to specify the location of points within the surface.

For example, the equation mapping out the surface of a sphere of radius R is $x^2 + y^2 + z^2 = R^2$. If we keep one of the variables constant, say z, while allowing x and y to vary (constrained by $x^2 + y^2 + (\text{chosen } z \text{ value})^2 = R^2$), we draw out circles around the z-axis with a radius determined by the value of z that we have chosen. Equally, if we fix a value for x while allowing y and z to change, we draw circles around the x-axis. The combination of these two sets of circles provides a valid co-ordinate system on the surface of the sphere (Figure 7.2).

On any 2D surface, a system made from two variables (s_1, s_2), with a fixed origin point, will be sufficient to denote any point.[1] That point will then have one set of co-ordinates within the surface, (s_1, s_2), and another from the external co-ordinate axes (x, y, z). Consequently, it must be possible to write the three external co-ordinates in terms of surface co-ordinates:

$$x = x(s_1, s_2) \quad y = y(s_1, s_2) \quad z = z(s_1, s_2)$$

and vice versa:

$$s_1 = s_1(x, y, z) \quad s_2 = s_2(x, y, z).$$

Take two co-ordinates mined from the sets of circles that we have drawn on the surface of a sphere (Figure 7.3):

$s_1 = $ fraction of a circumference measured down from a pole
$s_2 = $ fraction of a circle measured around at a longitude

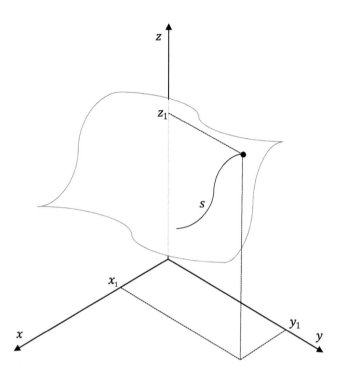

FIGURE 7.1 A curved surface described as a particular function of co-ordinates in 3D space with respect to (x, y, z) axes. Keeping one co-ordinate fixed and allowing the other two to vary will draw a line, s, on the surface. A collection of lines of this form can be used as a valid set of co-ordinates to specify any point on the surface.

then we have:

$$x = R \sin\left(\frac{s_1}{R}\right)\cos\left(\frac{s_2}{R\sin(s_1/R)}\right) \quad y = R\sin\left(\frac{s_1}{R}\right)\sin\left(\frac{s_2}{R\sin(s_1/R)}\right) \quad z = R\cos\left(\frac{s_1}{R}\right),$$

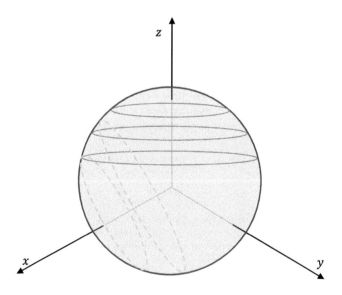

FIGURE 7.2   Sets of circles drawn on the surface of a sphere which can act as a co-ordinate system on the surface.

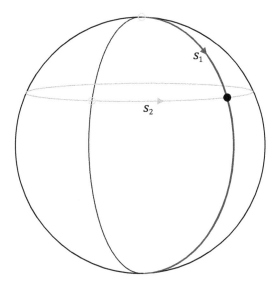

FIGURE 7.3 A curved co-ordinate system on the surface of a sphere uses two variables s_1 and s_2 to specify the position of a point on the surface. The same point can be specified using three co-ordinates in the embedding space.

where R is the radius of the sphere. The inverses are:

$$s_1 = R \cos^{-1}(z/R) \quad s_2 = R \sin\left(\cos^{-1}(z/R)\right) \tan^{-1}(y/x).$$

Think of this from the perspective of a (hypothetical) flat 2D physicist who was part of a race of flatters living on the surface of the sphere. Such beings are only able to slide around on the surface of the sphere (they have no dimensionality perpendicular to the surface) and can only 'see' due to light rays confined to the surface as well. In this analogy, light rays striking the surface from outside, would not be part of their universe.

A set of co-ordinates such as (s_1, s_2) could be measured within their universe with appropriate rulers, light rays, or similar. Locally, these would appear to be flat co-ordinates. On the other hand, (x, y, z) *cannot be directly measured* as that would require access to the third dimension, which these creatures do not have. They can *infer* the values from measurements made within their universe but are unable to have direct physical access to these dimensions.

Co-ordinates that can be directly measured from within a space are called *intrinsic co-ordinates*, while the external co-ordinates are *embedding co-ordinates* (see Section 9.1.1).

From the perspective of general relativity, intrinsic co-ordinates are crucial as they are *the only type that we have direct access to*. If our universe exists in an extended embedding co-ordinate space-time (and some versions of quantum gravity thinks in those terms) we do not have access to it, so any determination of the geometry of the universe needs to be done on the basis of measurements made within systems of intrinsic co-ordinates.

To go back to our 2D analogy, we need to know if and how our flat physicist could determine the curvature of a spherical surface from measurements made entirely within that surface. The answer is of course yes, but we need in due course to develop a specific way of doing that which will generalise to 3D or even the 3D+1 universe that we occupy.

Even then we will need to take great care, as ascribing co-ordinates can be a very tricky business. There is very rarely one 'correct' or 'most useful' set of co-ordinates for a given situation. General relativity's general covariance assures us that the chosen co-ordinates do not matter, but they can lead us

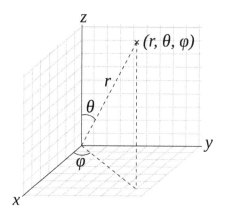

FIGURE 7.4   The spherical-polar co-ordinate system as used in the Schwarzschild metric.

astray due to our habit of thinking that the values for any set of co-ordinates must have direct physical meaning. We will need to develop this point later (see especially the discussion of co-ordinate time in Section 8.2), but for one quick example, we can pop back to the Schwarzschild metric from Chapter 4:

$$
g_{Sch} = \begin{pmatrix} \left(1 - \dfrac{2GM}{c^2 r}\right) & 0 & 0 & 0 \\ 0 & -\left(1 - \dfrac{2GM}{c^2 r}\right)^{-1} & 0 & 0 \\ 0 & 0 & -r^2 & 0 \\ 0 & 0 & 0 & -r^2 \sin^2(\theta) \end{pmatrix}
$$

written in spherical-polar co-ordinates (Figure 7.4).

In this metric, the gravitating mass is at the origin of the system and the metric applies in the free (unoccupied by masses or fields) surrounding space-time. The use of this co-ordinate system, supplemented by a *ct*-axis (which can't be rendered on such a diagram), suggests that we could 'walk' away from the mass at the origin along a radial line, *r*. In fact, the gravitational curvature applied to the space means that distance we would cover walking along that line is different to the change in the value of *r*. Worse, the time co-ordinate has no direct relationship to values shown on any clock that we are carrying. Co-ordinate time fades into clock time as you move further and further from the source of gravitation. Such issues probably lay behind Einstein's comment mentioned at the start of Chapter 6[2]:

> Why were another seven years required for the construction of the general theory of relativity? The main reason lies in the fact that it is not so easy to free oneself from the idea that coordinates must have an immediate metrical meaning.

7.3 Rates of Change for Generalised Co-ordinates

In normal co-ordinates, an expression like $\Delta E_x / \Delta y$ is the rate at which the *x* component of the electrical field changes as you move in the *y* direction. When general co-ordinates are introduced, things are no longer as simple as the *y*-axis at one point in the space-time *is no longer in the same direction as the y-axis somewhere nearby*. Even if the magnitude of the field itself has not changed, there will be a change in the components as we move along *y* due to the co-ordinate system rotating as we move. We must account for that as well.

Each point has its own local arrangement of axes and you need to flow from one set to another as you move around in a curved geometry. Given two points A and B in our curved system, the arrangement of axes at B will be a function of the co-ordinates of B, as measured from A.

It's easier to think of this via a concrete example: again, a sphere. For any two spots on the equator of a sphere, the lines connecting them to the North Pole will be pointing in different directions, as will be the lines pointing along the equator (in 3D).

In order to gain some mathematical purchase, we introduce four new vectors:

$$\epsilon^0 = \begin{pmatrix} 1 \\ 0 \\ 0 \\ 0 \end{pmatrix} \quad \epsilon^1 = \begin{pmatrix} 0 \\ 1 \\ 0 \\ 0 \end{pmatrix} \quad \epsilon^2 = \begin{pmatrix} 0 \\ 0 \\ 1 \\ 0 \end{pmatrix} \quad \epsilon^3 = \begin{pmatrix} 0 \\ 0 \\ 0 \\ 1 \end{pmatrix}$$

which point along each of the co-ordinate axes in turn. I have set these vectors to be unit length, which is convenient when it comes to specifying events or building other vectors from them. The four vectors[3] are distinguished by labelling them with *superscripts*, so that they do not get confused with the *components* of a vector, such as X_0, which I have been using *subscripts* to indicate. The bold font is intended to re-enforce the fact that these are vectors, and so ϵ^2 stands for vector 2, not a vector squared, $(\epsilon)^2$. When I wish to refer to a component of one of these vectors, I will be forced into a slightly clumsy notation such as ϵ_3^2 for the third component of the second vector, etc.

Now that we have these vectors, it is easy to see that any other vector can be assembled from them, with the right parts. For example, the 4-displacement of an event (ct, x, y, z) can be expressed as a vector:

$$X = ct\epsilon^0 + x\epsilon^1 + y\epsilon^2 + z\epsilon^3.$$

provided that the co-ordinate values are specified along the axis vectors employed.[4] Equally, a 4-velocity $U = (U_0, U_1, U_2, U_3)$ would be:

$$U = U_0\epsilon^0 + U_1\epsilon^1 + U_2\epsilon^2 + U_3\epsilon^3$$

and any other vector, A is:

$$A = A_0\epsilon^0 + A_1\epsilon^1 + A_2\epsilon^2 + A_3\epsilon^3.$$

If we are dealing with a flat space-time expressed in rectangular (Cartesian) co-ordinates (rather than say, spherical-polar co-ordinates), the axis vectors are constant, as the axes extend across the entire co-ordinate system without changing direction. However, with a general system this is no longer the case, and the axis vectors at point B are combinations of the axis vectors of a system based at A:

$$\left(\epsilon^0\right)'(ct,x,y,z) = C_{00}(ct,x,y,z)\epsilon^0 + C_{01}(ct,x,y,z)\epsilon^1 + C_{02}(ct,x,y,z)\epsilon^2 + C_{03}(ct,x,y,z)\epsilon^3.$$

The coefficients, like C_{01}, tell us how the zeroth vector at B is built from the 1-vector at A, the exact functions involved depending on the geometry that we are dealing with. We would need similar expressions for the other co-ordinate axes, $\left(\epsilon^1\right)', \left(\epsilon^2\right)', \left(\epsilon^3\right)'$ at B.

With a moving object, we would expect its co-ordinates to be functions of time. Unfortunately, moving through a curved space-time, the axis vectors would also be functions of co-ordinates. Consequently, we end up with something complicated looking:

$$X(\text{ct},x,y,z) = X_0(\text{ct},x,y,z)\epsilon^0(\text{ct},x,y,z) + X_1(\text{ct},x,y,z)\epsilon^1(\text{ct},x,\ y,\ z) + X_2(\text{ct},x,y,z)\epsilon^2(\text{ct},x,y,z)$$

$$+ X_3(\text{ct},x,y,z)\epsilon^3(\text{ct},x,y,z).$$

In the expression above, we have stuck with using our standard co-ordinates from the Lorentzian systems (ct,x,y,z), but in the general theory of relativity, any co-ordinate system goes. We could choose to use spherical-polar co-ordinates instead which would be (ct,r,θ,φ). It could be something more exotic. To provide a generic terminology for co-ordinates, I am going to use (x_0,x_1,x_2,x_3) to denote any general set of co-ordinates. In Cartesian terms $x_0 = ct, x_1 = x, x_2 = y, x_3 = z$; in spherical-polar $x_0 = ct, x_1 = r, x_2 = \theta, x_3 = \varphi$. In other systems, we would have to specify the appropriate relationship. Note the axis vectors would be different in each case as well. Our expression for the co-ordinates of an object becomes:

$$X(x_0,x_1,x_2,x_3) = X_0(x_0,x_1,x_2,x_3)\epsilon^0(x_0,x_1,x_2,x_3) + X_1(x_0,x_1,x_2,x_3)\epsilon^1(x_0,x_1,x_2,x_3)$$

$$+ X_2(x_0,x_1,x_2,x_3)\epsilon^2(cx_0,x_1,x_2,x_3) + X_3(x_0,x_1,x_2,x_3)\epsilon^3(x_0,x_1,x_2,x_3).$$

To avoid driving everyone mad (including the typesetter), I am not going to explicitly indicate that something is a function of the co-ordinates every time we see it. I will include the co-ordinates only if that helps make a specific point, so:

$$X = X_0\epsilon^0 + X_1\epsilon^1 + X_2\epsilon^2 + X_3\epsilon^3.$$

Now it is high time that we circled back to our original problem: how to deal with an expression like $\Delta E_x / \Delta y$ in general co-ordinates where the direction of the axes might be changing along with the components of the vector.

Let's state this in more general terms. We have a vector A with components (A_0, A_1, A_2, A_3) referred to some collection of axis vectors $(\epsilon^0, \epsilon^1, \epsilon^2, \epsilon^3)$, both of which are functions of our general co-ordinates. What happens to that vector when we move a small distance along one of the co-ordinate lines? In other words, what is the quantity $\Delta A_i / \Delta x_k$ where i and k could be 0, 1, 2, 3? We'll pick one example and look to see how the component A_2 changes with co-ordinate x_3.

As the component is effectively a function of more than one variable, the techniques we introduced in Section 3.9 apply. So, we write:

$$\frac{\Delta(A_2\epsilon^2)}{\Delta x_3} = \left(\frac{\Delta A_2}{\Delta x_3}\epsilon^2\right)_{\text{keeping axis vectors constant}} + \left(A_2\frac{\Delta\epsilon^2}{\Delta x_3}\right)_{\text{keeping components constant}}.$$

The first term is the normal expression for the change in a component. The second is telling us how things are changing across the geometry in which the vector components are embedded.

(In this calculation I am assuming that all the other co-ordinates (x_0, x_1, x_2) are not changing while we are twiddling with x_3.)

In a flat space-time geometry, *expressed in Cartesian co-ordinates*, the second bracket would vanish. The axis vectors are always pointing in the same direction, so they do not change with x_3, or any other co-ordinate for that matter. Remember that a geometry can be flat, and still use curved co-ordinates (e.g. spherical-polar co-ordinates), so in such a case the second bracket would not vanish. The axis vectors at any point on the surface of a sphere would be pointing in a different direction to those at a different point nearby.

Now, $A_2\Delta\epsilon^2/\Delta x_3$ is certainly a vector, so it must be possible to write it in terms of the axis vectors at the point (x_0, x_1, x_2, x_3).

(Do not jump to the conclusion that this vector is just pointing along ϵ^2. This is how ϵ^2 *changes* when x_3 changes by a small amount and the new axis vector $\epsilon^2(x_3 + \Delta x_3)$ is probably not pointing in the same direction as $\epsilon^2(x_3)$.)

Formally I write:

$$A_2 \frac{\Delta\epsilon^2}{\Delta x_3} = A_2\left(\Gamma^0_{23}\epsilon^0 + \Gamma^1_{23}\epsilon^1 + \Gamma^2_{23}\epsilon^2 + \Gamma^3_{23}\epsilon^3\right)$$

to fulfil this promise.

The funny Γ symbols are about to become very important to us. They are known as *connection coefficients* or *Christoffel symbols*[5] and are tagged with three indices:

- the superscript tells us which axis vector the symbol is the coefficient for,
- the two subscripts label the original axis vector that we are allowing to change and the co-ordinate over which it is changing.

So, specifically, the symbol Γ^k_{ij} links the axis vector ϵ^k to the way that the axis vector ϵ^i changes when we move along in x_j. In standard developments of general relativity, the bottom indices are symmetric: $\Gamma^k_{ij} = \Gamma^k_{ji}$.

Although this is a point that we will develop later, it is worth saying now that the Christoffel symbols are *not* tensors, they transform in very different manner (Section 7.5.3).

Flipping from our specific example to the general case we have:

$$A_i \frac{\Delta\epsilon^i}{\Delta x_k} = A_i\left(\Gamma^0_{ik}\epsilon^0 + \Gamma^1_{ik}\epsilon^1 + \Gamma^2_{ik}\epsilon^2 + \Gamma^3_{ik}\epsilon^3\right),$$

which we insert into our formula for the rate of change of a component:

$$\frac{\Delta\left(A_i\epsilon^i\right)}{\Delta x_k} = \left(\frac{\Delta A_i}{\Delta x_k}\epsilon^i\right)_{\text{keeping axis vectors constant}} + \left(A_i\frac{\Delta\epsilon^i}{\Delta x_k}\right)_{\text{keeping components constant}}$$

to give:

$$\frac{\Delta\left(A_i\epsilon^i\right)}{\Delta x_k} = \left(\frac{\Delta A_i}{\Delta x_k}\epsilon^i\right) + \left(A_i\frac{\Delta\epsilon^i}{\Delta x_k}\right) = \left(\frac{\Delta A_i}{\Delta x_k}\epsilon^i\right) + A_i\left(\Gamma^0_{ik}\epsilon^0 + \Gamma^1_{ik}\epsilon^1 + \Gamma^2_{ik}\epsilon^2 + \Gamma^3_{ik}\epsilon^3\right).$$

Once again, it is worth reminding ourselves how this breaks down:

rate of change of a vector as co-ordinates change $=$ (rate of change due to components changing)

$+$ (rate of change due to the geometry)

7.3.1 Summation, Metrics and Christoffel Symbols

As we are going to be using long expressions involving summations, it is useful to introduce a more compact notation. A standard sum over a sequence of terms such as:

$$\boldsymbol{U} = U_0\epsilon^0 + U_1\epsilon^1 + U_2\epsilon^2 + U_3\epsilon^3$$

can be expressed more neatly by:

$$U = \sum_i U_i \epsilon^i.$$

The large Σ, which is pronounced 'sigma', stands for 'sum' and the little i hanging about below means: sum the terms which involve an i, where i spans the values that are involved in the sum, in this case $i = 0,1,2,3$.

We will also have need of a 'double summation' $\Sigma\Sigma$, for example our space-time invariant is:

$$(\Delta s)^2 = \sum_{i,j} \eta_{ij} X_i X_j = \sum_i \sum_j \eta_{ij} X_i X_j = \sum_i \left(\eta_{i0} X_i X_0 + \eta_{i1} X_i X_1 + \eta_{i2} X_i X_2 + \eta_{i3} X_i X_3 \right).$$

Carrying out the second summation (for practice) gives:

$$(\Delta s)^2 = \left(\eta_{00} X_0 X_0 + \eta_{10} X_1 X_0 + \eta_{20} X_2 X_0 + \eta_{30} X_3 X_0 \right)$$

$$+ \left(\eta_{01} X_0 X_1 + \eta_{11} X_1 X_1 + \eta_{21} X_2 X_1 + \eta_{31} X_3 X_1 \right)$$

$$+ \left(\eta_{02} X_0 X_2 + \cdots \right) + \left(\eta_{03} X_0 X_3 + \cdots \right)$$

$$= \eta_{00} X_0 X_0 + \eta_{11} X_1 X_1 + \eta_{22} X_2 X_2 + \eta_{33} X_3 X_3.$$

The summation collapsing as the components of the metric, $\eta_{ij} = 0$, unless $i = j$:

$$(\Delta s)^2 = \sum_i \eta_{ii} X_i X_i.$$

Using our new notation, we render the rate of change of a vector in curved co-ordinates more compactly:

$$\frac{\Delta(A)}{\Delta x_k} = \frac{\Delta}{\Delta x_k} \left(\sum_i A_i \epsilon^i \right) = \sum_i \left\{ \frac{\Delta A_i}{\Delta x_k} \epsilon^i + A_i \frac{\Delta \epsilon^i}{\Delta x_k} \right\} = \sum_i \left\{ \frac{\Delta A_i}{\Delta x_k} \epsilon^i + A_i \sum_m \Gamma^m_{ik} \epsilon^m \right\}.$$

Or, splitting the terms up for the moment:

$$\frac{\Delta(A)}{\Delta x_k} = \sum_i \frac{\Delta A_i}{\Delta x_k} \epsilon^i + \sum_i A_i \sum_m \Gamma^m_{ik} \epsilon^m.$$

A summation such as Σ_i is really a shorthand for 'add up the terms involving i by letting i take the values 0, 1, 2, 3 in order'. In which case, *the letter used to denote the summation is not relevant* and we can switch letters without changing the result of the calculation. I'm going to pull this trick in the double sum by switching $i \rightarrow m$ and separately $m \rightarrow i$ (all will become clear):

$$\frac{\Delta(A)}{\Delta x_k} = \sum_i \frac{\Delta A_i}{\Delta x_k} \epsilon^i + \sum_m A_m \sum_i \Gamma^i_{mk} \epsilon^i = \sum_i \left\{ \frac{\Delta A_i}{\Delta x_k} + \sum_m A_m \Gamma^i_{mk} \right\} \epsilon^i.$$

Having done that, both terms now feature ϵ^i, which is why I was able to factor it out of the sum. So now we have:

$$\frac{\Delta(A)}{\Delta x_k} = \sum_i \frac{\Delta A_i}{\Delta x_k} \epsilon^i = \sum_i \left\{ \frac{\Delta A_i}{\Delta x_k} + \sum_m A_m \Gamma^i_{mk} \right\} \epsilon^i.$$

Or, if you prefer, in component form:

$$\frac{\Delta(A_i)}{\Delta x_k} = \frac{\Delta A_i}{\Delta x_k} + \sum_m A_m \Gamma^i_{mk},$$

which avoids having any axis vectors getting in the way.

To distinguish a 'normal' rate of change, which just involves the changes in the components, from a 'geometrical' rate of change, which involves the change in the axis vectors as well, I'm going to introduce the 'D' terminology and write:

$$\mathbb{D}_k(A_i) = \frac{D(A_i)}{\Delta x_k} = \frac{\Delta A_i}{\Delta x_k} + \sum_m A_m \Gamma^i_{mk}. \tag{7.1}$$

The Christoffel symbols are crucial, as without them we can't proceed to make any calculations. Their job is to stitch together parts of space-time by tracking how the co-ordinate grids, expressed in the metric, change as you move from point to point. They can be obtained directly from the metric $g_{\overline{ik}}$:

$$\Gamma^j_{ik} = \frac{1}{2} \sum_m (g^{-1})_{jm} \left\{ \frac{\Delta g_{\overline{mk}}}{\Delta x_i} + \frac{\Delta g_{\overline{im}}}{\Delta x_k} - \frac{\Delta g_{\overline{ik}}}{\Delta x_m} \right\},$$

where g^{-1} is the inverse of the metric. Note that here I am indicating that the components of the metric are *covariant* with a *bar over the indices*. This is something that we will discuss further in Section 7.5.2. Proving this relationship is fiddly and as there are more profitable proofs to spend our time on, I am not including this one. Most textbooks on general relativity do cover the argument if you are interested.[6]

Note that the symbols are constructed from terms such as $\Delta g_{\overline{im}}/\Delta x_k$, which makes sense when you consider their role in tracking co-ordinate changes. In ordinary Cartesian co-ordinates for a flat space, the metric is constant, which means all the terms like $\Delta g_{\overline{im}}/\Delta x_k$ are zero. Hence the Christoffel symbol is zero in this situation. However, the same flat space-time in spherical-polar co-ordinates would not have vanishing Christoffel symbols as the flat space-time metric in this system is:

$$\eta_{s-p} = \begin{pmatrix} 1 & 0 & 0 & 0 \\ 0 & -1 & 0 & 0 \\ 0 & 0 & -r^2 & 0 \\ 0 & 0 & 0 & -r^2 \sin^2(\theta) \end{pmatrix},$$

and the bottom two diagonal terms do show changes as we move around in the system.

This is very significant, so we should let it sink in for a moment.

The same space-time, in this case a flat Minkowskian one, can have vanishing Christoffel symbols in one choice of co-ordinates (Cartesian), but not vanishing symbols in another (spherical-polar). So, there is a cogent argument to say that the Christoffel symbols tell us about the co-ordinate systems used in the space-time, *not the structure of the space-time itself.*

We have been trained by experience not to separate these concepts out. We used to think of space as being geometrically flat (and time as an absolute), so the choice of Cartesian co-ordinates was natural and 'obvious'. This is no longer the case. General covariance allows us to choose any co-ordinate system and the complexities of curved space-time are such that the 'natural' choices are not so obvious anymore.

7.4 Acceleration in Curved Space-time

We have already seen how the 4-velocity is obtained from the 4-displacement by:

$$U = \frac{\Delta X}{\Delta \tau}.$$

In Minkowskian space-time, we would derive the *4-acceleration* by looking for the rate at which the 4-velocity changed with proper time:

$$A = \frac{\Delta U}{\Delta \tau}.$$

However, in curved co-ordinates things are not quite as simple. We start by writing the 4-velocity at a point in terms of the axis vectors at that point:

$$U = U_0 \epsilon^0 + U_1 \epsilon^1 + U_2 \epsilon^2 + U_3 \epsilon^3 = \sum_i U_i \epsilon^i$$

and then look for the rate of change of the velocity with proper time. This will be a collection of terms such as:

$$\frac{D\left(U_0 \epsilon^0\right)}{\Delta \tau} = \frac{\Delta U_0}{\Delta \tau} \epsilon^0 + U_0 \frac{\Delta \epsilon^0}{\Delta \tau}$$

or, including all of them together:

$$\frac{DU}{\Delta \tau} = \sum_i \left(\frac{\Delta U_i}{\Delta \tau} \epsilon^i + U_i \frac{\Delta \epsilon^i}{\Delta \tau} \right).$$

Now we pull a neat mathematical trick on the $U_i \dfrac{\Delta \epsilon^i}{\Delta \tau}$ terms. Looking specifically at ϵ^0, for example, this vector will be a function of the generalised co-ordinates (x_0, x_1, x_2, x_3). As we are dealing with an object that is accelerating, we know that these co-ordinates will be a function of the proper time, τ. So, ϵ^0 must *indirectly* be a function of τ as well:

$$\epsilon^0 = \epsilon^0 \left(x_0, x_1, x_2, x_3 \right) \quad x_0 = x_0(\tau), x_1 = x_1(\tau), x_2 = x_2(\tau), x_3 = x_3(\tau)$$

making:

$$\epsilon^0 = \epsilon^0 \left(x_0(\tau), x_1(\tau), x_2(\tau), x_3(\tau) \right).$$

If ϵ^0 were only a function of x_0, which is not realistic but I am only suggesting it as a simplification for the moment, then:

$$\frac{\Delta \epsilon^0}{\Delta \tau} = \frac{\Delta \epsilon^0 \left(x_0(\tau) \right)}{\Delta \tau} = \left(\frac{\Delta x_0}{\Delta \tau} \right) \left(\frac{\Delta \epsilon^0}{\Delta x_0} \right).$$

Think about what is happening here. When we change τ, it is bringing about a change in x_0, and that change is, in turn, producing a change in ϵ^0. It's like a leverage effect, or alternatively a scale factor. This is why the $\Delta \epsilon^0 / \Delta \tau$ has to be broken down into parts each of which takes the form:

$$\frac{\Delta x_0}{\Delta \tau} \times \frac{\Delta \epsilon^0}{\Delta x_0} = \left(\text{rate that } x_o \text{ is changing with } \tau\right) \times \left(\text{rate that } \epsilon^0 \text{ is changing with } x_0\right).$$

Once we bring in the other co-ordinates as well, we have:

$$\frac{\Delta \epsilon^0}{\Delta \tau} = \left(\frac{\Delta x_0}{\Delta \tau}\right)\left(\frac{\Delta \epsilon^0}{\Delta x_0}\right) + \left(\frac{\Delta x_1}{\Delta \tau}\right)\left(\frac{\Delta \epsilon^0}{\Delta x_1}\right) + \left(\frac{\Delta x_2}{\Delta \tau}\right)\left(\frac{\Delta \epsilon^0}{\Delta x_2}\right) + \left(\frac{\Delta x_3}{\Delta \tau}\right)\left(\frac{\Delta \epsilon^0}{\Delta x_3}\right) = \sum_j \left(\frac{\Delta x_j}{\Delta \tau}\right)\left(\frac{\Delta \epsilon^0}{\Delta x_j}\right).$$

Consequently, our rate of change is becoming a tad complicated:

$$\frac{DU}{\Delta \tau} = \sum_i \left(\frac{\Delta U_i}{\Delta \tau}\epsilon^i + U_i \frac{\Delta \epsilon^i}{\Delta \tau}\right) = \sum_i \left(\frac{\Delta U_i}{\Delta \tau}\epsilon^i + U_i \sum_j \left(\frac{\Delta x_j}{\Delta \tau}\right)\left(\frac{\Delta \epsilon^i}{\Delta x_j}\right)\right).$$

Noticing that $\Delta x_j/\Delta \tau = U_j$, makes things slightly more compact:

$$\frac{DU}{\Delta \tau} = \sum_i \left(\frac{\Delta U_i}{\Delta \tau}\epsilon^i + U_i \sum_j U_j \frac{\Delta \epsilon^i}{\Delta x_j}\right).$$

Each of the $\Delta \epsilon^i/\Delta x_j$ can be expressed in terms of the Christoffel symbols:

$$\frac{\Delta \epsilon^i}{\Delta x_j} = \sum_k \Gamma_{ij}^k \epsilon^k$$

so we can make things more complicated-looking again:

$$\frac{DU}{\Delta \tau} = \sum_i \left(\frac{\Delta U_i}{\Delta \tau}\epsilon^i + U_i \sum_j \sum_k U_j \Gamma_{ij}^k \epsilon^k\right).$$

Using $U_i = \Delta x_i/\Delta \tau$ we obtain:

$$\frac{DU}{\Delta \tau} = \sum_i \left(\frac{\Delta}{\Delta \tau}\left(\frac{\Delta x_i}{\Delta \tau}\right)\epsilon^i + U_i \sum_j \sum_k U_j \Gamma_{ij}^k \epsilon^k\right).$$

Splitting the right-hand side into two parts (does this look a familiar move?):

$$\frac{DU}{\Delta \tau} = \sum_i \frac{\Delta}{\Delta \tau}\left(\frac{\Delta x_i}{\Delta \tau}\right)\epsilon^i + \sum_i \left(U_i \sum_j \sum_k U_j \Gamma_{ij}^k \epsilon^k\right)$$

we can re-index the summation in the first term to be over k rather than i:

$$\frac{DU}{\Delta \tau} = \boxed{\sum_k \frac{\Delta}{\Delta \tau}\left(\frac{\Delta x_k}{\Delta \tau}\right)\epsilon^k} + \sum_i \left(U_i \sum_j \sum_k U_j \Gamma_{ij}^k \epsilon^k\right)$$

and change the order of summations in the second term:

$$\frac{DU}{\Delta\tau} = \sum_k \frac{\Delta}{\Delta\tau}\left(\frac{\Delta x_k}{\Delta\tau}\right)\epsilon^k + \boxed{\sum_k \sum_i \sum_j U_i U_j \Gamma_{ij}^k \epsilon^k}.$$

Now we can re-group to factor the ϵ^k out of the bracket:

$$\frac{DU}{\Delta\tau} = \sum_k \left(\frac{\Delta}{\Delta\tau}\left(\frac{\Delta x_k}{\Delta\tau}\right) + \sum_i \sum_j U_i U_j \Gamma_{ij}^k\right)\epsilon^k.$$

This allows us to write, on a component-by-component basis:

$$\frac{DU_k}{\Delta\tau} = \frac{\Delta}{\Delta\tau}\left(\frac{\Delta x_k}{\Delta\tau}\right) + \sum_i \sum_j U_i U_j \Gamma_{ij}^k,$$

which is one of the most important results that we have derived so far.

We started by trying to calculate the 4-acceleration in a curved co-ordinate system. This is worth a lot more than simply academic interest. Newton's second law of motion has the classical form:

$$\text{force applied} = \text{mass} \times \text{resulting acceleration}$$

$$F = ma$$

but the alternate form (for objects of constant mass);

$$\frac{\Delta p}{\Delta t} = m\frac{\Delta u}{\Delta t},$$

where p is the Newtonian momentum, mu. In the absence of an applied force, there is no acceleration and so the momentum remains constant. If the object is not moving, it remains stationary (a constant momentum of zero is still a constant momentum…). If the object is already moving, it will continue to move in a straight line at constant speed.

Transferring this framework to the general theory we write:

$$\frac{DP_k}{\Delta\tau} = m\frac{DU_k}{\Delta\tau} = m\left(\frac{\Delta}{\Delta\tau}\left(\frac{\Delta x_k}{\Delta\tau}\right) + \sum_i \sum_j U_i U_j \Gamma_{ij}^k\right).$$

If no external force is applied $\frac{DP_k}{\Delta\tau} = m\frac{DU_k}{\Delta\tau} = 0$, which is the general relativistic equivalent of Newton's first law of motion. Consequently, we have:

$$m\left(\frac{\Delta}{\Delta\tau}\left(\frac{\Delta x_k}{\Delta\tau}\right) + \sum_i \sum_j U_i U_j \Gamma_{ij}^k\right) = 0$$

or, simplifying things, provided $m \neq 0$:

$$\frac{\Delta}{\Delta\tau}\left(\frac{\Delta x_k}{\Delta\tau}\right) + \sum_i \sum_j U_i U_j \Gamma_{ij}^k = 0,$$

so that:

$$\frac{\Delta}{\Delta \tau}\left(\frac{\Delta x_k}{\Delta \tau}\right) = -\sum_i \sum_j U_i U_j \Gamma_{ij}^k. \tag{7.2}$$

The left-hand side of this equation is a component of the instantaneous 4-acceleration. The right-hand side is the curving geometry of space-time acting on the velocity 4-vector, via the Christoffel symbols.

This is the equation that determines how an object moves in curved space-time.

Notice that the mass we introduced (which was only ever an inertial mass) has dropped out of the equation. Hence all masses move in the same fashion through the same curved space-time…

The job is not completely done, however. The Christoffel symbols depend on the metric and we have to solve Einstein's field equations (which we have not come to as yet) to find that metric. However, once we have the metric, the motion of objects through space-time follows.

PONDER POINT

I think a period of quiet introspection is called for.

After all, this is quite a decisive moment. We have worked our way through the special theory and tackled the somewhat arcane mathematics of curved geometry (and to be honest there is a bit still to go), but here we have the promise of the general theory delivered. The curvature of space-time influences the motion of particles, without the intervention of a conventional force.

I, for one, think we all deserve a cup of tea.

7.5 General Co-ordinate Transformations

In order to demonstrate general covariance, we need to transform between sets of general co-ordinates. Such a transformation affects both the *components* of a vector and the *axis vectors* that the components are with reference to. Up to now, I have not been picky and just focussed on transforming components. So, as a sort of 'five-finger exercise' we will start by seeing how the axis vectors transform using the Lorentz transformations.

7.5.1 Lorentz Transformations of Axis Vectors

Recall that our axis vectors are:

$$\epsilon^0 = \begin{pmatrix} 1 \\ 0 \\ 0 \\ 0 \end{pmatrix} \quad \epsilon^1 = \begin{pmatrix} 0 \\ 1 \\ 0 \\ 0 \end{pmatrix} \quad \epsilon^2 = \begin{pmatrix} 0 \\ 0 \\ 1 \\ 0 \end{pmatrix} \quad \epsilon^3 = \begin{pmatrix} 0 \\ 0 \\ 0 \\ 1 \end{pmatrix}$$

and that the full Lorentz 4-matrix is:

$$L = \begin{pmatrix} \gamma & -\beta\gamma & 0 & 0 \\ -\beta\gamma & \gamma & 0 & 0 \\ 0 & 0 & 1 & 0 \\ 0 & 0 & 0 & 1 \end{pmatrix},$$

so that if we transform ϵ^0 we get:

$$\left(\epsilon^0\right)' = \begin{pmatrix} \gamma & -\beta\gamma & 0 & 0 \\ -\beta\gamma & \gamma & 0 & 0 \\ 0 & 0 & 1 & 0 \\ 0 & 0 & 0 & 0 \end{pmatrix} \begin{pmatrix} 1 \\ 0 \\ 0 \\ 0 \end{pmatrix} = \begin{pmatrix} \gamma \\ -\beta\gamma \\ 0 \\ 0 \end{pmatrix}$$

and for ϵ^1:

$$\left(\epsilon^1\right)' = \begin{pmatrix} \gamma & -\beta\gamma & 0 & 0 \\ -\beta\gamma & \gamma & 0 & 0 \\ 0 & 0 & 1 & 0 \\ 0 & 0 & 0 & 0 \end{pmatrix} \begin{pmatrix} 0 \\ 1 \\ 0 \\ 0 \end{pmatrix} = \begin{pmatrix} -\beta\gamma \\ \gamma \\ 0 \\ 0 \end{pmatrix}.$$

What were we expecting to see? I for one was hoping for the equations of the ct' and x' axes, to allow us to plot them in S. After all, back in Chapter 4 we saw that the gradient of the ct' axis viewed from S was $1/\beta$ and that of the x' was β.

That is not what we are seeing here, as the minus signs are getting in the way.

Our (deliberate) mistake was to assume that these vectors transform *contravariantly*. Actually, they are *covariant* which means that they transform by the *inverse* transformation, L^{-1}.

$$\left(\epsilon^0\right)' = \begin{pmatrix} \gamma & \beta\gamma & 0 & 0 \\ \beta\gamma & \gamma & 0 & 0 \\ 0 & 0 & 1 & 0 \\ 0 & 0 & 0 & 1 \end{pmatrix} \begin{pmatrix} 1 \\ 0 \\ 0 \\ 0 \end{pmatrix} = \begin{pmatrix} \gamma \\ \beta\gamma \\ 0 \\ 0 \end{pmatrix}$$

$$\left(\epsilon^1\right)' = \begin{pmatrix} \gamma & \beta\gamma & 0 & 0 \\ \beta\gamma & \gamma & 0 & 0 \\ 0 & 0 & 1 & 0 \\ 0 & 0 & 0 & 1 \end{pmatrix} \begin{pmatrix} 0 \\ 1 \\ 0 \\ 0 \end{pmatrix} = \begin{pmatrix} \beta\gamma \\ \gamma \\ 0 \\ 0 \end{pmatrix}.$$

Just to be clear, with a multiple of the vector ϵ^0 such as $5 \begin{pmatrix} \gamma \\ \beta\gamma \\ 0 \\ 0 \end{pmatrix}$ we would plot a point in S with

co-ordinates $\left(5\beta\gamma, 5\gamma\right)$ giving ϵ^0 a gradient of $5\gamma/5\beta\gamma = 1/\beta$. A similar argument shows that $\left(\epsilon^1\right)'$ is a line in S with gradient β: just as we expect.

By the way: this is the origin of the otherwise curious convention of calling a Lorentz vector that transforms using L *contravariant*. Its *contra* to the way the axis vectors transform[7]...

By the way 2: strictly it is not the Lorentz *vector* that transforms contravariantly, it is the *components* of the vector. If we are being punctilious about terminology, a vector *does not transform at all*. It is the same object viewed from any system. The *components* of the vector and the *axis vectors* are what transform.

Mathematically we would say:

$$\text{vector} = \left(\overleftarrow{\text{components}}\right)\left(\overrightarrow{\text{axis vectors}}\right) = \boldsymbol{A} = \overrightarrow{A_0}\overleftarrow{\epsilon^0} + \overrightarrow{A_1}\overleftarrow{\epsilon^1} + \overrightarrow{A_2}\overleftarrow{\epsilon^2} + \overrightarrow{A_3}\overleftarrow{\epsilon^3},$$

where I have used arrows pointing to the *right* to indicate *contravariance* and those pointing to the *left* to indicate *covariance*. Unsurprisingly, transforming axes and components in the opposite sense leaves the vector unchanged.

The same vector can be specified by many different sets of components, depending on the system being used for the axes. This is how space-time transcends the system used to delineate events within it, how the electromagnetic field transcends electrical and magnetic fields and how energy and momentum are merged into relativistic energy. To make the point visually, take a look at Figure 7.5 which shows a vector viewed from two different co-ordinate systems rotated with respect to each other.

7.5.2 General Transformations

Back in Chapter 3, we introduced a way of monitoring the change in a quantity if it were a function of more than one variable:

$$\Delta f = \left(\frac{\Delta f}{\Delta x}\right)\Delta x + \left(\frac{\Delta f}{\Delta y}\right)\Delta y + \left(\frac{\Delta f}{\Delta z}\right)\Delta z + \left(\frac{\Delta f}{c\Delta t}\right)c\Delta t. \tag{7.3}$$

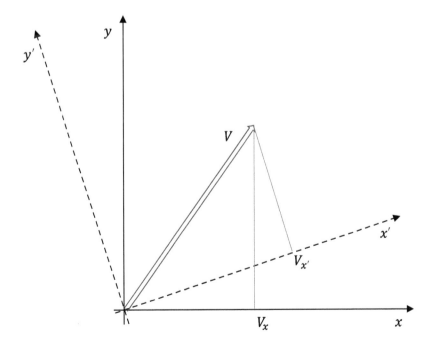

FIGURE 7.5   The same vector viewed from two co-ordinate systems has different components, but the vector has not changed.

Although we did not stress it at the time, this is the secret to transforming from one set of co-ordinates into another. In our Lorentz transformations, we have:

$$c\Delta t' = \gamma\left(c\Delta t - \beta\Delta x\right)$$

$$\Delta x' = \gamma\left(\Delta x - \beta c\Delta t\right).$$

So, we can regard $c\Delta t'$ as a function of Δx and $c\Delta t$ while $\Delta x'$ is a different function of Δx and $c\Delta t$. From this point of view:

$$c\Delta t' = \left(\frac{c\Delta t'}{c\Delta t}\right)_{\Delta x} c\Delta t + \left(\frac{c\Delta t'}{\Delta x}\right)_{c\Delta t} \Delta x \quad \text{and} \quad \Delta x' = \left(\frac{\Delta x'}{c\Delta t}\right)_{\Delta x} c\Delta t + \left(\frac{\Delta x'}{\Delta x}\right)_{c\Delta t} \Delta x.$$

From the Lorentz transformations we can see that:

$$\left(\frac{c\Delta t'}{c\Delta t}\right)_{\Delta x} = \gamma \qquad \left(\frac{c\Delta t'}{\Delta x}\right)_{c\Delta t} = -\beta\gamma$$

$$\left(\frac{\Delta x'}{c\Delta t}\right)_{\Delta x} = -\gamma\beta \qquad \left(\frac{\Delta x'}{\Delta x}\right)_{c\Delta t} = \gamma$$

and so:

$$c\Delta t' = \left(\frac{c\Delta t'}{c\Delta t}\right)_{\Delta x} c\Delta t + \left(\frac{c\Delta t'}{\Delta x}\right)_{c\Delta t} \Delta x = \gamma c\Delta t - \beta\gamma\Delta x = \gamma\left(c\Delta t - \beta\Delta x\right)$$

$$\Delta x' = \left(\frac{\Delta x'}{c\Delta t}\right)_{\Delta x} c\Delta t + \left(\frac{\Delta x'}{\Delta x}\right)_{c\Delta t} \Delta x = -\gamma\beta c\Delta t + \gamma\Delta x = \gamma\left(\Delta x - \beta c\Delta t\right).$$

Undoubtedly, there is a large loop in this argument, however I have spelt it out in order to demonstrate that the Lorentz transformations are just an example of the more general transformations illustrated by Equation 7.3.

Using general co-ordinates, the set (x_0, x_1, x_2, x_3) would transform into (x_0', x_1', x_2', x_3') with the transformation relationship being:

$$x_0' = \left(\frac{\Delta x_0'}{\Delta x_0}\right) x_0 + \left(\frac{\Delta x_0'}{\Delta x_1}\right) x_1 + \left(\frac{\Delta x_0'}{\Delta x_2}\right) x_2 + \left(\frac{\Delta x_0'}{\Delta x_3}\right) x_3$$

with similar transformations for the other components. We could write this in matrix form as:

$$\begin{pmatrix} x_0' \\ x_1' \\ x_2' \\ x_2' \end{pmatrix} = G \begin{pmatrix} x_0 \\ x_1 \\ x_2 \\ x_3 \end{pmatrix} = \begin{pmatrix} \dfrac{\Delta x_0'}{\Delta x_0} & \dfrac{\Delta x_0'}{\Delta x_1} & \dfrac{\Delta x_0'}{\Delta x_2} & \dfrac{\Delta x_0'}{\Delta x_3} \\ \dfrac{\Delta x_1'}{\Delta x_0} & \dfrac{\Delta x_1'}{\Delta x_1} & \dfrac{\Delta x_1'}{\Delta x_2} & \dfrac{\Delta x_1'}{\Delta x_3} \\ \dfrac{\Delta x_2'}{\Delta x_0} & \dfrac{\Delta x_2'}{\Delta x_1} & \dfrac{\Delta x_2'}{\Delta x_2} & \dfrac{\Delta x_2'}{\Delta x_3} \\ \dfrac{\Delta x_3'}{\Delta x_0} & \dfrac{\Delta x_3'}{\Delta x_1} & \dfrac{\Delta x_3'}{\Delta x_2} & \dfrac{\Delta x_3'}{\Delta x_3} \end{pmatrix} \begin{pmatrix} x_0 \\ x_1 \\ x_2 \\ x_3 \end{pmatrix}$$

or more compactly, using our summation notation:

$$x_i' = \sum_k G_{ik} x_k = \sum_k \frac{\Delta x_i'}{\Delta x_k} x_k.$$

The inverse transformation is:

$$x_i = \sum_k \left(G^{-1}\right)_{ik} x_k = \sum_k \frac{\Delta x_i}{\Delta x_k'} x_k.$$

Consequently, we should extend the definition of a contravariant vector to any quantity that transforms using G:

$$A_i' = \sum_k G_{ik} A_k,$$

whereas a covariant vector transforms using G^{-1}:

$$\overline{A_i'} = \sum_k \left(G^{-1}\right)_{ik} \overline{A_k}.$$

Tensor transformations can also be written using the summation notion. The transformation of a contravariant tensor is:

$$T_{ij}' = \sum_{k,l} G_{ik} G_{jl} T_{kl}$$

that of a covariant tensor:

$$T_{\overline{ij}}' = \sum_{k,l} \left(G^{-1}\right)_{ik} \left(G^{-1}\right)_{jl} T_{\overline{kl}}$$

and that of a mixed tensor:

$$T_{i\overline{j}}' = \sum_{k,l} G_{ik} \left(G^{-1}\right)_{jl} T_{k\overline{l}}.$$

Note that in the case of a tensor, I have put the bar over the indices that are covariant, rather than over the symbol as a whole.

With Lorentz 4-vectors, the metric serves to convert between contravariant and covariant, even at the index level. This makes the metric fully covariant, and the inverse of the metric contravariant. So, for a vector:

$$\overline{A_i} = \sum_k g_{\overline{ik}} A_k \quad A_i = \sum_k \left(g^{-1}\right)_{ik} \overline{A_k}$$

with the metric expressed in the general co-ordinate system as well as the vectors.

For a tensor, we can convert individual indexes like this:

$$T_{i\overline{j}} = \sum_k g_{\overline{jk}} T_{ik} \quad T_{ij} = \sum_k \left(g^{-1}\right)_{jk} T_{i\overline{k}}.$$

While we are on the subject, the product of a metric and its inverse forms the unit matrix, which in summation notation is:

$$\sum_k \left(g^{-1}\right)_{ik} g_{\overline{kj}} = \delta_{ij}$$

introducing the δ symbol, defined by $\delta_{ij} = 0$ if $i \neq j$ and $\delta_{ij} = 1$ if $i = j$.

In the special theory, we constructed invariants via the metric, or equivalently by multiplying covariant and contravariant vectors together. Exactly the same is true using general co-ordinates where:

$$\left(\Delta s\right)^2 = \sum_{i,k} g_{\overline{ik}} \Delta x_i \Delta x_k = \sum_i \Delta x_i \overline{\Delta x_i}$$

is an invariant, for example. However, the two vectors do not have to be the same. The process of multiplying a contravariant quantity by a covariant one is called *contraction*. When applied to vectors, this naturally produces an invariant. However, we can also contract tensors, for example:

$$C_i = \sum_k A_{\overline{ik}} B_k$$

provided the summation spans a contravariant and a covariant index. Indeed, the two indices do not have to be on separate tensors. We can always construct a new tensor out of a mixed one by adding appropriate components together. For example, starting with the mixed tensor $R_{\overline{ij}k}$, we can build:

$$C_i = \sum_k R_{\overline{ik}k}.$$

Note that contraction always reduces the number of indices involved by two, which follows from the combination of the covariant and contravariant aspects forming an invariant.

7.5.3 Transforming Christoffel Symbols

Christoffel symbols are a bit messier to transform than tensors. Their transformation properties derive from how they are used. Going back to Equation 7.1:

$$\mathbb{D}_k \left(A_i\right) = \frac{D\left(A_i\right)}{\Delta x_k} = \frac{\Delta A_i}{\Delta x_k} + \sum_m A_m \Gamma^i_{mk} \tag{7.1}$$

and requiring that $\mathbb{D}_k \left(A_i\right)$ be a two-index tensor, which is extremely convenient, allows us to deduce how the Christoffel symbols should transform. I won't take you through the complex demonstration, so please take it on trust that the transformation is:

$$\left(\Gamma^c_{ab}\right)' = \sum_i \frac{\Delta x'_c}{\Delta x_i} \left[\sum_{m,k} \left(\Gamma^i_{mk} \frac{\Delta x_m}{\Delta x'_a} \frac{\Delta x_k}{\Delta x'_b} + \frac{\Delta}{\Delta x'_a} \left(\frac{\Delta x_i}{\Delta x'_b} \right) \right) \right].$$

If you follow the subscripts and the superscripts around carefully you will:

- get dizzy
- see the logic of how the pieces fit together...

In essence, the m, k indices of the symbol transform like a covariant tensor and the i index transforms as a contravariant tensor. This also reflects in $\mathbb{D}_k \left(A_i\right)$ as i is going to be contravariant to match the vector that we are finding the rate of change for, and k is covariant which we might expect given that it comes from the $\Delta/\Delta x_k$ (in Section 3.10.1 we showed that such forms transform covariantly). From now on, we will write $\mathbb{D}_{\bar{k}}\left(A_i\right)$.

The extra $\dfrac{\Delta}{\Delta x'_a}\left(\dfrac{\Delta x_i}{\Delta x'_b}\right)$ term in the transformation comes about as Christoffel symbols are not tensors: it is sorting out the messy changes in the curving co-ordinates. This is why I am not putting a bar over the covariant indices in the symbols.

If we have a region of space-time with a particular metric in operation, we can calculate the Christoffel symbols for that region. According to the principle of equivalence (in one form or another), shifting to a locally co-moving system in free fall through that region negates any experience of gravitation.[8] Mathematically, we would see this once we transform the Christoffel symbols. It is always possible to find a system of local co-ordinates, y_k, in which the Christoffel symbols vanish.[9] In the new system, our equation for the acceleration $\dfrac{\Delta}{\Delta\tau}\left(\dfrac{\Delta x_k}{\Delta\tau}\right)=-\displaystyle\sum_i\sum_j U_iU_j\Gamma^k_{ij}$ would reduce to $\dfrac{\Delta}{\Delta\tau}\left(\dfrac{\Delta y_k}{\Delta\tau}\right)=0$, which is

what we would expect for an object floating freely alongside us.

7.6 Intrinsic Curvature

Now we have a handle on intrinsic co-ordinates and rates of change, we can develop a method to extract the curvature of space-time from measurements made solely within that space-time. The first person to explore geometry from this perspective was the mathematician G F B Riemann,[10] who now has an important tensor named after him.

Back in Section 4.7 I suggested that one way of telling the curvature of space was to measure the angles of a large triangle and see if they added up to 180°. An alternative method, which generalises more easily, is to see how a vector changes as you move it around a closed loop.

Imagine for a moment that we are living on a perfectly flat 2D plane. We start with a vector at A and move it around a small square loop (Figure 7.6).

We do this by sliding the vector along the line connecting A to B, and then slide it up from B to C, etc. When we get back to the starting point, the vector is still pointing in exactly the same direction.

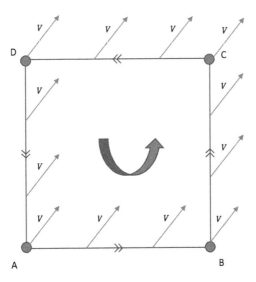

FIGURE 7.6 Moving a vector around a loop *A, B, C, D* in flat space.

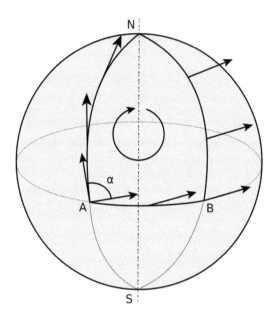

FIGURE 7.7   Parallel transporting a vector over the surface of a sphere.

In order to carry out a similar operation across a curved space, we need to specify exactly how the *transport* (sliding) of the vector is done.

When we moved the vector around the flat plane, we instinctively maintained a constant angle between the vector and the line along which it was being moved. This is called *parallel transport* and on a flat plane, it ensures that the vector arrives back at the starting point unchanged. However, that is a specific feature of the flatness of the plane. If we try and do the same thing on a curved surface, we end up with a very different result.

For a simple example, I'm going to parallel transport a vector in a loop over the surface of a sphere (Figure 7.7). Start with the vector at a point A, which is on the equator of the sphere and then move it along a path to the North Pole, N. If we keep a constant angle between the vector and the path, this inevitably causes the vector to change direction as the path is curving. Now we move it from N to B, keeping the angle between vector and path constant. Again, the vector changes direction as the path curves over the surface.

Finally, we slide the vector along BA to its starting point and we find that, overall, the vector has changed direction by an angle α. *This angle is a measure of the curvature of the surface.*

To establish this more formally, we define the *Riemann curvature tensor*, $R_{\overline{iabc}}$, in the following way (we will get to the contravariant/covariant nature of the indices).

Imagine we have a vector $V = (V_0, V_1, V_2, V_3)$ and we moved it in direction ϵ^b for a distance Δx_b and then along ϵ^c for a distance Δx_c. We then reverse the directions so that it goes back along a path parallel to ϵ^b at the end of ϵ^c and then back down parallel to ϵ^c to the start (Figure 7.8).

By the time we get the vector back to the starting point, each of its components will have changed by an amount ΔV_i depending on the paths, the vector's other components and the curvature of the space. In fact, we have a relationship that looks like this:

$$\Delta V_i = \Delta x_b \Delta x_c \sum_a R_{\overline{iabc}} V_a$$

with the Riemann tensor right in the middle, holding things together. Information about the curvature of the space is held in that tensor. Feed it with data such as the paths along which you are parallel

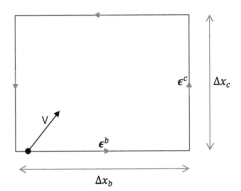

FIGURE 7.8 Parallel transporting a vector around a small loop. Although this is drawn as a rectangular loop, in practice this could be using curving co-ordinates in a non-flat geometry.

transporting, the vector you are transporting and how far you are going, and it will spit out the change in the components of the vector for you.

So that is what the tensor *does*: we still need to see how to *calculate* its components.

Well, as we move a vector along a distance Δx_b in direction ϵ^b, we are going to be interested in how the vector components change: $\dfrac{D(V_i)}{\Delta x_b}\Delta x_b$. Next, we move along ϵ^c for a distance Δx_c, keeping an eye on how the vector evolves $\dfrac{D}{\Delta x_c}\left(\dfrac{D(V_i)}{\Delta x_b}\right)\Delta x_b \Delta x_c$, and then we reverse the directions back to the starting point. In other words, we need to calculate:

$$\Delta V_i = \left\{ \frac{D}{\Delta x_c}\left(\frac{D(V_i)}{\Delta x_b}\right) - \frac{D}{\Delta x_b}\left(\frac{D(V_i)}{\Delta x_c}\right) \right\} \Delta x_b \Delta x_c.$$

Comparing this with our previous expression:

$$\Delta V_i = \Delta x_b \Delta x_c \sum_a R_{i\overline{abc}} V_a$$

shows us that:

$$\frac{D}{\Delta x_c}\left(\frac{D(V_i)}{\Delta x_b}\right) - \frac{D}{\Delta x_b}\left(\frac{D(V_i)}{\Delta x_c}\right) = \sum_a R_{i\overline{abc}} V_a.$$

The key to this is going to be the Christoffel symbols. Although I did not mention it when I introduced them in Section 7.3, they were specifically designed to facilitate parallel transport.

To see this, consider Figure 7.9 which shows how quantities like $\Delta V/\Delta x_j$ are actually defined by parallel transporting the vector back from B to A where the difference can be directly calculated. This is what is happening in the second term of the equation:

$$\frac{D(A_i \epsilon^i)}{\Delta x_k} = \left(\frac{\Delta A_i}{\Delta x_k}\epsilon^i\right) + A_i \sum_j \Gamma_{ik}^j \epsilon^j,$$

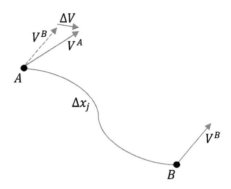

FIGURE 7.9 Calculating $\Delta V / \Delta x_j$ by parallel transporting the vector from B back to A and calculating the change ΔV.

and the Christoffel symbols are defined as:

$$\Gamma_{ik}^{j} = \frac{1}{2} \sum_{m} \left(g^{-1} \right)_{jm} \left\{ \frac{\Delta g_{\overline{mk}}}{\Delta x_i} + \frac{\Delta g_{\overline{im}}}{\Delta x_k} - \frac{\Delta g_{\overline{ik}}}{\Delta x_m} \right\}$$

to make this possible.

The actual calculation requires some dexterous manipulation of summations and the application of the following two rules:

$$\mathbb{D}_{\overline{k}} \left(\overline{A}_i \right) = \frac{D \left(\overline{A}_i \right)}{\Delta x_k} = \frac{\Delta \overline{A}_i}{\Delta x_k} - \sum_{m} \overline{A}_m \Gamma_{ik}^{m} \quad \text{covariant}$$

$$\mathbb{D}_{\overline{k}} \left(A_i \right) = \frac{D \left(A_i \right)}{\Delta x_k} = \frac{\Delta A_i}{\Delta x_k} + \sum_{m} A_m \Gamma_{mk}^{i} \quad \text{contravariant}. \tag{7.1}$$

These rules are derived in the online Appendix, where you can also find the full calculation of the Riemann tensor. Here I quote the important result:

$$R_{i\overline{abc}} = \frac{\Delta \left(\Gamma_{ac}^{i} \right)}{\Delta x_b} - \frac{\Delta \left(\Gamma_{ab}^{i} \right)}{\Delta x_c} + \sum_{n} \left(\Gamma_{ac}^{n} \Gamma_{nb}^{i} - \Gamma_{ab}^{n} \Gamma_{nc}^{i} \right).$$

7.6.1 Properties of the Riemann Tensor

So far, I have been coy about the covariant/contravariant nature of the Riemann tensor. Unsurprisingly, it is covariant in indices 3 and 4 (b, c), which follows from the $\Delta / \Delta x_b$, $\Delta / \Delta x_c$ terms. Given that we sum products of the tensor with components of a contravariant vector in the formula $\sum_{a} R_{i\overline{abc}} V_a$, we should

not be surprised to find that it is covariant in the second index, a, as well. That just leaves the first index, i. Christoffel symbols are contravariant in their 'upstairs' index, and the only time that is un-summed in the tensor is in the terms Γ^i, it follows that the first index of Riemann tensor must be contravariant. Hence, in total we have $R_{i\overline{abc}}$.

The Riemann tensor has a rich collection of symmetries and antisymmetries, most of which are expressed in the fully covariant version found by changing index one by using the metric. As one simple example, the tensor is antisymmetric in the third and fourth indices, so $R_{i\overline{abc}} = -R_{i\overline{acb}}$. The effect of these various symmetries/antisymmetries is to reduce the number of independent components. Given that R has four indices, each of which have four possible values (0, 1, 2, 3) we expect $4^4 = 256$ components. However, this is reduced to 20 independent components by the symmetries.

Obviously, the curvature tensor depends on the Christoffel symbols, which in turn depend on the metric. Hence the Riemann tensor depends, in the end, on the metric and changes in that metric along co-ordinate directions.

It is worth noting, in passing, that in a flat space-time, every component of the Riemann tensor is zero. This is true no matter what the co-ordinate system used.

Of course, even with a flat space-time we can use any co-ordinate system we choose and in that system the metric may look nothing like the simple Minkowskian one. It then becomes a challenge to separate out a fancy flat metric in funny co-ordinates from a metric in the presence of gravitation. If we just write down a metric, you could not necessarily tell, just by looking at it, if it were flat or gravitational.

The Riemann tensor is the answer. If it is zero, there is no gravitation and you can always find a transform to another set of co-ordinates which turns the funny metric into η *at every point in the space-time*. If the curvature tensor is *not* zero, transformations still exist that will turn the metric into something η-like, but *only in the local region of an event*, not across the whole space-time at once.

7.6.2 An Unfortunate Number of Indices

We now have the tensor that contains all the information there is regarding the curvature of a space-time, based on its intrinsic properties. So, you might think that we have reached our goal and that this tensor, or a multiple of it, is the correct factor to insert into the left-hand side of our prospective equation for gravitation:

$$\text{Something clever to do with metrics} = \kappa \mathbb{T}.$$

Unfortunately, the Riemann tensor will not do: it has the wrong number of components.

The energy density tensor $\mathbb{T}$ has 16 components $\mathbb{T}_{ij}$ where $i, j = 0, 1, 2, 3$. The Riemann tensor, $R_{i\overline{abc}}$, has 20 independent components. So, any equation we build from $R_{i\overline{abc}}$, will not balance. It can't work.

All is not lost, as the Riemann tensor happens to have one contravariant index and a range of covariant ones to choose from, hence we can contract it:

$$\mathbb{R}_{\overline{ab}} = \sum_i R_{i\overline{abi}} = R_{0\overline{ab0}} + R_{1\overline{ab1}} + R_{2\overline{ab2}} + R_{3\overline{ab3}}.$$

We have built a tensor, known as the *Ricci tensor*, with two indices, and hence 16 components, out of bits of the Riemann tensor.

I have chosen to contract over the first and the last index in the Riemann tensor, but clearly, I could have contracted over the first and second or the first and third. However, due to the various symmetries and antisymmetries among the components of the Riemann tensor, other contractions will either give zero or $-\mathbb{R}_{\overline{ab}}$. Hence, *the Ricci tensor is the only viable contraction of the Riemann tensor*. It is also symmetrical in the remaining indices, which makes it a good match for the energy density tensor.

It is worth knowing that this choice of contraction, along with my selection of the η_1 version of the Minkowski metric (back in Section 3.5) and the definition of the Riemann tensor, determine the final sign in the Einstein field equations. One ought to bear these sorts of things in mind when comparing the equations that appear in different presentations.

If we use the inverse metric to make one of the Ricci tensor's indices contravariant:

$$\mathbb{R}_{a}{}^{\overline{b}} = \sum_i \left(g^{-1}\right)_{ai} \mathbb{R}_{\overline{ib}},$$

we can then contract again, and end up with the *Ricci scalar*:

$$\mathcal{R} = \sum_k \mathbb{R}_{k\bar{k}} = \sum_k \sum_i \left(g^{-1}\right)_{ki} \mathbb{R}_{\overline{ik}}.$$

There is one final small detail that we need to attend to. The Ricci tensor is covariant in both its indices, which follows from being built from the Riemann tensor. The energy density tensor we have constructed is contravariant in both indices, so we need to tweak the Ricci tensor in order to get it to match:

$$\mathbb{R}_{ab} = \sum_{j,k} \left(g^{-1}\right)_{aj} \left(g^{-1}\right)_{bk} \mathbb{R}_{\overline{jk}}.$$

Relating this back to the Riemann tensor, we have:

$$\mathbb{R}_{ab} = \sum_{j,k} \left(g^{-1}\right)_{aj} \left(g^{-1}\right)_{bk} \left(\sum_i R_{i\overline{jki}}\right) = \sum_{i,j,k} \left(g^{-1}\right)_{aj} \left(g^{-1}\right)_{bk} R_{i\overline{jki}}.$$

7.6.3 And Finally, the Field Equations of General Relativity

In Chapter 6, we outlined the frantic weeks in November 1915 where all of Einstein's powers were deployed in fighting towards the fully covariant field equations. After some blind alleys and distractions, the first version that gave the correct figure for the deflection of light and which explained the odd quirk in Mercury's orbit was:

$$\mathbb{R}_{ab} = \kappa' \mathbb{T}_{ab}.$$

Even given the evident successes associated with these equations, they are ultimately incorrect, as Einstein discovered. The reason is subtle.

One of the properties of the energy density tensor is:

$$\sum_a \frac{\Delta \mathbb{T}_{ab}}{\Delta x_a} = 0$$

or expanding it out longhand:

$$\left\{\frac{\Delta \mathbb{T}_{0b}}{\Delta x_0} + \frac{\Delta \mathbb{T}_{1b}}{\Delta x_1} + \frac{\Delta \mathbb{T}_{2b}}{\Delta x_2} + \frac{\Delta \mathbb{T}_{3b}}{\Delta x_3}\right\} = 0,$$

which must hold no matter what value of b is chosen. To be completely obvious:

$$\left\{\frac{\Delta \mathbb{T}_{00}}{\Delta x_0} + \frac{\Delta \mathbb{T}_{10}}{\Delta x_1} + \frac{\Delta \mathbb{T}_{20}}{\Delta x_2} + \frac{\Delta \mathbb{T}_{30}}{\Delta x_3}\right\} = 0 \quad \text{and} \quad \left\{\frac{\Delta \mathbb{T}_{01}}{\Delta x_0} + \frac{\Delta \mathbb{T}_{11}}{\Delta x_1} + \frac{\Delta \mathbb{T}_{21}}{\Delta x_2} + \frac{\Delta \mathbb{T}_{31}}{\Delta x_3}\right\} = 0.$$

These relationships are an expression of the local (i.e. within a small region of space-time) conservation of energy and momentum. This is not instantly easy to see, but if we re-arrange one of these equations slightly:

$$\frac{\Delta \mathbb{T}_{0b}}{\Delta x_0} = -\left\{\frac{\Delta \mathbb{T}_{1b}}{\Delta x_1} + \frac{\Delta \mathbb{T}_{2b}}{\Delta x_2} + \frac{\Delta \mathbb{T}_{3b}}{\Delta x_3}\right\}$$

and recall the definitions of the energy density tensor's components from the end of Chapter 5, we see that the time rate of change is balanced by the spatial rates of change of flow in various component directions, which is an expression of energy/momentum conservation. This rough calculation has been done in a flat space-time, but it also holds in more general co-ordinates and metrics where the Christoffel symbols get involved as well. In that case the property of the tensor is written as:

$$\sum_a \frac{D\mathbb{T}_{ab}}{\Delta x_a} = 0.$$

Now, the problem is that the same calculation done to the Ricci tensor does not give zero. Whatever you do to one side of an equation has to balance when the same thing is done to the other side, so:

$$\sum_a \frac{D\mathbb{R}_{ab}}{\Delta x_a} = \kappa' \sum_a \frac{D\mathbb{T}_{ab}}{\Delta x_a}.$$

but $\sum_a \dfrac{D\mathbb{T}_{ab}}{\Delta x_a} = 0$ and $\sum_a \dfrac{D\mathbb{R}_{ab}}{\Delta x_a} \neq 0$ so, this field equation can't work.

Once Einstein had realised this, it was clear that he needed to find a tensor, $\mathbb{G}_{ab}$, to replace the Ricci tensor, that still had to be related to curvature, be symmetrical and have the property:

$$\sum_a \frac{D\mathbb{G}_{ab}}{\Delta x_a} = 0.$$

What he came up with was:

$$\mathbb{G}_{ab} = \mathbb{R}_{ab} - \frac{1}{2}\mathcal{R}\left(g^{-1}\right)_{ab},$$

which we now call the *Einstein tensor*.[11] It is the only tensor[12] in 4D that is a function of the metric, g, $\dfrac{\Delta g}{\Delta x_j}$ and $\dfrac{\Delta}{\Delta x_k}\left(\dfrac{\Delta g}{\Delta x_j}\right)$ for which $\sum_a \dfrac{D G_{ab}}{\Delta x_a} = 0$. Note that g^{-1} is present as that is the contravariant form of the metric.

Hence the field equations Einstein published at the end of November 1915 were:

$$\mathbb{R}_{ab} - \frac{1}{2}\mathcal{R}\left(g^{-1}\right)_{ab} = \kappa \mathbb{T}_{ab} \tag{7.4}$$

with the value of κ that we will get to in the next chapter.

7.7 What It All Means…

If we take the field equations and make everything covariant:

$$\mathbb{R}_{\overline{ab}} - \frac{1}{2}\mathcal{R}g_{\overline{ab}} = \kappa \mathbb{T}_{\overline{ab}},$$

then we can pop in the Riemann tensor contraction instead of the Ricci tensor:

$$\sum_i R_{\overline{iabi}} - \frac{1}{2}\mathcal{R}g_{\overline{ab}} = \kappa \mathbb{T}_{\overline{ab}}.$$

This formulation, although undoubtedly harder to read, drives home the point that *the curvature of space-time is only partly determined by matter and energy*. The Ricci tensor is constructed from some of the Riemann tensor's components, the rest of them do not get directly connected to the matter/energy density present as they are not in the field equations. Yet these components of the Riemann tensor cannot be redundant. As a whole, the Riemann tensor contains all the information there is to be had about the curvature of space-time (or any geometry for that matter) so clearly there are some aspects to space-time that are not directly influenced by matter/energy.

If there is no matter or energy present, then the field equations reduce to their *vacuum form*:

$$\sum_i R_{i\overline{ab}i} - \frac{1}{2}\mathcal{R}g_{\overline{ab}} = 0,$$

which in turn implies that the Ricci tensor vanishes. To see how this comes about, we need to step back to the field equations with matter/energy present, multiply by the inverse metric:

$$\sum_a \left\{ \left(g^{-1}\right)_{ma} \mathbb{R}_{\overline{ab}} - \frac{1}{2}\mathcal{R}\left(g^{-1}\right)_{ma} g_{\overline{ab}} \right\} = \kappa \sum_a \left(g^{-1}\right)_{ma} \mathbb{T}_{\overline{ab}}$$

giving us $\mathbb{R}_{m\overline{b}} - \frac{1}{2}\mathcal{R}\delta_{m\overline{b}} = \kappa\mathbb{T}_{m\overline{b}}$, which we then contract:

$$\sum_m \left\{ \mathbb{R}_{m\overline{m}} - \frac{1}{2}\mathcal{R}\delta_{m\overline{m}} \right\} = \kappa \sum_m \mathbb{T}_{m\overline{m}},$$

so that $\left\{ \mathcal{R} - \frac{1}{2}\mathcal{R} \right\} = \frac{1}{2}\mathcal{R} = \kappa\mathcal{T}$ with $\mathcal{T} = \sum_m \mathbb{T}_{m\overline{m}}$.

Now the original equations look like this:

$$\mathbb{R}_{\overline{ab}} - \kappa\mathcal{T}g_{\overline{ab}} = \kappa\mathbb{T}_{\overline{ab}}$$

which re-arranges to:

$$\mathbb{R}_{\overline{ab}} = \kappa\left(\mathbb{T}_{\overline{ab}} + \kappa\mathcal{T}g_{\overline{ab}}\right)$$

or equally:

$$\mathbb{R}_{ab} = \kappa\left(\mathbb{T}_{ab} + \kappa\mathcal{T}\left(g^{-1}\right)_{ab}\right).$$

In absence of any matter/energy $\mathbb{T}_{\overline{ab}} = \mathbb{T}_{ab} = 0$, which means that $\mathcal{T} = 0$ and so in the end $\mathbb{R}_{\overline{ab}} = 0$: the Ricci tensor vanishes in the vacuum situation. However, just because the Ricci tensor is zero does not mean that the space-time lacks curvature. After all, just because $\mathbb{R}_{\overline{ab}} = 0$, does not mean that the *whole* Riemann tensor vanishes, and that is the arbiter of curvature and hence gravitation. We appear to be saying that gravitation (curvature) can be present *even in the absence of mass/energy*.

From a Newtonian perspective, a gravitational field contains energy (gravitational potential energy), which according to the special theory has mass, and we know that mass is a source of gravitation. (We have seen that other aspects of the energy momentum tensor act as sources of gravitation as well). Hence, we are 'set up' to expect that a gravitational field will be a source of gravity. In the end, this idea leads to gravitational waves, which we know have been experimentally observed.

Unfortunately, the whole notion of 'gravitational potential energy' is tricky in general relativity as energy is co-ordinate system dependent (consider the problem generated by being able to transforming away the gravitational field as per the equivalence principle[13]). Equally, potential energy is intimately linked to force $\left(g_i = -\dfrac{\Delta\phi}{\Delta x_i} \right)$ and we have stressed how Einstein dispensed with the notion of a gravitational force and replaced it with the curvature of space-time. So that makes the problem worse…

In general relativity, the extra components of Riemann tensor, those that are not pinned down explicitly by the field equations, express how curvature determines curvature. Indeed, it is possible to form a new tensor (yippee!) called the *Weyl tensor* from the remains of the Riemann tensor. The Weyl tensor does not vanish in the vacuum and it governs how gravitational waves move through space-time in the absence of matter. There will be more to say on this in Chapter 11.

7.7.1 The Complexity of the Field Equations

When we look at Equation 7.4, we have to remember that there is a great deal of compactness in the notation. The elegant simplicity we see is disguising a fiendish complexity. The Riemann tensor is determined by the Christoffel symbols:

$$R_{i\overline{abc}} = \frac{\Delta\left(\Gamma^i_{ac}\right)}{\Delta x_b} - \frac{\Delta\left(\Gamma^i_{ab}\right)}{\Delta x_c} + \sum_n \left(\Gamma^n_{ac}\Gamma^i_{nb} - \Gamma^n_{ab}\Gamma^i_{nc}\right),$$

and the Christoffel symbols depend on the metric and its inverse:

$$\Gamma^j_{ik} = \frac{1}{2}\sum_m \left(g^{-1}\right)_{jm} \left\{ \frac{\Delta g_{\overline{mk}}}{\Delta x_i} + \frac{\Delta g_{\overline{im}}}{\Delta x_k} - \frac{\Delta g_{\overline{ik}}}{\Delta x_m} \right\}$$

so the best of luck if you intend to substitute in and write it all out…

The ultimate the aim, of course, is to solve these equations for the metric $g_{\overline{ab}}$. Given that $a, b = 0, 1, 2, 3$ there are 16 components that we need to find to fully specify $g_{\overline{ab}}$. The fact that metrics are always symmetric, reduces the number of independent components to ten, but that still means we have ten independent equations to solve.

Not only are these equations intricate, but as they depend on g, $\dfrac{\Delta g}{\Delta x_j}$, $\dfrac{\Delta}{\Delta x_k}\left(\dfrac{\Delta g}{\Delta x_j}\right)$ and $\left(\dfrac{\Delta g}{\Delta x_j}\right)^2$, they are *non-linear* as well. This mathematical structure gives rise to interesting physics. Generally, the laws of nature form a linear relationship between the source of a field and the field. In Maxwell's equations, the electric and magnetic fields are linear functions of the charges and currents. As a result, if we place one configuration of charges and currents on top of another, then the resulting fields just add together.[14] However, with gravitation as determined by Einstein's equations, that will not happen. The curvature produced by one mass/energy distribution will alter the curvature produced by the other, so that the combination is not the simple 'sum' of the separate situations. Again, this is down to gravitation producing gravitation.

You begin to see why exact solutions to the field equations are so rare and depend on highly symmetrical situations to allow us to simplify the calculations.

7.7.2 Solving the Field Equations

Broadly there are three approaches that you can take to solving these equations[15]:

- The T-approach: you can figure out what the energy density tensor is for the configuration of matter/energy that you are interested in, plug it into the equations and then pound away until you get a solution. This is where simplifications of symmetry come into play, or some method by which successively refined approximations are used. *Linear gravity theory* is an example of an approximation scheme. This will be discussed more in Chapter 11.

- The *g*-approach: you can pick a metric, plug it into the equations and then see what sort of energy density tensor comes out. This is a bit easier mathematically, as you are not 'solving' the equations in the same sense. However, it's hard to pick a metric that gives an energy density tensor that is physically plausible.
- The feedback approach: you can study the situation and 'guess' at the energy density tensor and metric, then use the field equations as a way to refine the guesses so that they are consistent with each other.

However, exact solutions are not the only story. In contemporary physics, the power of computing is so great that numerical simulations can provide a great deal of insight where exact mathematical calculations would not be able to tread.

7.7.3 The Gravitational Field in General Relativity

… nowhere has a precise definition of the term 'gravitational field' been given --- nor will one be given. Many different mathematical entities are associated with gravitation; the metric, the Riemann curvature tensor, the curvature scalar… Each of these plays an important role in gravitation theory, and none is so much more central than the others that it deserves the name 'gravitational field'.

C Misner, K Thorne and J Wheeler[16]

Einstein believed that the Christoffel symbols represented the gravitational field in general relativity. His thinking can be attributed, in part, to the hypnotic attraction of the principle of equivalence: that you can transform away a (local) gravitational field by moving to a system in free fall. The vanishing of the Christoffel symbols in a properly chosen set of free fall co-ordinates reflects this. Equally, the gravitational field in Newtonian mechanics is expressed as the rate of change of the potential over distance $g_i = -\dfrac{\Delta\phi}{\Delta x_i}$

and the Christoffel symbols are also expressed as rates of change $\Gamma^j_{ik} = \dfrac{1}{2}\sum_m \left(g^{-1}\right)_{jm}\left\{\dfrac{\Delta g_{\overline{mk}}}{\Delta x_i} + \dfrac{\Delta g_{\overline{im}}}{\Delta x_k} - \dfrac{\Delta g_{\overline{ik}}}{\Delta x_m}\right\}$,

which appears to make them analogous to the gravitational field.

However, modern relativists tend not to think in these terms. As the Christoffel symbols can vanish in one system but not in another, they are not tensors. These days we prefer to allocate physical quantities to tensors, vectors and scalars. So, to modern eyes, Christoffel symbols are disqualified from carrying any direct physical content.

Also, the Christoffel symbols tell us how to knit together different parts of the co-ordinate system chosen, not the geometry of the space-time. Once again, it is easy to fall into a trap and think of the co-ordinate system as directly disclosing the space-time structure. Many different systems can be used in the same space-time. The Christoffel symbols are as much to do with that arbitrary choice as they are any fundamental aspect of the geometry.

I noted earlier that in a flat space-time $R_{i\overline{abc}} = 0$, which is true no matter what co-ordinate system is used. Equally, if there is curvature to the space-time $R_{i\overline{abc}}$ is not zero, which will also be true in any co-ordinate system. Having a non-zero curvature tensor is indicative of the presence of gravitation, but this does not mandate a formal identification of the gravitational field with that tensor.

In the end, Newtonian theory allows us to calculate (or estimate) how gravitating masses move around in the vicinity of each other, at least in most elementary situations. To do this, we work with a gravitational field which is generated by the masses and which causes a force to act on other masses.

General relativity achieves the same end: determining how gravitating masses move in the vicinity of each other. However, it does so from a completely different standpoint, that of mass/energy distorting the geometry of space-time and objects moving through that space-time. This is such a different perspective on the problem, it is not too surprising to find that aspects of the old ideas are not compatible with the new. So, why should we be bothered if there is no representation of the Newtonian gravitational field in general relativity?

The only constraint is this: in the correct context, general relativity must reduce to Newtonian theory. This is the *weak field approximation*, which is the subject of the next chapter.

Notes

1. I'm using *s* here to refer to co-ordinates, not the space-time interval – sometimes you run out of letters…
2. A Einstein to P Ehrenfest, 26 December 1915 Collected Papers of Albert Einstein, 8, Doc. 173, as quoted in Janssen, M., 2004. *Einstein's First Systematic Exposition of General Relativity.* http://philsci-archive.pitt.edu/id/eprint/2123.
3. NOT 4-vectors, four OF the vectors…this comes dangerously close to a *Two Ronnies* sketch, for those who understand the reference.
4. The axis vectors are designed to be of unit length to facilitate this.
5. Elwin Bruno Christoffel, 1829–1900, German mathematician and physicist whose work on differential geometry paved the way for the manipulation of tensors and hence the general theory of relativity. These are the things that Weinberg referred to as affine connections, in the quote we used in Chapter 6.
6. For a treatment broadly similar to the one I am following, I recommend Lambourne, R.J.A., 2010. *Relativity, Gravitation and Cosmology.* Cambridge University Press, ISBN-13: 978-0521131384.
7. Still think it's a bit silly…but the mathematicians seem to like it and I can (sort) of see their point.
8. In truth, shifting to any system in relative inertial motion compared to the co-moving system in free fall will do the job.
9. Careful here, even in flat space the Christoffel symbols do not vanish. So, vanishing symbols does not imply an absence of gravitation. The principle of equivalence implies that you can always find a system of local co-ordinates in which they do vanish.
10. Georg Friedrich Bernhard Riemann 1826–1866 who made great contributions to analysis (especially integration), number theory and differential geometry.
11. The property $\sum_a \dfrac{D\mathbb{G}_{ab}}{\Delta x_a} = 0$ was discovered independently by Aurel Voss in 1880, Ricci in 1889, and by Luigi Bianchi in 1902. The tensor is named due to Einstein's procuring it for the field equations.
12. Lovelock, D, 1971. The Einstein tensor and its generalizations. *Journal of Mathematical Physics*, **12** (3), pp. 498–502.
13. Specifically, if we could write an energy density tensor for the gravitational field, then in a freely falling inertial system the principle of equivalence would tell us that this tensor would be zero. The problem is, that if a tensor is zero in one co-ordinate system, then it must be zero in all of them (consider the way tensors transform). Hence, there is no meaningful energy momentum tensor for the gravitational field. In part, because there is no gravitational field in general relativity.
14. Provided we can do this without one collection of charges and currents disturbing the other. In truth, things become very messy quiet easily, even with linear laws of Nature. If a charge generates and electric field, that will exert a force on another charge, but that other charge can push back. If the source charge is not nailed in place, then it will move as a result changing the electric field, which in turn changes the force on the charge etc…

15. Here I follow the outline presented in *Relativity, Gravitation and Cosmology* by R.J.A. Lambourne (see ref. 6 above) and the more detailed exposition in *Relativity: The General Theory*, Synge, J.L., North-Holland, Dordrecht, (1960) ASIN: B01A0CY2OW.

16. Misner, C.W., Thorne, K.S. and Wheeler, J.A. (1973) *Gravitation*. W.H. Freeman and Company, New York, ISBN-10: 0691177791, page 399.

8

Weak Field Gravitation

8.1 Linearising the Field Equations

Exploring the weak field limit[1] of general relativity allows us to demonstrate how Einstein's field equations reduce to Poisson's equation in the right circumstances, and in the process obtain a value for the constant κ. It also enables us to explore the application of general relativity to the gravitational conditions that we find on Earth and as we mooch about in the solar system.

We start by constructing a metric, $g_{\overline{ab}}$, which is a small shift away from the Minkowski metric:

$$g_{\overline{ab}} = \eta_{\overline{ab}} + h_{\overline{ab}}$$

where $h_{\overline{ab}}$ and $\Delta h_{\overline{ab}}/\Delta x_j$ (for any x_j) are assumed to be small enough that the products $\overline{h} \times \overline{h}, \overline{h} \times \dfrac{\Delta \overline{h}}{\Delta x_j}, \dfrac{\Delta \overline{h}}{\Delta x_i} \times \dfrac{\Delta \overline{h}}{\Delta x_j}$ can all be ignored. In fact, as $h \ll \eta$ and all the components of $|\eta| = 1$ we can also write $h \ll 1$.

Grabbing the definition of the Christoffel symbols:

$$\Gamma_{ik}^{j} = \frac{1}{2} \sum_{m} \left(g^{-1}\right)_{jm} \left\{ \frac{\Delta g_{\overline{mk}}}{\Delta x_i} + \frac{\Delta g_{\overline{im}}}{\Delta x_k} - \frac{\Delta g_{\overline{ik}}}{\Delta x_m} \right\}$$

and putting in our weak field metric gives:

$$\Gamma_{ik}^{j} = \frac{1}{2} \sum_{m} \left(\left(\overline{\eta} + \overline{h}\right)^{-1}\right)_{jm} \left\{ \frac{\Delta\left(\eta_{\overline{mk}} + h_{\overline{mk}}\right)}{\Delta x_i} + \frac{\Delta\left(\eta_{\overline{im}} + h_{\overline{im}}\right)}{\Delta x_k} - \frac{\Delta\left(\eta_{\overline{ik}} + h_{\overline{ik}}\right)}{\Delta x_m} \right\}.$$

As every component of $\overline{\eta}$ is constant, all their rates of change are zero; hence, we can reduce this sum to:

$$\Gamma_{ik}^{j} = \frac{1}{2} \sum_{m} \left(\left(\overline{\eta} + \overline{h}\right)^{-1}\right)_{jm} \left\{ \frac{\Delta h_{\overline{mk}}}{\Delta x_i} + \frac{\Delta h_{\overline{im}}}{\Delta x_k} - \frac{\Delta h_{\overline{ik}}}{\Delta x_m} \right\}.$$

A useful approximation is $\left(\overline{\eta} + \overline{h}\right)^{-1} \approx (\eta - h)$, where $h \sim \overline{h}$. Also, as $\eta^2 = I$, $\eta = \eta^{-1}$ so that $(\eta^{-1})_{ab} = \eta_{\overline{ab}} = \eta_{ab}$.

In this approximation, h is the fully contravariant version of $\overline{h}$, and *not* the inverse, $\overline{h}^{-1}$. The steps needed to confirm this are slightly tricky, so I have put the derivation into the online Appendix.

Substituting $(\eta - h)$, we now have:

$$\Gamma_{ik}^{j} = \frac{1}{2} \sum_{m} \left(\eta_{jm} - h_{jm}\right) \left\{ \frac{\Delta h_{\overline{mk}}}{\Delta x_i} + \frac{\Delta h_{\overline{im}}}{\Delta x_k} - \frac{\Delta h_{\overline{ik}}}{\Delta x_m} \right\}.$$

The h_{jm} in the first bracket multiplies the terms in $\Delta h_{\overline{ab}}/\Delta x_j$, generating products that we have already decided can be neglected. It is easier to drop the h_{jm}, so we don't create such terms in the first place, reducing the Christoffel symbols to:

$$\Gamma^j_{ik} = \frac{1}{2}\sum_m \eta_{jm}\left\{\frac{\Delta h_{\overline{mk}}}{\Delta x_i} + \frac{\Delta h_{\overline{im}}}{\Delta x_k} - \frac{\Delta h_{\overline{ik}}}{\Delta x_m}\right\}.$$

So far, the steps that we have taken correspond to *linear gravity theory*, a simplification of the full theory of general relativity that is used for many important calculations (as discussed later in Chapter 11). The additional move that we need to take, in order to morph into the weak field limit, is to argue for the temporal curvature of space-time having the only significant impact at Newtonian velocities. The spatial curvature we can neglect.

8.1.1 Newtonian Gravity

We can justify this in two ways. The first is more 'hand-wavy' and focusses on the temporal co-ordinates having the extra factor of c. For example, an object dropped from a height of 15 m on Earth would take $\approx 1.75\,$s to reach the ground: that's a 'distance' of $ct = 525,000\,$km along the temporal co-ordinate axis. Hence, the space-time world lines for objects of domestic speeds are mostly along the temporal axis, and so temporal distortion has a much bigger effect on their path than spatial distortion. At relativistic speeds, the spatial components are just as significant (see Section 8.1.4).

The second justification comes from the energy density tensor. In the Newtonian regime, that tensor is dominated by the mass density term, ρc^2, as the relativity low velocities in the other terms make them negligible. As the mass density term is the $\mathbb{T}_{00}$ component, it follows that we should focus on the g_{00} component of the metric as well. Hence, in our linear case, h_{00} is crucial.

In summary, linear gravity theory has the metric in the form $g_{\overline{ab}} = \eta_{\overline{ab}} + h_{\overline{ab}}$ with $\overline{h} \ll 1$. Weak field gravity has, in addition, $v \ll c$, leading to a focus on the temporal components of the metric.

We can see the first effect of this further simplification by recalling the equation from Section 7.4 which determines how an object moves in curved space-time when no force is acting:

$$\frac{\Delta}{\Delta\tau}\left(\frac{\Delta x_k}{\Delta\tau}\right) = -\sum_i\sum_j U_i U_j \Gamma^k_{ij}. \tag{7.2}$$

In our weak field scenario, objects are not moving at relativistic speeds which means $\tau \to t$ for starters. As a result, Equation 7.2 becomes:

$$\frac{\Delta}{\Delta t}\left(\frac{\Delta x_k}{\Delta t}\right) = -\sum_i\sum_j U_i U_j \Gamma^k_{ij} = -\left\{U_0 U_0 \Gamma^k_{00} + U_0 U_1 \Gamma^k_{01} + U_1 U_0 \Gamma^k_{10} + \ldots + U_1 U_1 \Gamma^k_{11} + \ldots\right\},$$

where I have only illustrated enough terms in the sum (which remember has $i,j = 0,1,2,3$) to make the next point.

From our definition of the 4-velocity, $U_0 = \gamma_u c$ and $U_1 = \gamma_u u_x$, etc. As $u \ll c$ we are entitled to take $\gamma_u \approx 1$ so that the sum becomes:

$$\frac{\Delta}{\Delta t}\left(\frac{\Delta x_k}{\Delta t}\right) = -\sum_i\sum_j U_i U_j \Gamma^k_{ij} \approx -\left\{c^2\Gamma^k_{00} + cu_x\Gamma^k_{01} + u_x c\Gamma^k_{10} + \ldots + u_x^2\Gamma^k_{11} + \ldots\right\}$$

which will be dominated by the first term,[2] to such an extent that we can approximate the whole sum by just taking the value of the first term. Hence, we end up with:

$$\frac{\Delta}{\Delta t}\left(\frac{\Delta x_k}{\Delta t}\right) \approx -c^2 \Gamma_{00}^k$$

considerably reducing the set of Christoffel symbols that we have to worry about. In fact, all we need are:

$$\Gamma_{00}^k = \frac{1}{2}\sum_m \eta_{km}\left\{\frac{\Delta h_{\overline{m0}}}{\Delta x_0} + \frac{\Delta h_{\overline{0m}}}{\Delta x_0} - \frac{\Delta h_{\overline{00}}}{\Delta x_m}\right\}.$$

We also take it that the metric's components are not changing over time. If the masses acting as the source of gravitation are not doing anything funny, like varying from moment to moment, then the Newtonian field is static. Although there is no direct analogue of the Newtonian gravitational field in general relativity, it is reasonable, given the field equations, to say that a static source of gravitation is going to give rise to metric components that are not functions of time. This being the case, any terms like $\Delta \overline{h}/\Delta x_0 = 0$ (remembering that in general co-ordinates systems, x_0 is the temporal co-ordinate).

The implication is $\Delta h_{\overline{m0}}/\Delta x_0 = 0$ and $\Delta h_{\overline{0m}}/\Delta x_0 = 0$, leaving us with:

$$\Gamma_{00}^k = -\frac{1}{2}\sum_m \eta_{km}\frac{\Delta h_{\overline{00}}}{\Delta x_m}.$$

To get further, we take a specific example, say $k = 0$:

$$\Gamma_{00}^0 = -\frac{1}{2}\sum_m \eta_{0m}\frac{\Delta h_{\overline{00}}}{\Delta x_m} = -\frac{1}{2}\left\{\eta_{00}\frac{\Delta h_{\overline{00}}}{\Delta x_0} + \eta_{01}\frac{\Delta h_{\overline{00}}}{\Delta x_1} + \eta_{02}\frac{\Delta h_{\overline{00}}}{\Delta x_2} + \eta_{03}\frac{\Delta h_{\overline{00}}}{\Delta x_3}\right\}.$$

The term $\eta_{\overline{00}}\frac{\Delta h_{\overline{00}}}{\Delta x_0} = 0$ as the field is static and all the other terms vanish as $\eta_{01}, \eta_{02}, \eta_{03} = 0$, so $\Gamma_{00}^0 = 0$.

Now, let's try $k = 1$:

$$\Gamma_{00}^1 = -\frac{1}{2}\sum_m \eta_{1m}\frac{\Delta h_{\overline{00}}}{\Delta x_m} = -\frac{1}{2}\left\{\eta_{10}\frac{\Delta h_{\overline{00}}}{\Delta x_0} + \eta_{11}\frac{\Delta h_{\overline{00}}}{\Delta x_1} + \eta_{12}\frac{\Delta h_{\overline{00}}}{\Delta x_2} + \eta_{13}\frac{\Delta h_{\overline{00}}}{\Delta x_3}\right\}.$$

Three of these terms drop out, as the metric components are zero. The remaining one, $\eta_{11} = -1$, which means:

$$\Gamma_{00}^1 = \frac{1}{2}\frac{\Delta h_{\overline{00}}}{\Delta x_1}.$$

A similar argument applies to $k = 2, 3$, so we can say that $\Gamma_{00}^k = \frac{1}{2}\frac{\Delta h_{\overline{00}}}{\Delta x_k}$ where $k = 1, 2, 3$. Physically this is telling us that the rate of change of a vector along a spatial axis, k, is connected with the rate of change of the temporal component of the distortion metric, $h_{\overline{00}}$, along that axis; something to keep in mind for later.

8.1.2 The First Newtonian Limit

Just a few paragraphs ago, we reduced equation 7.2 to:

$$\frac{\Delta}{\Delta t}\left(\frac{\Delta x_k}{\Delta t}\right) \approx -c^2 \Gamma_{00}^k.$$

Plugging in the expression we have just derived for the Christoffel symbols we end up with:

$$\frac{\Delta}{\Delta t}\left(\frac{\Delta x_k}{\Delta t}\right) \approx -c^2 \frac{1}{2} \frac{\Delta h_{\overline{00}}}{\Delta x_k}$$

with $k = 1, 2, 3$ due to the restrictions on the surviving Christoffel symbols.

In Newtonian theory, free fall acceleration is due to a gravitational force, and we have seen how the gravitational field strength, g, is related to the potential, ϕ:

$$g_x = -\frac{\Delta \phi}{\Delta x} \qquad g_y = -\frac{\Delta \phi}{\Delta y} \qquad g_z = -\frac{\Delta \phi}{\Delta z} \qquad \text{or} \qquad g_k = -\frac{\Delta \phi}{\Delta x_k}.$$

As gravitational field strength is the same as gravitational acceleration, we can set:

$$g_k = -\frac{\Delta \phi}{\Delta x_k} = \frac{\Delta}{\Delta t}\left(\frac{\Delta x_k}{\Delta t}\right) \approx -c^2 \frac{1}{2} \frac{\Delta h_{\overline{00}}}{\Delta x_k}$$

from which we can extract the conclusions[3]:

$$g_k \approx -c^2 \frac{1}{2} \frac{\Delta h_{\overline{00}}}{\Delta x_k}$$

$$\frac{\Delta \phi}{\Delta x_k} \approx c^2 \frac{1}{2} \frac{\Delta h_{\overline{00}}}{\Delta x_k} \quad \text{so that} \quad \phi = \frac{1}{2} c^2 h_{\overline{00}}.$$

The first statement clearly relates to our previous assertion that, in the weak field limit, the everyday effects of gravity are down to the curvature of time. The second connects the gravitational potential to the temporal distortion of the weak field metric.[4]

Originally, we set up the metric as $g_{\overline{ab}} = \eta_{\overline{ab}} + h_{\overline{ab}}$, so now we can write:

$$g_{\overline{00}} = \eta_{\overline{00}} + h_{\overline{00}} \approx \eta_{\overline{00}} + \frac{2\phi}{c^2} = 1 + \frac{2\phi}{c^2}.$$

While this is an extremely pretty result, we should not get carried away and attribute more physical meaning to it than is warranted, given the approximations that we had to go through to get here. In any case, the thought should be "ah, so the Newtonian potential was actually always a metrical distortion in disguise", rather than "ah, so metrical distortions are actually the Newtonian potential at work". In the latter case, we might be tempted to bring a penumbra of other physical interpretations associated with Newtonian potential and try and shoe-horn them into general relativity, where they do not belong and cannot be supported.

8.1.3 The Second Newtonian Limit

Now we turn to establishing the value of κ in:

$$\mathbb{R}_{ab} - \frac{1}{2}\mathcal{R}\left(g^{-1}\right)_{ab} = \kappa \mathbb{T}_{ab}.$$

In order to do this, we apply our weak field metric to the contravariant Ricci tensor:

$$\mathbb{R}^{ab} = \sum_{i,j,k} \left(g^{-1}\right)_{aj}\left(g^{-1}\right)_{bk} R_{i\overline{jki}} = \sum_{i,\,j,\,k} \left(\eta_{aj} - h_{aj}\right)\left(\eta_{bk} - h_{bk}\right) R_{i\overline{jki}}.$$

The components of the Riemann tensor are as follows:

$$R_{i\overline{jki}} = \frac{\Delta\left(\Gamma^i_{ji}\right)}{\Delta x_k} - \frac{\Delta\left(\Gamma^i_{jk}\right)}{\Delta x_i} + \sum_n \left(\Gamma^n_{ji}\Gamma^i_{nk} - \Gamma^n_{jk}\Gamma^i_{ni}\right);$$

so, inserting them gets us to:

$$\mathbb{R}_{ab} = \sum_{i,j,k}\left(\eta_{aj} - h_{aj}\right)\left(\eta_{bk} - h_{bk}\right)\left\{\frac{\Delta\left(\Gamma^i_{ji}\right)}{\Delta x_k} - \frac{\Delta\left(\Gamma^i_{jk}\right)}{\Delta x_i} + \sum_n \left(\Gamma^n_{ji}\Gamma^i_{nk} - \Gamma^n_{jk}\Gamma^i_{ni}\right)\right\}.$$

The products $\Gamma^n_{ji}\Gamma^i_{nk}$ and $\Gamma^n_{jk}\Gamma^i_{ni}$ can be dropped, as each symbol has terms like $\Delta\overline{h}/\Delta x_j$, and we have licence to disregard anything containing $\frac{\Delta\overline{h}}{\Delta x_i} \times \frac{\Delta\overline{h}}{\Delta x_j}$. Also, h can be removed from the multiplying metrics, to save generating products such as $h \times \frac{\Delta\overline{h}}{\Delta x_j}$, which we would neglect in any case:

$$\mathbb{R}_{ab} = \sum_{i,j,k}\eta_{aj}\eta_{bk}\left\{\frac{\Delta\left(\Gamma^i_{ji}\right)}{\Delta x_k} - \frac{\Delta\left(\Gamma^i_{jk}\right)}{\Delta x_i}\right\}.$$

Earlier, I argued that in Newtonian situations, the only term in the energy density tensor that would matter was $\mathbb{T}_{00}$. Now I take that argument as a justification for concentrating on $\mathbb{R}_{00}$, especially if I remember that an alternative form of the field equations is as follows (Section 7.7):

$$\mathbb{R}_{ab} = \kappa\left(\mathbb{T}_{ab} + \kappa T\left(g^{-1}\right)_{ab}\right).$$

Focussing on $\mathbb{R}_{00}$ gives me $a = b = 0$ hence:

$$\mathbb{R}_{00} = \sum_{i,j,k}\eta_{0j}\eta_{0k}\left\{\frac{\Delta\left(\Gamma^i_{ji}\right)}{\Delta x_k} - \frac{\Delta\left(\Gamma^i_{jk}\right)}{\Delta x_i}\right\},$$

which in turn shows us that we can collapse the sums over j,k as the only non-zero metric component of the form η_{0j} or η_{0k} is $\eta_{00} - 1$; hence, $j = k = 0$:

$$\mathbb{R}_{00} = \sum_i \left\{\frac{\Delta\left(\Gamma^i_{0i}\right)}{\Delta x_0} - \frac{\Delta\left(\Gamma^i_{00}\right)}{\Delta x_i}\right\}.$$

The first term can be neglected as the field is static, resulting in a pleasing simplicity:

$$\mathbb{R}_{00} = -\sum_i \frac{\Delta\left(\Gamma^i_{00}\right)}{\Delta x_i}.$$

This also exposes the wisdom of focussing on Γ^i_{00} earlier, as we can insert our previous result (switching k into i) $\Gamma^i_{00} = \frac{1}{2}\frac{\Delta h_{\overline{00}}}{\Delta x_i}$ to give us:

$$\mathbb{R}_{00} = -\sum_i \frac{\Delta\left(\Gamma^i_{00}\right)}{\Delta x_i} = -\frac{1}{2}\sum_i \frac{\Delta}{\Delta x_i}\left(\frac{\Delta h_{\overline{00}}}{\Delta x_i}\right).$$

Back in Section 5.4.1, we learned to write the expression $\frac{\Delta}{\Delta x}\left(\frac{\Delta\phi}{\Delta x}\right)+\frac{\Delta}{\Delta y}\left(\frac{\Delta\phi}{\Delta y}\right)+\frac{\Delta}{\Delta z}\left(\frac{\Delta\phi}{\Delta z}\right)$ as $\nabla^2\phi$.

The same abbreviation will apply to $\frac{\Delta}{\Delta x_1}\left(\frac{\Delta h_{\overline{00}}}{\Delta x_1}\right)+\frac{\Delta}{\Delta x_2}\left(\frac{\Delta h_{\overline{00}}}{\Delta x_2}\right)+\frac{\Delta}{\Delta x_3}\left(\frac{\Delta h_{\overline{00}}}{\Delta x_3}\right)$, which means:

$$\mathbb{R}_{00} = -\frac{1}{2}\nabla^2\left(h_{\overline{00}}\right) = \nabla^2\left(-\frac{1}{2}h_{\overline{00}}\right).$$

Now we switch tack (don't worry, we're getting there…) and consider the Einstein field equations in the form:

$$\mathbb{R}_{ab} - \frac{1}{2}\mathcal{R}\left(g^{-1}\right)_{ab} = \kappa\mathbb{T}_{ab}.$$

Since $\mathbb{R}_{ab}$ and $\mathcal{R}$ are going to involve terms in h or its rate of change, we do not need to use the full metric $\left(g^{-1}\right)_{ab} = \eta_{ab} - h_{ab}$ in the field equations. If we did, the terms involving h multiplied into $\mathbb{R}_{ab}$ and $\mathcal{R}$ would be small and we could neglect them. We can get away with:

$$\mathbb{R}_{ab} - \frac{1}{2}\mathcal{R}\eta_{ab} = \kappa\mathbb{T}_{ab}.$$

To progress further, we multiply both sides by η_{ba} so that:

$$\eta_{ba}\mathbb{R}_{ab} - \frac{1}{2}\mathcal{R}\eta_{ba}\eta_{ab} = \kappa\eta_{ba}\mathbb{T}_{ab}$$

and then sum over a:

$$\sum_a \eta_{ba}\mathbb{R}_{ab} - \frac{1}{2}\sum_a \mathcal{R}\eta_{ba}\eta_{ab} = \kappa\sum_a \eta_{ba}\mathbb{T}_{ab}.$$

This is the same as:

$$\mathbb{R}_{\overline{bb}} - \frac{1}{2}\sum_a \mathcal{R}\eta_{ba}\eta_{ab} = \kappa\sum_a \eta_{ba}\mathbb{T}_{ab}.$$

Summing over b forms contractions:

$$\sum_b \mathbb{R}_{\overline{bb}} - \frac{1}{2}\sum_b\sum_a \mathcal{R}\eta_{ba}\eta_{ab} = \kappa\sum_b\sum_a \eta_{ba}\mathbb{T}_{ab}.$$

The left-hand side is:

$$\mathcal{R} - \frac{1}{2}\mathcal{R}\left(\eta_{00}^2 + \eta_{11}^2 + \eta_{22}^2 + \eta_{33}^2\right) = \mathcal{R} - \frac{1}{2}\mathcal{R}\left(1+1+1+1\right) = -\mathcal{R}$$

and the right-hand side is going to be:

$$\kappa\left\{\eta_{00}\mathbb{T}_{00} + \eta_{01}\mathbb{T}_{01} + \ldots + \eta_{10}\mathbb{T}_{10} + \eta_{11}\mathbb{T}_{11} + \ldots + \eta_{43}\mathbb{T}_{43} + \ldots\right\}.$$

The off-diagonal terms of the metric are zero, which kills off everything but:

$$\kappa \left\{ \eta_{00} \mathbb{T}_{00} + \eta_{11} \mathbb{T}_{11} + \eta_{22} \mathbb{T}_{22} + \eta_{33} \mathbb{T}_{33} \right\}.$$

Now we bring back the argument I mentioned earlier regarding the energy density tensor. From Section 5.5.1, $\mathbb{T}_{00} = (n\gamma_u)(\gamma_u mc^2) = \rho c^2$, where n is the number density of particles in the rest system. Hence, $\eta_{00} \mathbb{T}_{00} = \eta_{00} \rho c^2$.

The other terms in the tensor are of the form $(n\gamma_u)(\gamma_u mu_i^2)$ and, as they contain u_i^2, in our Newtonian regime is going to render them much smaller than the first term, so again we can approximate this whole sum as just the value of the first term:

$$\kappa \left\{ \eta_{00} \mathbb{T}_{00} + \eta_{11} \mathbb{T}_{11} + \eta_{22} \mathbb{T}_{22} + \eta_{33} \mathbb{T}_{33} \right\} \approx \kappa \eta_{00} \rho c^2 = \kappa \rho c^2$$

as $\eta_{00} = 1$. Inserting our calculations for the right-hand and left-hand sides transforms the equation:

$$\sum_b \mathbb{R}_{\overline{bb}} - \frac{1}{2} \sum_b \sum_a \mathcal{R} \eta_{ba} \eta_{ab} = \kappa \sum_b \mathbb{T}_{\overline{bb}}$$

into:

$$-\mathcal{R} = \kappa \rho c^2.$$

Putting this value of $\mathcal{R}$ back into the field equations we get:

$$\mathbb{R}_{ab} + \frac{1}{2} \kappa \rho c^2 \eta_{ab} = \kappa \mathbb{T}_{ab} \quad \text{or} \quad \mathbb{R}_{ab} = \kappa \left(\mathbb{T}_{ab} - \frac{1}{2} \rho c^2 \eta_{ab} \right).$$

Restricting our focus to just $\mathbb{T}_{00}$ and consequently $\mathbb{R}_{00}$:

$$\mathbb{R}_{00} = \kappa \left(\mathbb{T}_{00} - \frac{1}{2} \rho c^2 \eta_{00} \right) = \kappa \left(\rho c^2 - \frac{1}{2} \rho c^2 \right) = \frac{1}{2} \kappa \rho c^2.$$

Dropping in our expression for $\mathbb{R}_{00}$ we produce (tension building):

$$\nabla^2 \left(-\frac{1}{2} h_{\overline{00}} \right) = \frac{1}{2} \kappa \rho c^2.$$

In the previous section, we established that $\phi = \frac{1}{2} c^2 h_{\overline{00}}$ which we turn around to give $\frac{1}{2} h_{\overline{00}} = \frac{\phi}{c^2}$

and insert into the remains of the field equations:

$$\nabla^2 \left(-\frac{\phi}{c^2} \right) = \frac{1}{2} \kappa \rho c^2 \quad \text{or} \quad \nabla^2 (\phi) = -\frac{1}{2} \kappa \rho c^4.$$

Thinking back to Poisson's equation:

$$\nabla^2 (\phi) = 4\pi G \rho$$

and, comparing the two shows us that:

$$\kappa = -\frac{8\pi G}{c^4}.$$

So, we now have the complete Einstein field equations:

$$\mathbb{R}_{ab} - \frac{1}{2}\mathcal{R}\left(g^{-1}\right)_{ab} = -\frac{8\pi G}{c^4}\mathbb{T}_{ab}$$

and have demonstrated that they reduce to Poisson's equation in the weak field limit.

8.1.4 Weak Field Schwarzschild Metric

There are very few exact solutions to the Einstein field equations, which is not surprising given their complex and non-linear structure (Section 7.7.1). Possibly the most famous one is the *Schwarzschild metric* that we have already mentioned on a couple of occasions and which takes the form:

$$g_{Sch} = \begin{pmatrix} \left(1-\dfrac{2GM}{c^2 r}\right) & 0 & 0 & 0 \\ 0 & -\left(1-\dfrac{2GM}{c^2 r}\right)^{-1} & 0 & 0 \\ 0 & 0 & -r^2 & 0 \\ 0 & 0 & 0 & -r^2\sin^2(\theta) \end{pmatrix}$$

in spherical-polar co-ordinates. This metric describes the empty regions of space-time surrounding certain gravitating masses, which makes it a solution to the *vacuum field equations*:

$$\mathbb{R}_{ab} - \frac{1}{2}\mathcal{R}\left(g^{-1}\right)_{ab} = 0.$$

Of course, the Schwarzschild metric would have to blend smoothly into the space-time applicable within the volume of the gravitating mass, which would be a solution to the non-vacuum equations with an appropriate energy density tensor.

Schwarzschild obtained this metric by making some assumptions about the space-time surrounding a non-rotating, spherically symmetrical, constant mass object. This enabled him to produce an educated guess about the structure of the metric:

$$g_{Sch} = \begin{pmatrix} F_1(r,t)c^2 & 0 & 0 & 0 \\ 0 & -F_2(r,t) & 0 & 0 \\ 0 & 0 & -r^2 & 0 \\ 0 & 0 & 0 & -r^2\sin^2(\theta) \end{pmatrix}$$

with the functions $F_1(r,t)$ and $F_2(r,t)$ to be determined. The 22 and 33 components of the metric look just like those of the flat spherical-polar metric due to the assumption of symmetry in the gravitating mass.

Plugging this guess into the Einstein field equations reduced them to a set of equations that could be solved for F_1 and F_2, which is how the final metric is derived.

The Schwarzschild metric can be used to understand the properties of simpler (non-charged, non-rotating, spherically symmetrical) black holes, but it can also be applied to ordinary stars, which fall under our weak field approximation if the objects near to the gravitating mass are moving at speeds much slower than light.

In these situations, we only need consider the temporal term as the space-time interval in the vicinity of an event would be:

$$(\Delta s)^2 = \left(1 - \frac{2GM}{c^2 r}\right)c^2(\Delta t)^2 - \frac{(\Delta r)^2}{\left(1 - \frac{2GM}{c^2 r}\right)} - r^2(\Delta \theta)^2 - r^2 \sin^2(\theta)(\Delta \varphi)^2$$

which is dominated by the term $c^2(\Delta t)^2$. To illustrate this, let's try a couple of calculations.

Our Sun masses $\sim 2.0 \times 10^{30}$ kg and $G = 6.674 \times 10^{-11}$ m^3/kgs^2, so the term $2GM/c^2 r \sim 4 \times 10^{-6}$, for just outside the Sun's radius. That makes the interval:

$$(\Delta s)^2 = \left(1 - 4 \times 10^{-6}\right)c^2(\Delta t)^2 - \frac{(\Delta r)^2}{\left(1 - 4 \times 10^{-6}\right)} - r^2(\Delta \theta)^2 - r^2 \sin^2(\theta)(\Delta \varphi)^2$$

$$= 0.999996 c^2(\Delta t)^2 - 1.000004(\Delta r)^2 - r^2(\Delta \theta)^2 - r^2 \sin^2(\theta)(\Delta \varphi)^2.$$

If we evaluate this interval for two events along the world line of an object that is moving radially at non-relativistic speeds, the first term will dominate. For the spatial term to be comparable to the temporal one we would have to be moving briskly, as to make $0.999996 c^2(\Delta t)^2 = 1.000004(\Delta r)^2$ when $\Delta r = v\Delta t$, implies:

$$V = \sqrt{\frac{0.999996}{1.000004}} c = 0.999996c.$$

On the other hand, the supermassive black hole at the centre of our galaxy, known as Sagittarius A^* or Sgr A^*, has a mass of $\sim 4 \times 10^6$ solar masses and an estimated event horizon[5] radius of 1.18×10^{10} m. For such an object, the smallest stable orbit is $\sim 4.5 \times$ the event horizon radius, making $2Gm/c^2 r \sim 0.22$ and the interval:

$$(\Delta s)^2 = \left(1 - 0.22\right)c^2(\Delta t)^2 - \frac{(\Delta r)^2}{(1 - 0.22)} - r^2(\Delta \theta)^2 - r^2 \sin^2(\theta)(\Delta \varphi)^2.$$

In this instance, for the spatial term to be comparable to the temporal one:

$$0.78 c^2(\Delta t)^2 = 1.29(\Delta r)^2 = 1.29 v^2(\Delta t)^2,$$

which happens if $v^2 = (0.78/1.29)c^2$ or $v = 0.78c$, which is fast, but not ridiculously so. For example, many supermassive black holes found at the centres of galaxies emit streams of material called *jets* which move at relativistic speeds[6]. One of the closest powerful radio source galaxies is Centaurus A, which can be seen in the composite image Figure 8.1. This well-studied active galaxy has a black hole which is 55 million times more massive than our Sun. The giant jets visible in the image extend for thousands of light years in X-ray and can be detected over millions of light years in radio emissions.

The inner portions of the X-ray jets are moving at an estimated speed of 0.5 c, so clearly in a situation where spatial and temporal curvature would be relevant to them.

So, once again we see that in domestic situations gravitational effects are determined by temporal curvature.

FIGURE 8.1 Centaurus A, showing the lobes and jets emanating from the active galaxy's central black hole. This composite image is built with data from three instruments operating at very different wavelength from 870-micron submillimetre through to X-ray. The visible light component shows the stars and the galaxy's characteristic dust lane. (Image credit: ESO/WFI [Optical]; MPIfR/ESO/APEX/A. Weiss et al. [Submillimetre]; NASA/CXC/CfA/R.Kraft et al. [X-ray].)

It is also worth noting, in passing, that the Schwarzschild metric smoothly transforms into the (spherical-polar) Minkowski metric as we get further and further from the gravitating mass. Formally:

$$\left(1-\frac{2GM}{c^2 r}\right) \to 1 \quad \text{as} \quad r \to \infty.$$

Equally, as the mass declines in value, so the metric folds into the (spherical-polar) Minkowski metric at all distances from the diminishing mass:

$$\left(1-\frac{2GM}{c^2 r}\right) \to 1 \quad \text{as} \quad M \to 0.$$

Both of these are sensible conditions which were part of the assumptions used to set up the metric. Finally, as the functions F_1 and F_2 turn out not to depend on time, the metric has an interesting property: even if the gravitating mass is changing with time (e.g. a star pulsating) the metric does not change, provided these pulsations remain spherically symmetrical.[7] As a result, such pulsations cannot act as sources of gravitational waves. Also, and significantly, if the star is steadily expanding or contracting in a symmetrical manner, then the Schwarzschild metric holds good as well. Consequently, Schwarzschild space-time holds outside simply collapsing stars that form black holes (Chapter 10).

Returning to our weak field, $g_{\overline{ab}} = \eta_{\overline{ab}} + h_{\overline{ab}}$, as the temporal component of the Schwarzschild metric is $1-\frac{2GM}{c^2 r}$ and $\eta_{\overline{00}} = 1$, it is evident that:

$$g_{\overline{00}} = 1 + h_{\overline{00}} = 1 - \frac{2GM}{c^2 r}$$

making:

$$h_{\overline{00}} = -\frac{2GM}{c^2 r}.$$

We have already found that the Newtonian gravitational potential relates to the metric by:

$$\phi = \frac{1}{2}c^2 h_{\overline{00}} = \frac{1}{2}c^2\left(-\frac{2GM}{c^2 r}\right) = -\frac{GM}{r},$$

which is exactly what we expect to see for the Newtonian potential of a point mass M.

The equation that determines the motion of an object in curved space-time, reduced to:

$$\frac{\Delta}{\Delta t}\left(\frac{\Delta x_k}{\Delta t}\right) \approx -c^2 \frac{1}{2}\frac{\Delta h_{\overline{00}}}{\Delta x_k}$$

in the weak field limit, so if we insert our value for $h_{\overline{00}}$ and set $x_k = r$, so the object is falling radially towards the gravitating mass, we get:

$$\frac{\Delta}{\Delta t}\left(\frac{\Delta r}{\Delta t}\right) = -c^2 \frac{1}{2}\frac{\Delta h_{\overline{00}}}{\Delta x_r} = -c^2 \frac{1}{2}\frac{\Delta}{\Delta r}\left(-\frac{2GM}{c^2 r}\right).$$

Calculating this rate of change is a mathematical manipulation that many of you may already be familiar with. For those who have not seen this before, the calculation is shown in the online Appendix. Either way, the upshot is:

$$\frac{\Delta}{\Delta r}\left(-\frac{2GM}{c^2 r}\right) = \frac{2}{c^2}\frac{GM}{r^2}.$$

Putting this into the equation of motion gives:

$$\frac{\Delta}{\Delta t}\left(\frac{\Delta r}{\Delta t}\right) = -c^2 \frac{1}{2}\frac{\Delta}{\Delta r}\left(-\frac{2GM}{c^2 r}\right) = -\frac{c^2}{2} \times \frac{2}{c^2} \times \frac{GM}{r^2}$$

which, when tidied up results in:

$$\frac{\Delta}{\Delta t}\left(\frac{\Delta r}{\Delta t}\right) = -\frac{GM}{r^2}.$$

Once again, this is exactly the result we would have expected, given the Newtonian gravitational force law.

8.1.5 Free Fall in the Schwarzschild Weak Field

Classically, an object falling from rest at infinity would have achieved a velocity, v, by the time it had reached a distance r from the centre of a gravitating mass where:

$$\frac{1}{2}mv^2 = \frac{GMm}{r} \quad \text{or} \quad v^2 = \frac{2GM}{r}.$$

Now, imagine that we are watching this from a system that is placed at infinity and stationary. In order to boost into a system that is co-moving with the falling object when it reaches r, we need a boost factor:

$$\gamma = \frac{1}{\sqrt{1 - v^2/c^2}} = \frac{1}{\sqrt{1 - 2GM/c^2 r}}.$$

Comparing this with the Schwarzschild space-time interval at the same r:

$$(\Delta s)^2 = \left(1 - \frac{2GM}{c^2 r}\right) c^2 (\Delta t)^2 - \frac{(\Delta r)^2}{\left(1 - \dfrac{2GM}{c^2 r}\right)} - r^2 (\Delta \theta)^2 - r^2 \sin^2(\theta)(\Delta \varphi)^2$$

we can see that:

$$(\Delta s)^2 = \frac{c^2 (\Delta t)^2}{\gamma^2} - \gamma^2 (\Delta r)^2 - r^2 (\Delta \theta)^2 - r^2 \sin^2(\theta)(\Delta \varphi)^2.$$

If we observe two events at the same position, Δr, $\Delta \theta$ and $\Delta \varphi$ are all zero, so $\Delta s = c \Delta t / \gamma$. As we will discuss further in Section 8.2, we define *gravitational proper time* in the general theory analogously to that in the special theory and write $c \Delta \tau = \Delta s$ so that $\Delta \tau = \Delta t / \gamma$ or $\Delta t = \gamma \Delta \tau$. Hence, Δt, known as the *co-ordinate time*, is dilated compared to the proper time.

Equally, if the two events are simultaneous, so $\Delta t = 0$, but separated in radial distance only, $\Delta \theta$ and $\Delta \varphi$ are both zero and we have $\Delta s = -\gamma \Delta r$. The *proper distance*, σ, is defined by $\Delta \sigma = \sqrt{-(\Delta s)^2}$ so that $\Delta \sigma = \gamma \Delta r$, which re-arranges to $\Delta r = \Delta \sigma / \gamma$: the *co-ordinate distance*, Δr, is contracted compared to the proper distance.

These comparisons show us that there is something to be gained in thinking of the Schwarzschild space-time as being analogous to that of an object falling at a rate given by $v^2 = 2GM/r$, which has some resonance with the principle of equivalence. However, we should not make too much of this as co-ordinate distance and co-ordinate time turn out to have a tenuous physical reality.

8.2 Gravitational Time Dilation

In the special theory of relativity, with its flat space-time metric, the space-time interval is:

$$(\Delta s)^2 = (c \Delta t)^2 - (\Delta x)^2 - (\Delta y)^2 - (\Delta z)^2$$

in Cartesian co-ordinates. In this expression, and others like it, Δt is a duration of *co-ordinate time*. In flat space-time, the co-ordinate time is simply the physical time recorded on clocks stationed around the co-ordinate system. The specific physical time recorded on clocks in a co-moving system is singled out as *proper time*.

When we discussed time dilation in the context of the special theory, we were comparing clocks sitting in a co-moving system to those in other systems. The time dilation was symmetrical: observers in each system regarded their time as normal and that of the other system as dilated. This ties in with the relativity of inertial motion.

Time dilation near to a gravitating mass is not relative. We can absolutely see that one clock is near to the gravitating mass and is experiencing dilated time. However, dilated compared to what? Where is the reference for this comparison?

Once more, the space-time interval for the Schwarzschild metric is:

$$(\Delta s)^2 = \left(1 - \frac{2GM}{c^2 r}\right) c^2 (\Delta t)^2 - \frac{(\Delta r)^2}{\left(1 - \dfrac{2GM}{c^2 r}\right)} - r^2 (\Delta \theta)^2 - r^2 \sin^2(\theta)(\Delta \varphi)^2.$$

For events taking place at the same spatial location $\Delta r = \Delta\theta = \Delta\varphi = 0$, hence:

$$(\Delta s)^2 = \left(1 - \frac{2GM}{c^2 r}\right) c^2 (\Delta t)^2$$

giving:

$$\Delta s = c\Delta\tau = \sqrt{\left(1 - \frac{2GM}{c^2 r}\right)} c\Delta t$$

using τ as the *gravitational proper time*, which in this context refers to the time recorded on a clock that is stationary at the same spatial co-ordinates as the event. This means that we have two subtly different usages of the term proper time, so we need to keep a close eye on the context to be clear about which is which.

Just absorb the physics of this for a moment, because there is something non-intuitive about it. If the proper time is the time recorded on a stationary clock at the location of an event, then what is the co-ordinate time interval, Δt?

Instinct, and our experience from special relativity, prompts us to think that the co-ordinate time ought to be the time recorded on a clock at the location of an event, otherwise what is the use of co-ordinate time? However, in the context of general relativity instincts need to be re-trained. *No clock in the vicinity of a gravitating mass ever records co-ordinate time.* All they can show is the appropriate gravitational proper time. Co-ordinate time loses any physical meaning if the metric is not flat.

To make the rest of our discussion a little easier, we introduce the *Schwarzschild radius*:

$$R_s = \frac{2GM}{c^2}$$

so that $\Delta\tau = \sqrt{\left(1 - \frac{R_s}{r}\right)}\Delta t$. Straight away we see that when $r = R_s$ strange things happen (even stranger when we think of the spatial component of the metric), but that is a subject for later (Sections 9.1.3 and 10.2).

For the moment, consider the ratio between gravitational proper time and co-ordinate time at a distance r from the centre of our mass:

$$\frac{\Delta\tau}{\Delta t} = \sqrt{\left(1 - \frac{R_s}{r}\right)}.$$

Figure 8.2 shows how this ratio develops along a radial line from Sgr A*. For radial distance, I have used multiples of the Schwarzschild radius for that black hole, which gives a more convenient scale than metres or any other distance unit.

Clearly $\Delta\tau/\Delta t \to 1$ the further we get from the black hole, without ever actually reaching that value. This of course follows directly from the equation as $\frac{\Delta\tau}{\Delta t} = \sqrt{\left(1 - \frac{R_s}{r}\right)} \to 1$ as $r \to \infty$ because $R_s/r \to 0$ as $r \to \infty$, but it is useful to see it visually in the graph.

The gravitational proper time would correspond with co-ordinate time if we were able to get the gravitating mass to fade to zero, as $\Delta\tau = \sqrt{\left(1 - \frac{R_s}{r}\right)}\Delta t \to \Delta t$ as $M \to 0$ because $R_s \to 0$ as $M \to 0$.

From both of these arguments we conclude that *co-ordinate time is the same as the gravitational proper time read on a clock outside the influence of the gravitating mass*. In other words, if the

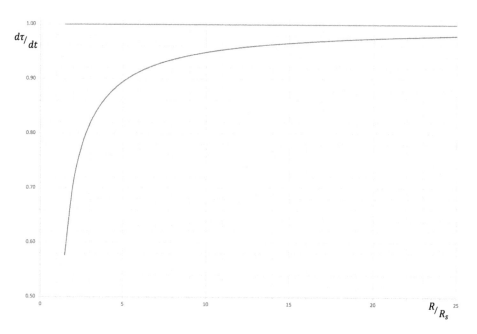

FIGURE 8.2 Comparing co-ordinate time to proper time as we move away from a Sgr A^*. The distance axis is plotted in multiples of the Schwarzschild radius. The horizontal line across the top of the graph is the line $\Delta\tau/\Delta t = 1$, which is there to guide the eye.

Schwarzschild metric faded into a flat metric, then the time recorded on clocks would be the same as the co-ordinate time.

Some authors summarise this by saying that co-ordinate time is the clock time (gravitational proper time) measured by a clock that is stationary at infinity. However, as there are all sorts of physical issues involved with this notion,[8] I prefer a different prescription.

While co-ordinate time is never recorded on clocks in a curved metric, we can make the difference as small as we need by moving further from the gravitating mass. Once we have set the level of precision we require, we can find the distance, R_Z, from the mass where $1 - \dfrac{R_s}{R_Z} \approx 1$ to that degree of precision. Clocks placed at $r > R_Z$ will then be declared as outside the zone of influence of the gravitating mass and their time will be considered effectively equal to co-ordinate time. This gives us an experimental scenario by which approximate co-ordinate time values can be ascribed to events.

To put this in concrete terms, in Figure 8.3 the clock far, far away from the black hole is recording a gravitational proper time that is effectively co-ordinate time. An observer at this clock sees a pair of events at $r = 2R_s$ and times 10 s between them. This is the approximate co-ordinate time interval for these events. However, the gravitational proper time interval recorded by the clock at $r = 2R_s$ is 7.1 s. An observer at $r = 2R_s$ will look towards the clock at $r > R_s$ and see it racing ahead with 1.4 s to every second of their clock.

8.2.1 Global Positioning Systems

Compared to Sgr A^*, the gravitation in the vicinity of the Earth can definitely be considered as a weak field. Nevertheless, the Earth's mass distorts time significantly enough for its effects to be critical to the working of the global positioning system (GPS) extensively used for navigation.

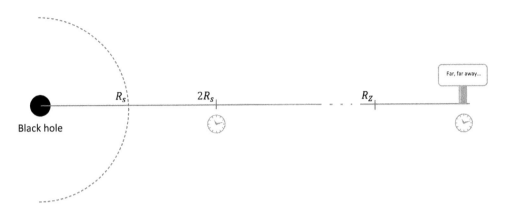

FIGURE 8.3 Comparing the time recorded on two clocks: one at $r = 2R_s$ and the other at $r > R_Z$.

In order to triangulate its position on Earth, a GPS receiver compares microwave signals from a mesh of orbiting satellites. Each signal contains information about the time it left the satellite, and the satellite's position in space at the moment of transmission. Inside the signal is a time-ordered sequence of ones and zeros following a pattern known to the receiver. By aligning the received pattern with its internally generated version, the receiver can figure out how long the signal took to travel from the satellite. Then, by knowing where the satellite is, the receiver starts to triangulate its own location relative to that satellite. Four such triangulations are generally sufficient to locate the receiver accurately.

If we want the precision of this location to be better than 10 m (say), then we need a timing accuracy for the satellite clocks $\sim 10\,\mathrm{m}/c = 33$ ns.

GPS satellites orbit at a height of roughly 20,000 km, at which they are moving with a velocity given by (Newtonian calculation):

$$\frac{mv^2}{r} = \frac{GMm}{r^2} \quad \text{or} \quad v = \sqrt{\frac{GM}{r}} = \sqrt{\frac{6.67 \times 10^{-11} \times 5.97 \times 10^{24}}{\left(6.37 \times 10^6 + 2 \times 10^7\right)}} = 3{,}890\,\mathrm{m/s}$$

using values for the Earth's mass and radius.

At such speeds, the time dilation factor, γ, is tiny, so it is best to calculate its cumulative effect across a whole day:

$$\Delta\mathrm{GPS} = \frac{24 \times 60 \times 60}{\gamma} - 24 \times 60 \times 60 = 7.26\,\mu\mathrm{s/day}.$$

This is huge compared to the time precision needed for 10 m physical location accuracy. (Actually, with the Earth's rotation, only the satellite's velocity relative to the surface counts, which brings the error down to $\sim 6\mu\mathrm{s}/$ day, but this factor will then depend on latitude, so we are not going to consider it in our simple estimations.)

Time dilation due to speed of travel is not the only factor. We also have to consider the relative gravitational time dilation between a clock on the surface and one in orbit. As $\Delta\tau = \sqrt{\left(1 - \dfrac{2GM}{c^2 r}\right)}\Delta t$ we have:

$$\Delta\tau_{\mathrm{surface}} = \sqrt{\left(1 - \frac{2GM}{c^2 R_e}\right)}\Delta t \quad \text{and} \quad \Delta\tau_{\mathrm{orbit}} = \sqrt{\left(1 - \frac{2GM}{c^2 \left(R_e + h\right)}\right)}\Delta t,$$

so that:

$$\frac{\Delta\tau_{\text{orbit}}}{\Delta\tau_{\text{surface}}} = \sqrt{\frac{\left(1 - \dfrac{2GM}{c^2(R_e + h)}\right)}{\left(1 - \dfrac{2GM}{c^2 R_e}\right)}} = \sqrt{\frac{R_e\left(c^2(R_e + h) - 2GM\right)}{\left(c^2 R_e - 2GM\right)(R_e + h)}}$$

which equates to 46 μs per day. The special relativistic effect slows the clock in orbit down compared to the ground, but the gravitational effect speeds it up, so the net error over the course of a day is $46 - 7 = 39$ μs per day. That is equivalent to a spatial error of $c \times 39 \times 10^{-6} = 11.7$ km! Don't forget, this error would accumulate each day, so such a system would be completely useless. Of course, the software is designed to correct for these relativistic effects given the known orbital parameters of the satellites. Nevertheless, the fact that such a commonly used system as the GPS network has to take general relativity into account shows both the reality of time dilation and its significance, even in day-to-day life.

8.3 Gravitational Frequency Shift

Another observable consequence of time dilation near to a gravitating mass comes from the impact it has on the frequency of an electromagnetic wave. To set this up, consider some form of laser beam mounted on the surface of the Earth and directed to fire its beam vertically upwards. The laser is monitored by two observers: one of whom is stationary on the surface and next to the laser and the other is posted far from the Earth at $r > R_Z$ for our chosen precision of co-ordinate time.

   Although this is not a practical possibility, imagine that the observer next to the beam were able to count the number of complete cycles of the wave, n, in proper time $\Delta\tau$. For this observer, the frequency of the wave, $f_r = n/\Delta\tau$. Our second, much more distant, observer watching the light being emitted will count the same number of cycles, but in a time interval that they have as $\Delta\tau_{R_z}$; hence, their view of the frequency is $f_{R_z} = n/\Delta\tau_{R_z}$. Comparing the two frequencies, we see that:

$$\frac{f_{R_z}}{f_r} = \frac{\Delta\tau}{\Delta\tau_{R_z}}.$$

The gravitational time dilation formula from the last section is $\Delta\tau = \sqrt{1 - \dfrac{2GM}{c^2 R_e}}\,\Delta t$, so if we use the gravitational proper time measured by our distant observer as the best substitute for co-ordinate time, we can write $\Delta t = \Delta\tau_{R_z}$ and $\Delta\tau = \sqrt{1 - \dfrac{2GM}{c^2 R_e}}\,\Delta\tau_{R_z}$ making:

$$\frac{\Delta\tau}{\Delta\tau_{R_z}} = \sqrt{1 - \frac{2GM}{c^2 R_e}}$$

hence:

$$\frac{f_{R_z}}{f_r} = \frac{\Delta\tau}{\Delta\tau_{R_z}} = \sqrt{1 - \frac{2GM}{c^2 R_e}} \qquad f_{R_z} = f_r\sqrt{1 - \frac{2GM}{c^2 R_e}}.$$

As $\sqrt{1 - \dfrac{2GM}{c^2 R_e}} < 1$, the distant observer will see a *lower frequency* than the one declared by the observer on the surface.

There is a subtle aspect to this calculation that needs thought. We have just worked out the frequency that the distant observer would see watching the photon being emitted. In principle, *we have not calculated the frequency they will detect when the photon arrives at them*. However, if the co-ordinate time taken for n cycles does not change during the flight of the photon (which can be shown to be the case by a suitable ancillary calculation), the two frequencies will be the same. The co-ordinate frequency $= n/\Delta t$ does not change in flight.

In the context of quantum theory, observations and interactions with photons need more careful consideration than this simple argument, but the essential physics is not changed.

Different observers at different heights will see shifted frequencies due to the relative time dilation between them and the observer at the surface. At some height H above the surface, a third observer would record a frequency $f_1 = n/\Delta \tau_1$. With $n = f_r \Delta \tau$, we see that:

$$f_1 = f_r \frac{\Delta \tau}{\Delta \tau_1} = f_r \sqrt{\frac{\left(1 - \dfrac{2GM}{c^2 R_e}\right)}{\left(1 - \dfrac{2GM}{c^2 (R_e + H)}\right)}}.$$

Now it is interesting to see if this shift in frequencies could be measured over the height of, say, a building. In that case $H \ll R_e$.

First, we apply the $(1 + x)^n \approx 1 + nx$ approximation assuming $\dfrac{2GM}{c^2 R_e} \ll 1$:

$$\sqrt{\left(1 - \frac{2GM}{c^2 R_e}\right)} \approx 1 - \frac{GM}{c^2 R_e} \quad \text{and} \quad \frac{1}{\sqrt{\left(1 - \dfrac{2GM}{c^2 (R_e + H)}\right)}} \approx 1 + \frac{GM}{c^2 (R_e + H)}.$$

So now we have:

$$f_1 \approx f_r \left(1 - \frac{GM}{c^2 R_e}\right)\left(1 + \frac{GM}{c^2 (R_e + H)}\right) \approx f_r \left(1 + \frac{GM}{c^2 (R_e + H)} - \frac{GM}{c^2 R_e}\right)$$

dropping the term in $1/c^4$. Carrying out the subtraction for the fractions:

$$f_1 \approx f_r \left(1 + \frac{GMR_e - GM(R_e + H)}{c^2 (R_e + H) R_e}\right) = f_r \left(1 - \frac{GMH}{c^2 (R_e + H) R_e}\right) \approx f_r \left(1 - \frac{GMH}{c^2 R_e^2}\right),$$

where the last step has used $H \ll R_e$. Next, we substitute $g = GM/R_e^2$ to give us:

$$f_1 \approx f_r \left(1 - \frac{gH}{c^2}\right).$$

This can be more usefully re-arranged by expanding:

$$f_1 - f_r = -f_r \frac{gH}{c^2} \quad \text{or even better} \quad \frac{f_1 - f_r}{f_r} = -\frac{gH}{c^2}.$$

Given a reasonably tall building, say roughly 22.5 m, $\dfrac{gH}{c^2} \approx 2.45 \times 10^{-15}$, a figure that was experimentally confirmed in 1960 (see Section 8.3.4).

8.3.1 Gravitational Frequency Shift from the Equivalence Principle

It's interesting to see how we can get the same equation for frequency shift by using the equivalence principle and the relativistic Doppler effect from Section 5.2.1. To do this, we use a different combination of observers. At the top of the tower, we park two observers. One remains there for the duration of the experiment and the other steps off the tower at the moment that the photon is emitted from the ground. This Tom Cruise-like observer is in free fall, so according to the equivalence principle their observations are immune from the effect of gravity. At the moment the photon is emitted, Tom is momentarily stationary so would observe the emitted frequency to be f_r. While the photon is climbing to the top of the tower, from Tom's point of view the top, with his partner observer, is moving away at increasing speed (the ground is also approaching, but Tom does not worry about such things...). By the time the photon gets to the top of the tower, Tom's speed has reached $g\dfrac{H}{c}$ where g is the acceleration due to gravity. From Tom's perspective the observer at the top of the tower is moving away from him at $v = g\dfrac{H}{c}$, so Tom predicts that this observer will see a Doppler shifted frequency, f_1 where:

$$f_1 = \gamma f_r (1 - \beta)$$

from Section 5.2. As $\gamma = 1/\sqrt{1 - \beta^2} \approx 1$ at this level of approximation, this reduces to:

$$f_1 \approx f_r (1 - \beta) = f_r \left(1 - \frac{v}{c}\right) = f_r \left(1 - \frac{gH}{c^2}\right)$$

as before.

Tom of course is perfectly safe due to the large cushions positioned to arrest his fall.

8.3.2 Quantum Approach to Frequency Shift

As the relationship between energy and frequency for a photon is $E = hf$, if our laser, of mass M_1, stationary at ground level, emits a photon of frequency f_1, the mass of the laser must decrease by an amount $\Delta M_1 = -hf_1$. This photon is then absorbed by another device, of mass M_2, which is also stationary at height H above the laser. If the frequency absorbed is f_2, the increase of mass in the absorber $\Delta M_2 = hf_2$. If we factor in the gravitational potential, ϕ, for the two devices then at the moment just before emission the total energy of the laser plus absorber is:

$$E_1 = M_1 c^2 + M_1 \phi_1 + M_2 c^2 + M_2 \phi_2.$$

After the emission and absorption, we have:

$$E_2 = \left\{ M_1 c^2 - hf_1 + \left(M_1 - \frac{hf_1}{c^2}\right)\phi_1 \right\} + \left\{ M_2 c^2 + hf_2 + \left(M_2 + \frac{hf_2}{c^2}\right)\phi_2 \right\}.$$

As $E_1 = E_2$ (viewing both from the same system):

$$\cancel{M_1 c^2} + \cancel{M_1 \phi_1} + \cancel{M_2 c^2} + \cancel{M_2 \phi_2} = \left\{ \cancel{M_1 c^2} - hf_1 + \left(\cancel{M_1} - \frac{hf_1}{c^2}\right)\phi_1 \right\} + \left\{ \cancel{M_2 c^2} + hf_2 + \left(\cancel{M_2} + \frac{hf_2}{c^2}\right)\phi_2 \right\}$$

$$hf_1 + \frac{hf_1}{c^2}\phi_1 = hf_2 + \frac{hf_2}{c^2}\phi_2.$$

Factorising gives us:

$$hf_1\left(1+\frac{\phi_1}{c^2}\right)=hf_2\left(1+\frac{\phi_2}{c^2}\right) \quad \text{or} \quad \frac{f_1}{f_2}=\frac{\left(1+\dfrac{\phi_2}{c^2}\right)}{\left(1+\dfrac{\phi_1}{c^2}\right)}.$$

Using our $(1+x)^n\approx 1+nx$ approximation, and dropping the term in $1/c^4$:

$$\frac{f_1}{f_2}\approx 1+\frac{\phi_2-\phi_1}{c^2}=1+\frac{gH}{c^2} \quad \text{or} \quad hf_1=hf_2+g\left(\frac{hf_2}{c^2}\right)H.$$

In a subtle way, this derivation relies on the principle of equivalence as the inertial mass of the emitter and receiver has changed due to the photon's emission and absorption bringing about a change in their internal energies and we have assumed an equivalent change in the gravitational mass.

Although nothing in this discussion has mentioned any mass for the photon, the result certainly looks as if the frequency shift has come about due to some of the photon's energy turning into gravitational energy. However, we should not place too much emphasis on this interpretation for a variety of reasons. As we have said before, energy is not that simple in general relativity. Also, we have had to go through a number of approximations to get to this point (not the least of which is the stationary and hence non-relativistic laser and absorber), which restrict the validity of the conclusion. Finally, this gives the impression that the photon is continually losing energy as it climbs, whereas our earlier argument suggests that the frequency difference, and hence the energy difference, really comes from the different perspectives of observers at different heights and the relative time dilations that produces.

8.3.3 Gravitational Redshift

The relationship between frequency and wavelength is the same for any wave:

$$v=c \ (\text{in this case})=f\lambda.$$

Hence, the effect of a gravitating mass is to change the wavelength as if:

$$f_{R_z}=f_r\sqrt{\left(1-\frac{2GM}{c^2R_e}\right)} \quad \text{then} \quad \frac{c}{\lambda_{R_z}}=\frac{c}{\lambda_r}\sqrt{\left(1-\frac{2GM}{c^2R_e}\right)}$$

giving:

$$\lambda_{R_z}=\frac{\lambda_r}{\sqrt{\left(1-\dfrac{2GM}{c^2r}\right)}}.$$

The observed wavelength at $r>R_z$ is *longer* than the emitted wavelength, an effect commonly known as *gravitational redshift*. In Section 5.2.1, we defined redshift as:

$$z=\frac{\Delta\lambda}{\lambda}=\frac{\lambda'-\lambda}{\lambda}=\frac{\lambda_{R_z}-\lambda_r}{\lambda_r}$$

in the context of the relativistic Doppler effect. The redshift due to a gravitating mass is down to time dilation, not a Doppler effect, but applying the same definition we get:

$$z = \frac{1}{\sqrt{\left(1 - \dfrac{2GM}{c^2 r}\right)}} - 1 \approx 1 + \frac{GM}{c^2 r} - 1 = \frac{GM}{c^2 r}.$$

Redshift will take up a crucially significant role when we come to discuss cosmology, as observing the light from distant galaxies and the extent to which it is redshifted provides evidence for the expansion of the universe.

8.3.4 Confirmation of Gravitational Frequency Shift

The first laboratory-based experimental test of general relativity was conducted in 1960 when Robert Pound and his graduate student, Glen A. Rebka Jr., confirmed gravitational time dilation via its effect on electromagnetic waves.

In their experiment, a sample of the γ ray emitting source ^{57}Fe was placed in the centre of a loudspeaker cone near the roof of the Jefferson laboratory at Harvard University. With a 10-Hz signal fed to it, the loudspeaker cone vibrated back and forth, moving the source vertically up and down. A duplicate sample of ^{57}Fe was placed in the basement, the distance between the two being 22.5 m (hence my choice of height before). Scintillating material, which emits visible light when it absorbs γ radiation, was placed under the receiving source to detect waves that had not been absorbed.

Normally, a source emitting electromagnetic radiation will absorb radiation of the same frequency. However, due to gravitational time dilation, the frequency emitted by the source at the top the building is different to that received in the basement.

Measurements were achieved via the moving loudspeaker cone, which introduced an additional Doppler shift into the radiation's frequency. With the cone moving away from the basement, the resulting Doppler shift reduced the frequency; with it moving towards the basement the Doppler shift increased the frequency. At some stage during this cycle, the Doppler shift reduced the frequency by the same amount as the gravitational shift increased it, then the source in the basement absorbed the radiation and the scintillation counts dropped. At other times during the cone's cycle of movement, the Doppler effect was either insufficient to counter the gravitational shift, or acted to increase the total frequency shift. In either case, the scintillations counts were higher during these times.[9] Hence, being able to match the scintillations counts to the movement of the cone enabled them to find the phase of the cone's cycle when the counts where maximum, which in turn gave the resulting Doppler shift and finally the gravitational frequency shift.

The experimental result gave $\dfrac{f_i - f_r}{f_r} = 2.57 \pm 0.26 \times 10^{-15}$, in good agreement with the predicted 2.45×10^{-15} (see the calculation in Section 8.3).

Pound and Synder then improved the precision to 1% in 1964.[10]

The results of a most intriguing experiment were published in 2010 by Holder Müller, Achim Peters and Steven Chu.[11] They exploited the quantum mechanics of particles and used the wave nature of a beam of caesium atoms to create interference effects, similar to those in the Michelson-Morley experiment discussed in Chapter 2. In essence, the caesium atoms were projected horizontally, but some of them were kicked 0.1 mm into a higher path by a laser beam. A second laser then pushed them down to recombine with the others. The team was then able to measure quantum interference effects because the quantum frequency of the atoms in the higher path had been gravitationally shifted. The end result confirmed the general relativistic effect to a precision of 7×10^{-9}. What makes this experiment significant, aside from the much higher precision than any previous result, is the small distance over which gravitational redshift was confirmed, as well as the extension to matter waves rather than the more traditional electromagnetic waves. It is reassuring, although unsurprising, to know that the effect applies to all wave forms.

Impressive as these experimental confirmations are, they still deal with relatively weak gravitational effects. In June 2018, further confirmation of the gravitational frequency shift came from examining the light emitted by a star in orbit around the supermassive black hole (Sgr A*) at the centre of our galaxy. This companion star, known as S2, orbits the black hole with a period of 16 years. In its highly elliptical orbit, S2 approaches to ~1,400 Schwarzschild radii of the black hole at which distance it is travelling at $7,650\,\text{km/s}\,(\approx 0.03c)$, bringing both special and general relativistic effects within the measurement capabilities of modern instruments. Over 26 years, a group of scientists[12] measured the orbital parameters of S2 including during the closest approaches in April 2002 and May 2018. This data confirmed that S2 was orbiting a compact gravitating object with mass $\sim 4 \times 10^6$ solar masses. At closest approach, the light from S2 was affected by the relativistic Doppler effect and the general relativistic frequency shift, with a predicted total shift of $z \approx 6.7 \times 10^{-4}$. The analysis of the data is subtle and complex, but the team employed a parameter, f, to characterise the results with $f = 0$ indicating pure Newtonian physics at work and $f = 1$ the exact general relativistic predictions. The reported value of the parameter from a best first to the data was $f = 0.9 \pm 0.17$.

Notes

1. Given our discussion at the end of Chapter 7, the terminology *weak field limit* is slightly unfortunate, but it is the common parlance...
2. Unless the Christoffel symbols $\sim c$, which would not fit in with our assumption that $\Delta\bar{h}/\Delta x_j$ is very small.
3. Note that the gravitational field strength/acceleration is a Newtonian vector, so covariance/contravariance is not applicable.
4. Strictly, we can only say that $\phi = \frac{1}{2}c^2 h_{\overline{00}} + \text{constant}$. However, if the metric is to become Minkowskian as we move very far from the gravitating mass, then $h_{\overline{00}} \to 0$ as we get further away. The gravitational potential is also defined so that $\phi = 0$ at infinity, hence we can say that for consistency, the constant must be zero.
5. We will discuss event horizons when we turn to black holes in Chapter 10. For the moment think of them as the best estimate of a physical radius for a black hole.
6. The exact physical cause of such jets is still under active research. However, the material is not being emitted directly from the black hole itself, that is just a convenient shorthand expression. It is most likely that accelerating material in orbit around the black hole is the cause of the jet emission.
7. This is known as Birkhoff's theorem and was first proven in 1923 by the mathematician George Birkhoff (1884–1944). The theorem actually states that any spherically symmetric solution to the vacuum field equations can always be transformed into the Schwarzschild metric.
8. Such as: can you place a clock at infinity in a spatially finite universe? How can you observe events from the vantage point of infinity? Etc.
9. Whenever a nucleus emits a γ ray photon it will recoil, which takes some of the photon's energy. In order to prevent this from masking the effect, Pound & Rebka had to prepare their source and detector carefully. In 1958, Rudolf Mössbauer had demonstrated that in a solid lattice at low temperature, the recoil was absorbed by all the atoms in the lattice, hence the emitted photon's energy was largely unaffected. The source that Pound & Rebka actually used was Cobalt-57 fused with iron to create the rigid lattice. This isotope of Cobalt decays by electron capture to produce an excited state of Iron-57, which then decays to give the γ ray. Without exploiting the Mössbauer effect, the expected recoil would have been five orders of magnitude larger than the gravitational effect.

10. Pound, R.V. and Snider J.L., 1964. Effect of gravity on nuclear resonance. *Physical Review Letters*, **13**(18), pp. 539–540.
11. Müller, H., Peters, A. and Chu, S., 2010. *Nature*, **463**, p. 926.
12. *Detection of the Gravitational Redshift in the Orbit of the Star S2 Near the Galactic Centre Massive Black Hole*, Genzel, R. et al., 2018. The gravity collaboration. *Astronomy & Astrophysics*, **615**, p. L15.

9

Space-time in the General Theory

As you see, the war treated me kindly enough, in spite of the heavy gunfire, to allow me to get away from it all and take this walk in the land of your ideas.

K Schwarzschild[1]

9.1 Metrics and Dimensionality

In our 3D+1 space-time, a metric has 16 components as it is a 4×4 tensor. However, as the metric is *symmetrical*, only 10 of these components are independent. In general, for an n-dimensional space, the number of free components in the metric is not n^2 but $\frac{1}{2}n(n+1)$. Think of it like this: the number of components in the metric $= n^2$, with n of them being along the diagonal and hence independent. This leaves $n^2 - n$ components that are not on the diagonal. As the metric is symmetrical, only half of the non-diagonal components are independent, i.e. $\frac{1}{2}(n^2 - n)$. Hence, the total number of independent components in the metric is $= n + \frac{1}{2}(n^2 - n) = \frac{1}{2}n(n+1)$.

The values of the metrical components depend on the co-ordinate system chosen. If you recall, the flat Minkowski metric looks rather different in Cartesian and spherical-polar co-ordinates while expressing exactly the same geometry. As we have complete freedom to choose the co-ordinates system we wish, we can always find co-ordinates in our 3D+1 space-time that make four of the metrical components zero. In general, in n dimensions, we can make n components zero by the right choice of co-ordinate system. This system may be of no practical purpose, but the *idea* illustrates that of the ten components that we have, only six of them actually carry any physics, four are just due to co-ordinate choices (it does not matter *which* six and *which* four). In higher dimensions, the number of free components would be down to $\frac{1}{2}n(n+1) - n = \frac{1}{2}n(n-1)$.

This freedom over co-ordinate choice will help us to extend the weak field calculations from Chapter 8 into a full *linear gravity theory*, when we come to look at gravitational waves in Chapter 11.

Moving on, the energy-density tensor is subject to the condition:

$$\sum_a \frac{D\mathbb{T}_{ab}}{\Delta x_a} = 0.$$

Given the field equations, the same must be true for the Einstein tensor:

$$\sum_a \frac{D\mathbb{G}_{ab}}{\Delta x_a} = 0$$

and this, in turn, presents a set of four equations that have to be satisfied by the Einstein tensor's components. As this tensor depends on the metric, this is tantamount to four equations determining metrical components that must be obeyed in all situations. Hence, the number of *genuinely free* metrical components, with no constraints of symmetry, conservation physics or co-ordinate choice acting on them is 2... Their values come from the field equations.

Interestingly, our groping around looking for a quantum theory of gravity (via linear gravity) has revealed that the hypothetical particle which acts as the granularity of space-time, the *graviton*, has two 'degrees of freedom' associated with it, which matches the 'freedom' in the metric.

9.1.1 Embedding Space and Embedding Diagrams

Embedding is a process by which a space-time is 'encased' into some higher dimensions that have no physical significance, but allow us to show the curvature of the space-time.

In Section 7.2, I talked about a race of flat-physicists living on the 2D surface of a 3D sphere. Their 2D universe is embedded in our 3D space, but the third dimension has no physical meaning for them, as they do not have the dimensionality to access it. However, it allows us to visualise the curvature of their universe.

You might think that you can embed by using just one more dimension than that of the space-time, so we could embed our 3D+1 universe into five dimensions. However, it is not at all that simple.

As we argued earlier, in n dimensions we have $\frac{1}{2}n(n-1)$ free metrical components. If we wish to embed this geometry, we are going to need N dimensions in total, n being needed for the space-time leaving $N-n$ to display the curvature. In our flat-physicists case, the co-ordinates working on the surface of the space (s_1, s_2) need to be mapped into (x, y) out of the embedding (x, y, z) space, leaving z free to be the direction that shows how the surface is curved. The upshot is that $N-n \geq \frac{1}{2}n(n-1)$ which, when you re-arrange, gives us $N \geq \frac{1}{2}n(n+1)$. To fully display the curvature of 3D+1 space we would need to embed into $N = \frac{1}{2} \times 4 \times (4+1) = 10$ dimensions *at least*. Even if we abandoned that and decided to display the curvature of 3-space at a moment in time, we would still need $N = \frac{1}{2} \times 3 \times (3+1) = 6$ dimensions, which would still be something of a struggle. Dropping our ambitions further and trying a 2D slice through space-time at a fixed moment in time, we would need $N = \frac{1}{2} \times 2 \times (2+1) = 3$ dimensions: that we can work with!

This explains why books show 2D slices of space-time for illustrations and why the 'rubber sheet' analogy for mass causing space-time curvature has become so popular, even if it can be somewhat misleading (Figure 9.1).

By the way, implicit in all this has been the assumption that the N-space has a flat geometry, but as the N-space is our choice, that does not represent a constraint.

9.1.2 The Embedding Diagram for Schwarzschild Space-time

Our aim is to illustrate the spatial curvature of the Schwarzschild metric.

Given the limitations of what is possible, we have to work with a snapshot at a fixed moment in co-ordinate time, so that $\Delta t = 0$. For convenience we will also take a slice through the equatorial plane, which corresponds to $\theta = \pi/2$ and hence $\Delta\theta = 0$ (see Figure 9.2).

With these constraints in place, the interval becomes:

$$(\Delta s)^2 = -\frac{(\Delta r)^2}{\left(1 - \dfrac{2GM}{c^2 r}\right)} - r^2 \sin^2\left(\frac{\pi}{2}\right)(\Delta\varphi)^2 = -\left(\frac{(\Delta r)^2}{\left(1 - \dfrac{2GM}{c^2 r}\right)} + r^2 (\Delta\varphi)^2\right).$$

FIGURE 9.1 Why the rubber sheet analogy for space-time curvature can be misleading. (Image credit: Randall Munroe, http://www.xkcd.com/.)

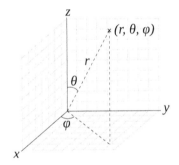

FIGURE 9.2 The spherical-polar co-ordinate system used in the Schwarzschild metric.

As this is a space-like interval, it is more convenient to set:

$$\left(\Delta\sigma\right)^2 = \left|\left(\Delta s\right)^2\right| = \frac{\left(\Delta r\right)^2}{\left(1 - \dfrac{2GM}{c^2 r}\right)} + r^2 \left(\Delta\varphi\right)^2.$$

Our embedding space is going to be 3D, and to make the co-ordinate translation easier, we use a system of *cylindrical co-ordinates* (r, θ, z) (Figure 9.3) in that space:

In such a system, the interval is as follows:

$$\left(\Delta s\right)^2 = \left(\Delta r\right)^2 + \left(\Delta z\right)^2 + r^2 \left(\Delta\varphi\right)^2$$

so we can map the (r, φ) co-ordinates from the Schwarzschild space directly into the similar co-ordinates for the embedding space.

If we are going to set up a 2D surface that has the same geometry as our fixed co-ordinate time, equatorial snapshot of the Schwarzschild space-time, we need to find a relationship $z = z(r, \varphi)$ which determines that surface. As the Schwarzschild metric is spherically symmetrical, we can reduce our surface's

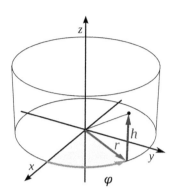

FIGURE 9.3 A cylindrical co-ordinate system to help with transferring our space-time interval into the embedding space. On this diagram, r and φ are to be mapped over from our space-time and z is the non-physical direction of the curvature. Here h is a specific z value. (Image credit: Svjo via Wikimedia commons.)

equation to $z = z(r)$. If we knew what this relationship was, we could use it to calculate $\Delta z / \Delta r$ which we could then insert into the interval:

$$(\Delta s)^2 = \left(1 + \left(\frac{\Delta z}{\Delta r}\right)^2\right)(\Delta r)^2 + r^2(\Delta \varphi)^2.$$

Comparing this to the Schwarzschild snapshot interval:

$$\left|(\Delta s)^2\right| = \frac{(\Delta r)^2}{\left(1 - \dfrac{2GM}{c^2 r}\right)} + r^2(\Delta \varphi)^2$$

shows us that:

$$1 + \left(\frac{\Delta z}{\Delta r}\right)^2 = \frac{1}{\left(1 - \dfrac{2GM}{c^2 r}\right)}.$$

This is an example of a particular type of equation (a *differential equation*[2]) that comes up often in physics. In this case, the equation has the solution:

$$z = \sqrt{\frac{8GM}{c^2}\left(r - \frac{2GM}{c^2}\right)} \quad \text{for} \quad r > \frac{2GM}{c^2}$$

which looks like Figure 9.4.

This 'goblet-shaped' structure is often used to represent a gravitating mass curving space-time. However, as we have been stressing, it needs careful consideration otherwise it can be misleading.

What would be the experience of an observer moving across this surface? How would the spatial curvature make itself known to them? Well first, this diagram shows the spatial curvature at a moment in co-ordinate time, so we can't really use it to describe an observer *moving* across the surface. In any case, elapsed time for the observer would be their proper time and hence related to how fast they are moving. What we can do is compare the experience of different observers posted at different locations across the surface. Importantly, an observer sitting near to the Schwarzschild radius does not feel 'lower' in any

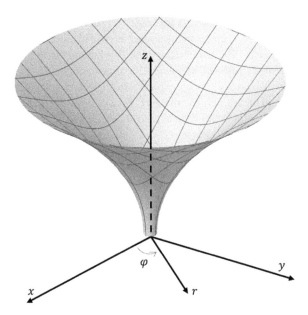

FIGURE 9.4 The embedding diagram for a single co-ordinate time, equatorial slice through Schwarzschild space-time for *r* > Schwarzschild radius. The lines across the curved surface simply act as 'eye lines' to help demonstrate the shape. They do not refer to co-ordinate lines. The Schwarzschild radius would be around the 'throat' at the base of the goblet's stem.

sense than another observer further away. On this diagram the 'height' of the observer is measured in the *z* direction, and this is a non-physical co-ordinate.

A collection of observers posted around a circle in the equatorial plane of the gravitating mass and all at the same co-ordinate distance, r_1, from the centre of the mass would measure the circumference of that circle to be $2\pi r_1$. This follows from the interval:

$$\left|(\Delta s)^2\right| = \frac{(\Delta r_1)^2}{\left(1 - \dfrac{2GM}{c^2 r_1}\right)} + r_1^2 (\Delta \varphi)^2$$

which reduces to:

$$\left|(\Delta s)^2\right| = r_1^2 (\Delta \varphi)^2$$

as $\Delta r = 0$. Moving around a closed loop of constant r_1 sweeps out a total angle $\varphi = 2\pi$ so as we add up little intervals around the loop we end up with:

$$\Sigma^2 = \sum_i (\Delta s_i)^2 = \sum_i r_1^2 (\Delta \varphi_i)^2 = 4\pi^2 r_1^2 \quad \text{or} \quad \Sigma = 2\pi r_1.$$

This, by the way, is how observers can determine their co-ordinate radius. All they need do is measure the circumference of this circle (or the sphere in the full 3D space) and divide by 2π. However, the co-ordinate radius is very different to the physical radius.

A collection of observers posted *radially* could not extend right to the centre of the mass: even if we discount their inability to stand inside a dense material, once they get within the region of the gravitating object, the Schwarzschild metric is no longer appropriate, as the vacuum equations are not applicable. We can post radial observers from the circumference of one circle inwards to that of another at a smaller co-ordinate radius r_2. The circumference of this second circle is $2\pi r_2$, so the difference in circumference between the two circles is $2\pi(r_2 - r_1)$. One immediately jumps to the conclusion that the radial distance between the two circles is $r_2 - r_1$, which is indeed the *co-ordinate distance* between them, but not the *physical distance*. This is where the metric comes in. Indeed, the reason that these objects are called *metrics*[3] is that they relate the non-physical co-ordinate distances, times and angles to the physical measurements of distance, time and angle.

The best way to understand how this difference comes about is to see the embedded surface in cross section, as in Figure 9.5.

In this graph, the co-ordinate distance between radial locations equal to 10 Schwarzschild radii and 15 Schwarzschild radii are shown. However, a collection of observers along a radial direction would connect up measuring rods from one to another and come up with a distance between them equal to the length of the *curved part of the surface*. Again, this does not mean that they would 'see' curvature – as if one of them were 'lower down' in the goblet from the other.

The physical distance, or *proper distance*, between the two radii has to be calculated carefully as we can't simply use the relationship:

$$(\Delta s)^2 = -\frac{(\Delta r_1)^2}{\left(1 - \dfrac{2GM}{c^2 r_1}\right)} \quad \text{or} \quad \Delta\sigma = \frac{\Delta r_1}{\sqrt{1 - \dfrac{2GM}{c^2 r_1}}}.$$

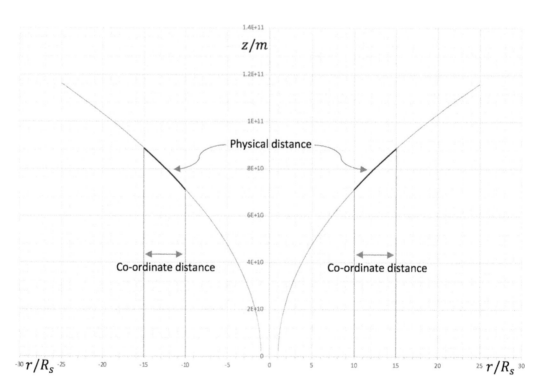

FIGURE 9.5 A vertical cross section through the embedding surface for the Schwarzschild metric. In this case, all the calculations have been made for the black hole Sgr A*. The radial distance, r, is measured in multiples of the Schwarzschild radius for that black hole (negative values of r simply indicate pointing to the left). The vertical height, z, is in metres, but remember that this is not a physical distance in the actual space-time.

The problem stems from the $2GM/c^2r_1$ part of the formula, which is changing as r_1 changes. As long as Δr_1 is small, we can assume that $2GM/c^2r_1$ stays reasonably constant between r_1 and $r_1 + \Delta r_1$, which is why the formula works for small (infinitesimal) distances. In order to calculate a real, macroscopic, distance we need to break it up into a collection of small intervals and add them together. Inside each small interval, we assume that $2GM/c^2r_n$ is constant and then jump to the next value in the next interval. The procedure looks something like this:

$$\Sigma \approx \sum_n \Delta\sigma_n = \sum_n \left\{ \frac{\Delta r_n}{\sqrt{1 - \dfrac{2GM}{c^2 r_n}}} \right\}.$$

The smaller we make the intervals, the more accurate the result will be, but the more of them we have to add up. The full mathematical procedure involves doing this again and again for smaller and smaller intervals and longer and longer sums. If the whole calculation is 'well behaved', then the results get closer and closer to a value, and that value is taken as the actual distance in this case. The whole process is called *integration*. Over the centuries, mathematicians who enjoy this sort of thing have calculated lists of standard integrals, so all we have to do in many cases is to look up the standard result, or to spend time massaging the problem until it looks sufficiently like a standard result to read off the answer. On-line calculation tools have made this task much easier. In its correct mathematical form, our calculation looks like this:

$$\Sigma = \int_{r_1}^{r_2} \frac{dr}{\sqrt{1 - \dfrac{2GM}{c^2 r}}} = \int_{r_1}^{r_2} \frac{dr}{\sqrt{1 - \dfrac{R_s}{r}}}$$

which is a standard integral, but unfortunately, the result is rather horrid:

$$\Sigma = \left[\sqrt{r(r - R_s)} + \frac{1}{2} R_s \log\left(2\sqrt{r(r - R_s)} + 2r - R_s \right) \right]_{r_1}^{r_2}.$$

Using this formula to calculate the physical distance between $r_1 = 10R_s$ and $r_2 = 15R_s$ you get $\Sigma \approx 5.2R_s$, rather than $r_2 - r_1 = 5R_s$. That may not sound like much of a difference, but don't forget that having a difference *at all* means that something is up with space, but also as $0.2R_s \approx 2.4 \times 10^6$ km, that's actually quite a big physical difference.

Most dramatically, the physical circumference of a circle centred on the gravitating mass is less than 2π times the physical radius. The geometry of space is 'wrong'. However, the observers would still this as an equatorial *plane*. They would not *see* curvature. The Schwarzschild geometry shows in the radial scale being distorted. It is hard to visualise this, still less draw it. Figure 9.6 shows concentric circles at $2R_s, 3R_s, 4R_s, 5R_s$, labelled by their proper distances from R_s on the radial scale. Hence, a circle of co-ordinate radius $2R_s$ is in fact a proper distance of $1.3R_s$ from the Schwarzschild radius. The circles have the correct scale circumferences for their co-ordinate radii.

As we might expect, as the metric becomes Minkowskian when we are a long distance from the gravitating mass, the difference between the proper distance and the co-ordinate distance diminishes when $r_1, r_2 \gg R_s$.

9.1.3 The Ghost of Zeno

So far, our treatment of the Schwarzschild radius has been limited to its definition:

$$R_s = \frac{2GM}{c^2}$$

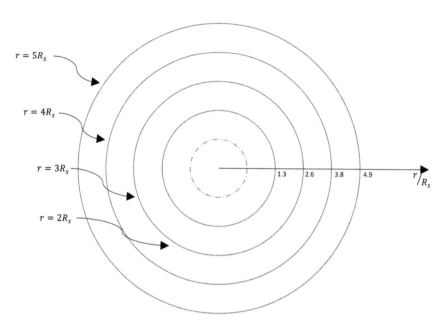

FIGURE 9.6 The curvature of the equatorial Schwarzschild plane shows up in the radial scale. The proper distances from the Schwarzschild radius are shown on the *r* scale as well as labels for each circle depending on their co-ordinate radius.

and its use as a distance scale for our calculations, to give a more tractable set of values to include on graphs, etc. Yet, it is clear that there is rather more to this specific radius, as evidenced by the behaviour of the space-time interval for the Schwarzschild metric:

$$\left(\Delta s\right)^2 = \left(1 - \frac{R_s}{r}\right)c^2\left(\Delta t\right)^2 - \frac{\left(\Delta r\right)^2}{\left(1 - \frac{R_s}{r}\right)} - r^2\left(\Delta\theta\right)^2 - r^2\sin^2\left(\theta\right)\left(\Delta\varphi\right)^2.$$

If $r = R_s$ then the first term in the interval vanishes. This would imply that an observer sitting at $r = R_s$ would experience no passage of proper time, even as co-ordinate time ticked away. That might be palatable as an extreme form of time dilation, but even worse things happen in the second term, which appears to be infinite! Strictly speaking, we have to write:

$$\frac{\left(\Delta r\right)^2}{\left(1 - \frac{R_s}{r}\right)} \to \infty \quad \text{as} \quad r \to R_s.$$

In other words, intervals of physical distance climb without limit as *r* approaches the Schwarzschild radius. You can also see this in the increasing steepness of the curve in Figure 9.5. If we are too brisk with our conclusions, we might think that this implies an infinite distance between any observer and the Schwarzschild radius.

From the point of view of someone approaching the Schwarzschild radius, they move from one collection of observers to another. Each observer measures the incremental distance to the next observer, and the one after does the same. Consequently, to get the distance covered by the person approaching the Schwarzschild radius, we need to add up all the incremental distances from one observer to the next. This is what we did by integration in the previous section, ending up with the result:

$$\Sigma = \left[\sqrt{r(r-R_s)} + \frac{1}{2} R_s \log \left(2\sqrt{r(r-R_s)} + 2r - R_s \right) \right]_{r_1}^{r_2}$$

which is finite, even when $r_1 = R_s$:

$$\Sigma = \left(\begin{array}{l} \sqrt{r_2(r_2-R_s)} + \frac{1}{2} R_s \log \left(2\sqrt{r_2(r_2-R_s)} + 2r_2 - R_s \right) - \cancel{\sqrt{R_s(R_s-R_s)}} \\ - \frac{1}{2} R_s \log \left(2\cancel{\sqrt{R_s(R_s-R_s)}} + 2R_s - R_s \right) \end{array} \right),$$

so that:

$$\Sigma = \sqrt{r_2(r_2-R_s)} + \frac{1}{2} R_s \log \left(2\sqrt{r_2(r_2-R_s)} + 2r_2 - R_s \right) - \frac{1}{2} R_s \log(R_s).$$

How does this square with the tendency of the radial co-ordinate to 'blow up' at $r = R_s$?

The apparent paradox is reminiscent of sort of argument that Zeno (495–430 BC) used to discuss: in order to cross a certain distance, you must first travel half way, and then you need to travel half the next distance, and then the next ... ad infinitum. As there is always half the remaining distance still to go, you can never actually cross any distance. The resolution is the same: Zeno did not know about integration, nor that infinite sums can add to finite values.

There is much more to say about the Schwarzschild radius and what happens when you approach and then cross this rubicon, but first we have to develop a more detailed understanding about space-time in the context of the general theory.

9.2 Space-time Intervals in the General Theory

So far, we have used at least two versions of the space-time interval: that related to the Minkowski metric (in Cartesian co-ordinates):

$$(\Delta s)^2 = c^2 (\Delta t)^2 - (\Delta x)^2 - (\Delta y)^2 - (\Delta z)^2$$

and that for the Schwarzschild metric:

$$(\Delta s)^2 = \left(1 - \frac{2GM}{c^2 r} \right) c^2 (\Delta t)^2 - \frac{(\Delta r)^2}{\left(1 - \frac{2GM}{c^2 r} \right)} - r^2 (\Delta \theta)^2 - r^2 \sin^2 (\theta)(\Delta \varphi)^2$$

both of which are examples of the general space-time interval for any metric $g_{\overline{ab}}$:

$$(\Delta s)^2 = \sum_{a,b} g_{\overline{ab}} \Delta x_a \Delta x_b.$$

As with Minkowskian intervals, these general intervals are:

- time-like if $(\Delta s)^2 > 0$ (followed by sub-light particles)
- space-like if $(\Delta s)^2 < 0$ (non-physical)
- null if $(\Delta s)^2 = 0$ (followed by light rays).

Time-like intervals are crucial as they relate to well-defined proper times. Any path that is a solution of the acceleration equation:

$$\frac{\Delta}{\Delta \tau}\left(\frac{\Delta x_k}{\Delta \tau}\right) = -\sum_i \sum_j U_i U_j \Gamma_{ij}^k$$

will track a time-like interval. By definition and common sense, any event on such a path will be causally linked to all other events on the same path, so they are possible physical paths of particles. Space-like intervals would require a particle to travel faster than light in order to cross the spatial interval in the time allowed, so they are clearly not physical.

9.2.1 Geodesics

Oftentimes, a set of co-ordinates can be written in terms of a single variable known as a *parameter*, so that:

$$x_0 = x_0(\lambda), x_1 = x_1(\lambda), x_2 = x_2(\lambda), x_3 = x_3(\lambda).$$

For example, in classical mechanics, if we launched a particle with speed v at an angle ϑ to the horizontal, it will follow a path $y(x) = vx\tan(\vartheta) - \dfrac{gx^2}{2(v\cos\vartheta)^2}$. This can be parameterised in terms of the time elapsed since we launched the particle, t:

$$y = y(t) = vt\sin(\vartheta) - \frac{1}{2}gt^2$$

$$x = x(t) = vt\cos(\vartheta).$$

When the co-ordinates refer to events on the path of an object, the most obvious parameter is the proper time of the object. However, other parameters are often possible, for example, the proper distance along the path or any linear function of the proper time will do. However, the parameter *does not have to be a physical quantity at all*, as long as it uniquely determines co-ordinates.

Consider some world line C along which the co-ordinates are parameterised by α so that $x_i = x_i(\alpha)$. The tangent to this curve, T has components:

$$T_i(\alpha) = \frac{\Delta x_i(\alpha)}{\Delta \alpha} = \sum_j \left(\frac{\Delta x_i(\alpha)}{\Delta x_j}\right)\frac{\Delta x_j}{\Delta \alpha}.$$

Now, in a perfectly flat space, the tangent to a straight line at any point is parallel to the tangent at any other point, but that is not necessarily the case if the path is curved or the geometry is not flat (Figure 9.7). In flat space, the condition for a straight line can be expressed mathematically as:

$$T_i(\alpha + \Delta\alpha) = k(\alpha)T_i(\alpha)$$

i.e. the tangent at $\alpha + \Delta\alpha$ is proportional (and so parallel) to the tangent at α.

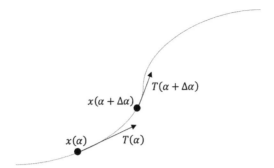

FIGURE 9.7 The tangent to a curve at two nearby points. A straight line, or geodesic, is best described as a curve where the tangent at any point is parallel to that at another point.

In a curved geometry, the only way to compare vectors from two points (and the tangent is simply a vector) is to parallel transport one back to the location of the other. So, in this terminology we must have (using Equation 7.1):

$$\frac{D\big(T_i(\alpha)\big)}{\Delta\alpha}\Delta\alpha = \sum_k \frac{D\big(T_i(\alpha)\big)}{\Delta x_k}\frac{\Delta x_k}{\Delta\alpha}\Delta\alpha = \sum_k \left\{\frac{\Delta T_i}{\Delta x_k} + \sum_m T_m\Gamma^i_{mk}\right\}\frac{\Delta x_k}{\Delta\alpha}\Delta\alpha$$

$$= \left(\frac{\Delta T_i}{\Delta\alpha}\right)\Delta\alpha + \sum_k\sum_m T_m\Gamma^i_{mk}\left(\frac{\Delta x_k}{\Delta\alpha}\right)\Delta\alpha$$

as the difference between a tangent vector and a parallel transported version of another. Applying the same definition of straightness, that the transported vector should be parallel (proportional) to the original vector, we have:

$$\left(\frac{\Delta T_i}{\Delta\alpha}\right)\Delta\alpha + \sum_k\sum_m T_m\Gamma^i_{mk}\left(\frac{\Delta x_k}{\Delta\alpha}\right)\Delta\alpha = T_i(\alpha+\Delta\alpha) - T_i(\alpha) = \{k(\alpha)-1\}T_i(\alpha).$$

If we search for a parameter, $\alpha = \lambda$, that has $k(\lambda)=1$, then this equation becomes:

$$\left\{\left(\frac{\Delta T_i}{\Delta\lambda}\right) + \sum_k\sum_m T_m\Gamma^i_{mk}\left(\frac{\Delta x_k}{\Delta\lambda}\right)\right\}\Delta\lambda = 0$$

and as the equation has to be true for all possible $\Delta\lambda$, this reduces to:

$$\left(\frac{\Delta T_i}{\Delta\lambda}\right) + \sum_k\sum_m T_m\Gamma^i_{mk}\left(\frac{\Delta x_k}{\Delta\lambda}\right) = 0.$$

Remembering that $T_i(\lambda) = \Delta x_i(\lambda)/\Delta\lambda$ gives:

$$\frac{\Delta}{\Delta\lambda}\left(\frac{\Delta x_i}{\Delta\lambda}\right) + \sum_k\sum_m \frac{\Delta x_m}{\Delta\lambda}\frac{\Delta x_k}{\Delta\lambda}\Gamma^i_{mk} = 0. \tag{9.1}$$

If we pull up the acceleration equation from Section 7.4:

$$\frac{\Delta}{\Delta\tau}\left(\frac{\Delta x_k}{\Delta\tau}\right)+\sum_i\sum_j U_i U_j \Gamma_{ij}^k = \frac{\Delta}{\Delta\tau}\left(\frac{\Delta x_k}{\Delta\tau}\right)+\sum_i\sum_j \frac{\Delta x_i}{\Delta\tau}\frac{\Delta x_j}{\Delta\tau}\Gamma_{ij}^k = 0$$

we can see the similarity. Our new equation (Equation 9.1) becomes the older one if $\lambda = \tau$. Solutions to Equation 9.1 are referred to as *geodesics*, and they represent the straightest lines possible in a curved space.

There is a physically intuitive definition of a geodesic that is more rigorous than 'the straightest line possible in a curved geometry', namely that a geodesic is a line where the tangent at any point is parallel to the tangent at any other point. Furthermore, if the geodesic is time-like, then it can be parameterised by proper time, and the tangent becomes the 4-velocity. In this case, the definition is equivalent to saying that *the 4-velocity remains constant along a geodesic*. However, remember that the definition of 'constant' is wrapped up with the notion of parallel transport in curved space-time, so the action of gravity is folded into this. Any curvature apparent in a geodesic is down to the geometry of the space-time, and is not characterised as an acceleration (rate of change of 4-velocity). The restriction to $k(\lambda)=1$ ensures that the length of the tangent is the same at each point.[4]

Back in Section 2.2.7, I defined *inertial systems* as systems of co-ordinates that are not in relative accelerated motion. In the context of the general theory, *any object following a geodesic is classed as undergoing inertial motion*.

9.2.2 Geodesic Deviation

In our discussion of the equivalence principle, back in Section 6.3.1, we noted that in practice it is quite easy to distinguish uniform acceleration from the effect of a gravitating mass, as the mass would pull objects radially towards its centre. This is why the equivalence principle is best expressed as a local equivalence.

In Newtonian physics, phenomena that come about due to any non-uniformity in the gravitational field are known as *tidal effects*, the tides we experience with the Earth's oceans being one example. In that case, the Moon's pull on the water at the side of the Earth facing the Moon is greater than the pull on the water at the opposite side of the Earth. Combined with the influence of the Earth's own gravity, this causes the distortions in the surface of the water that lead to the regular tides.

In the general theory, the notion of tidal effects needs to be incorporated, but the language of forces is inappropriate. Instead, we consider *geodesic deviation*.

In Figure 9.8, two geodesics have been drawn passing close to each other through the same region of space-time. I have assumed that both of them can be parameterised by the same variable, λ.

(Note that we are not assuming that these are time-like geodesics.) A specific choice for the value of λ will uniquely pinpoint co-ordinates on each geodesic (as shown in the diagram) and allow us to connect the points with a vector, ξ, which has components:

$$\xi_i(\lambda)= x_i^D(\lambda)- x_i^C(\lambda).$$

If this is being done in a flat region of space-time, then the two geodesics will be parallel straight lines, in which case ξ will be constant in length. However, in curved space-time, geodesics which might be parallel in one region can diverge in another, in which case ξ will not be constant. It then becomes interesting to inspect how ξ changes for different values of λ, which will tell us how the geodesics converge or diverge across space-time. This is the basis for tidal effects in the general theory.

The appropriate expression for the rate of change of ξ with λ in general co-ordinates is:

$$\frac{D\xi_i(\lambda)}{\Delta\lambda}=\frac{\Delta\xi_i(\lambda)}{\Delta\lambda}+\sum_{j,k}\Gamma_{j,k}^i\xi_j(\lambda)\frac{\Delta x_k}{\Delta\lambda}.$$

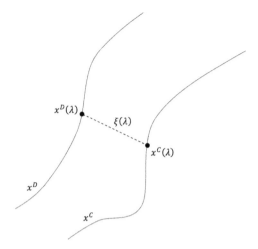

FIGURE 9.8 Geodesic deviation.

However, this equation is of limited usefulness, partly because Christoffel symbols are co-ordinate system dependent, but also as it does not pin down how geodesics deviate. Consider the situation of parallel lines in flat space. In that instance, ξ is constant in size *but changes with λ* as it moves from point to point along the geodesics (Figure 9.9).

With a flat space like this, ξ has a constant rate of change with λ, but that is *not* geodesic deviation. Hence, $D\xi_i(\lambda)/\Delta\lambda$ is not telling us what we really want to know.

However, if the geodesics diverge as in a curved space (Figure 9.10), then ξ has a rate of change of rate of change with λ.

So, we need the second rate of change of ξ (i.e. the rate of change of the rate of change) in order to expose true geodesic deviation. After a calculation similar to the derivation of the Riemann tensor, we get:

$$\frac{D}{\Delta\lambda}\left(\frac{D\xi_i(\lambda)}{\Delta\lambda}\right) = -\sum_{j,k,l} R_{i\overline{jkl}}\xi_l(\lambda)\frac{\Delta x_j}{\Delta\lambda}\frac{\Delta x_k}{\Delta\lambda} \tag{9.2}$$

which is co-ordinate system independent, as tensor properties are independent of the co-ordinate system chosen. This is a (complicated) equation that can be solved for $\xi_i(\lambda)$ (presumably by some painful algebraic or numerical method) and hence tells us how geodesics converge or diverge in a region of

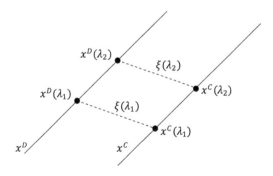

FIGURE 9.9 Two parallel geodesics in flat space-time. In this case, ξ is constant in size and direction. However, it does change with λ: it changes location. Effectively it slides along between the geodesics as λ changes.

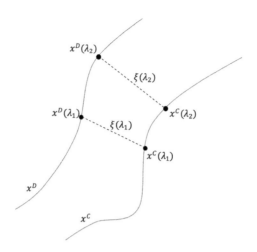

FIGURE 9.10   Geodesics in curved space. Here ξ is changing in both size and direction as we sweep through λ, which gives it a rate of change of rate of change, something like an 'acceleration'.

space-time described by the Riemann tensor. It is interesting to see that the whole tensor is involved, not just the Ricci tensor. This is a clear indication that geodesic paths are not simply and solely determined by the density of matter/energy in the vicinity: the curvature of space-time also depends on the curvature of space-time in the region, as expressed by the (still slightly mysterious to us) Weyl tensor.

Equation 9.2 provides an alternative physical interpretation of the Riemann tensor. We defined it in Chapter 7 in terms of the change in a vector that has been parallel transported around a closed loop. Here we see that it governs the deviation of two geodesics mapped through space-time. Clearly if the space-time is flat, $R_{i\overline{jkl}} = 0$ and the two geodesics run parallel to each other.

There is an interesting consequence of this. When Einstein first set down his principle of equivalence, he compared acceleration with the presence of a uniform gravitational field. In such a (Newtonian) field, if we released two particles they would fall vertically, but parallel to each other: they would not converge as they would if we dropped them on a planet. This implies that their geodesics do not converge. However, according to Equation 9.2, that can only happen if $R_{i\overline{jkl}} = 0$, which is generally interpreted as signalling the absence of any gravitational field (if anything plays that role in the general theory). As a result, there is no genuinely uniform gravitation in the general theory! Gravitation is synonymous with tidal effects, of some form.

9.2.3 The Ricci Tensor Rides Again

Following on from the role of the Riemann tensor in geodesic deviation, we can progress to an interesting relationship involving the Ricci tensor.

In a flat space-time, an infinitesimal volume element takes the form:

$$\delta V = c\Delta t \Delta x \Delta y \Delta z.$$

In general co-ordinates, the intuitive thing to do would be to replace this with:

$$\delta V = \sqrt{|g_{00}|}\Delta x_0 \sqrt{|g_{11}|}\Delta x_1 \sqrt{|g_{22}|}\Delta x_2 \sqrt{|g_{33}|}\Delta x_3$$

which is sometimes written as:

$$\delta V = \sqrt{|g|}\Delta x_0 \Delta x_1 \Delta x_2 \Delta x_3.$$

However, this is only true if the co-ordinate axes at all at right angles to each other. In the general case, the volume element is a bit more complicated. If we just assume that the volume element can be calculated in some fashion, then it becomes interesting to see how this element changes with geodesic motion. Physically, we can think of the volume as being filled with dust particles following geodesic paths, and then see how the volume varies over proper time (or how it evolves with any parameter). In order to do this, we need the rate of change of the rate of change of the volume:

$$\frac{D}{\Delta\tau}\left(\frac{D}{\Delta\tau}(\delta V)\right)$$

for the same reason as with geodesic deviation – it evolves over proper time in flat space, just by simply changing position.

Calculating this formally would involve introducing chunks of mathematics that we would only need for this specific task. So, I will simply quote the result:

$$\frac{D}{\Delta\tau}\left(\frac{D}{\Delta\tau}(\delta V)\right) = -\delta V \sum_{i,j} R_{ij}\frac{\Delta x_i}{\Delta\tau}\frac{\Delta x_j}{\Delta\tau} + FT,$$

where $x_i(\tau)$ marks the centre of the volume element and FT is the term that governs how volumes change in flat space-times. It might be surprising to see the Ricci tensor pop up in this context, but consider the geodesic deviation (using the proper time as the parameter):

$$\frac{D}{\Delta\tau}\left(\frac{D\xi_i(\tau)}{\Delta\tau}\right) = -\sum_{j,k,l} R_{i\overline{jkl}}\xi_l(\tau)\frac{\Delta x_j}{\Delta\tau}\frac{\Delta x_k}{\Delta\tau}.$$

If we interpret $\xi_l(\tau)$ as being one edge, Δx_l, of the volume element, δV, then geodesic deviation of that edge will bring about volume change. We can then see how the calculation would involve summing over all the edges, and that symmetry will force $i = k$ at some point along the way, collapsing the Riemann tensor into the Ricci tensor.

The upshot is a nice physical interpretation of the *Ricci tensor as controlling how the volume of an element evolves during geodesic flow.* How the shape distorts is controlled by the Weyl tensor, which is something that we will discuss in Chapter 11 when we get to gravity waves.

9.2.4 Light Cones in the General Theory

In its more general form, with a parameter λ, rather than proper time, the geodesic equation can be used to find the path followed by a photon moving through a curved space-time. As the photon is travelling at the speed of light, it has no elapsed proper time on its journey, its generalised space-time interval is null and the path it follows can't be parameterised by anything to do with proper time. However, other parameters are possible,[5] so this more general equation can be solved.

In Section 4.3.4, we introduced the light cone as a boundary between space-like and time-like world lines, in particular that the causal future, or past, of an event had to lie within the span of a light cone originating at that event. The edges of the light cone are drawn out by a hypothetical sphere of light originating at the event, or contracting in to the event. Hence, they are the null geodesics of individual particles of light.

In Minkowski diagrams of ct vs. x, the edges of the light cones are always at 45°, as the space-time is flat. In a curved space-time, the angle is not constrained in the same way, nor is the direction in which the light cone is facing. We will shortly come across a space-time where the cones can tip over. Indeed inside the Schwarzschild radius of a gravitating mass, the light cones are such that the *singularity* (to be defined in Section 10.2) lies in the future of all objects falling through that region of space-time.

9.2.5 Exotic Geometries

In 1949, the Princeton mathematician and close friend of Einstein's, Kurt Gödel, derived a contrived but fascinating solution to the field equations.[6] Gödel was impressed with Einstein's theory, but set out to demonstrate that it was not consistent with traditional philosophical views regarding time. Using the modified field equations that Einstein had promoted in 1917 and then subsequently abandoned after the discovery of the expanding universe, Gödel set up a model universe, where the matter comprised a uniform distribution of swirling dust particles, with density $\rho = 1/8\pi G a^2$ and with a cosmological constant $\Lambda = -1/2a^2$. The resulting metric generates the space-time interval:

$$(\Delta s)^2 = a^2 \left\{ (\Delta x_0)^2 - (\Delta x_1)^2 + \frac{1}{2}e^{2x_1}(\Delta x_2)^2 - (\Delta x_3)^2 + 2e^{x_1}\Delta x_0 \Delta x_2 \right\}$$

which is unusual in that it mixes spatial and temporal terms in a manner that we have not seen before. The geometry of this universe has several interesting, if not bizarre, features:

- It is singularity free: singularities are the banes of the general theory; they are predicted and unavoidable points of space-time where the curvature becomes infinite and the laws of physics as we understand them consequently break down. We will have much to say about them in Chapter 10. The lack of singularities in this model universe inclines us to trust the mathematics involved.
- Any observer within the Gödel universe would observe the matter to be rotating about an axis through their position with angular velocity $1/\sqrt{2}a$.
- Any two events in the universe can be connected with a *closed time-like world line*.

This last point is more of a bomb-shell that it might at first appear. A closed world line is one that loops around and returns to its starting event. A time-like world line has a well-defined proper time: every event on the world line lies within the light cone of an earlier event. So, a closed time-like world line implies that a particle, steered appropriately, continually advances into its proper time future, but ends up at the space-time event where it started (or before). The particle would loop back through an external observer's time while continually 'ageing' through its proper time (Figure 9.11).

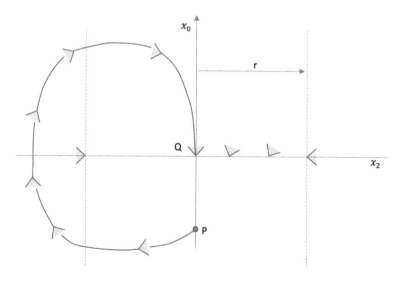

FIGURE 9.11 In a Gödel universe, light cones tip over the further they are from an event. At a critical radius, r, the cones become parallel to the distance axis, and beyond that they tip below the axis. Hence, it is possible to find a closed time-like curve that loops from event Q round to event P and hence back to Q again.

A detailed analysis shows that it is very hard to define *any* global measure of time in this universe and local time measures are troublesome as well. Coupled with the lack of any expansion or contraction of the space-time in this Gödel universe, a better interpretation might be that this universe, while being four dimensional, does not contain a dimension we would think of as time at all. That, in itself, throws up fascinating philosophical and physical questions about the nature of time, if Einstein's theory allows up a physically possible, if highly unlikely, solution which is 'timeless'.

More recent work, notably by Stephen Hawking, George Ellis and Kip Thorne, among others, has explored time travel and closed time-like curves in the context of *worm holes*, which are themselves rather exotic objects within the general theory. Hawking proposed a *chronology protection conjecture* whereby a quantum theory of gravity would, in some manner, physically forbid closed time-like curves.

9.3 Proper Time Intervals

In Section 8.2, we discussed the relationship between proper time for a stationary clock in the vicinity of a gravitating mass and co-ordinate time, defined in terms of the proper time of a clock outside the gravitational influence. Of course, when we refer to a clock being stationary, we have to remember that it has a world line that is running parallel to the appropriate time axis. Now we need to extend the idea of proper time for other world lines. We also need to consider the difference between an infinitesimal interval of time, $\Delta\tau$, and something more macroscopic, $\Delta\Gamma = \tau_2 - \tau_1$.

In the context of the special theory, proper time is defined as the time recorded by a clock in a co-moving system. The interval takes the form:

$$(\Delta s)^2 = c^2 (\Delta t)^2 - (\Delta x)^2 - (\Delta y)^2 - (\Delta z)^2 = \sum_{u,v} \eta_{\overline{uv}} \Delta x_u \Delta x_v,$$

so that $\Delta s = \sqrt{\sum_{u,v} \eta_{\overline{uv}} \Delta x_u \Delta x_v}$ and, as before, we define proper time by $\Delta s = c\Delta\tau$ or $\Delta\tau = \Delta s/c$. In order to get a finite time interval, we need to add up infinitesimal time intervals. We did something similar for tiny distance jumps in Section 9.1.2 by integrating. In this case, we need:

$$\Delta\Gamma = \tau_2 - \tau_1 = \frac{1}{c} \int_{t_1}^{t_2} \sqrt{\sum_{u,v} \eta_{\overline{uv}} \Delta x_u \Delta x_v}.$$

In order to make this more pliable, we carry out some mathematical massaging:

$$\Delta\Gamma = \frac{1}{c} \int_{t_1}^{t_2} \sqrt{\sum_{u,v} \eta_{\overline{uv}} \Delta x_u \Delta x_v} = \frac{1}{c} \int_{t_1}^{t_2} \sqrt{\sum_{u,v} \eta_{\overline{uv}} \frac{\Delta x_u}{c\Delta t} \frac{\Delta x_v}{c\Delta t} (c\Delta t)^2} = \frac{1}{c} \int_{t_1}^{t_2} \sqrt{\sum_{u,v} \eta_{\overline{uv}} \frac{\Delta x_u}{c\Delta t} \frac{\Delta x_v}{c\Delta t}} c\, dt.$$

The sum over u,v is taking the values $u,v = 0,1,2,3$ so we can split this up ($\eta_{\overline{00}} = 1$):

$$\Delta\Gamma = \frac{1}{c} \int_{t_1}^{t_2} \sqrt{1 + \sum_{i,j} \eta_{\overline{ij}} \frac{\Delta x_i}{c\Delta t} \frac{\Delta x_j}{c\Delta t}} c\, dt,$$

where $i,j = 1,2,3$. Remembering that $\eta_{\overline{ij}} = -\delta_{ij}$ gives us:

$$\Delta\Gamma = \frac{1}{c} \int_{t_1}^{t_2} \sqrt{1 + \sum_{i,j} \eta_{\overline{ii}} \frac{\Delta x_i}{c\Delta t} \frac{\Delta x_j}{c\Delta t}} c\, dt = \frac{1}{c} \int_{t_1}^{t_2} \sqrt{1 - \sum_i \left(\frac{\Delta x_i}{c\Delta t}\right)^2} c\, dt$$

which, with a little more tidying becomes:

$$\Delta\Gamma = \frac{1}{c}\int_{t_1}^{t_2}\sqrt{1 - \frac{1}{c^2}\sum_i\left(\frac{\Delta x_i}{\Delta t}\right)^2}\,cdt = \frac{1}{c}\int_{t_1}^{t_2}\sqrt{1 - v^2/c^2}\,cdt = \int_{t_1}^{t_2}\frac{1}{\gamma(t)}\,dt.$$

You may be wondering how the world line comes into this calculation, in which case recall that for a time-like path, the world line can be parameterised by time (Section 9.2), which is why I have shown $\gamma(t)$ as a function of time. The integral is then trotting along the world line from point to point picked out by the values of t; different world lines having different functions $\gamma(t)$. Without that specific function, we can't do anything more than show how the calculation might be done. However, we can use a simple example to draw an important conclusion.

In Figure 9.12, one object is stationary in the co-ordinate system and ages from Event A to Event B. Reading from the graph, we see that the proper time for this object is $c\Delta t = c\Delta\tau = 6$ units. By comparison, another object starts at Event A and moves to Event C, where it reverses course to end up at Event B as well. For this particle, $\beta = 2/3$ during the outward phase and $\beta = -2/3$ for the inward phase. The co-ordinate time elapsed for this object is $c\Delta t_1 = 3$ on the outwards journey and $c\Delta t_2 = 3$ during the flight back. This means that the proper time elapsed for the object is:

$$c\Delta\Gamma = \sqrt{1 - \beta^2}\,c\Delta t_1 + \sqrt{1 - \beta^2}\,c\Delta t_2 = 2 \times 3 \times \sqrt{1 - (2/3)^2} = 4.47$$

which is clearly less elapsed proper time than for the object that remained at rest. We can extend the argument by moving Event C to a position x at time $ct = 3$ and then calculate how proper time for the moving object changes for different values of x. The results are shown in Figure 9.13.

This figure makes it clear that the longest proper time occurs when $x = 0$, so that both particles share the same world line from Event A to Event B: that is to say, aging at rest.

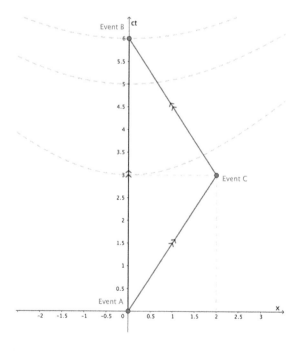

FIGURE 9.12 Comparison of the world lines of two particles: one which is stationary and the other that is moving. Both world lines end with Event B.

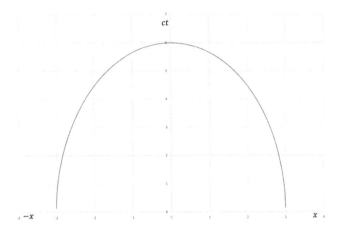

FIGURE 9.13 Elapsed proper time for different paths routed via an event C which is a distance x from the origin.

In this calculation we have assumed that the contribution due to the act of changing direction is negligible. The object moving along the indirect path has to use some manner of propulsion to change course, during which time the flat space-time metric no longer applies. We can wave our hands and explain that the contribution can be ignored by 'rounding off' the acceleration (rather than a sharp kink as in the figure) and doing it as gently as we like. As long as the time we take to change course is small compared to the length of the world lines, we can get away with ignoring this phase of the motion.

To generalise the calculation algebraically, assume that Event A is (x_1, ct_1), Event B (x_2, ct_2) and Event C (x, ct). Then the proper time for the moving object is:

$$c\Gamma = \frac{ct - ct_1}{\gamma_1} + \frac{ct_2 - ct}{\gamma_2}$$

and $\beta_1 = (x - x_1)/(ct - ct_1)$, $\beta_2 = (x_2 - x)/(ct_2 - ct)$.

Therefore:

$$c\Gamma = \frac{(x - x_1)}{\beta_1 \gamma_1} + \frac{(x_2 - x)}{\beta_2 \gamma_2}.$$

Now we move x slightly to the side, to $x + \Delta x$:

$$c(\Gamma + \Delta\Gamma) = \frac{(x + \Delta x - x_1)}{\beta_1 \gamma_1} + \frac{(x_2 - x - \Delta x)}{\beta_2 \gamma_2}$$

making:

$$c\Delta\Gamma = \frac{\Delta x}{\beta_1 \gamma_1} - \frac{\Delta x}{\beta_2 \gamma_2}.$$

So that:.

$$c\frac{\Delta\Gamma}{\Delta x} = \frac{1}{\beta_1 \gamma_1} - \frac{1}{\beta_2 \gamma_2}.$$

At the peak of a curve, such as $(0, 6)$ in Figure 9.13, the rate of change is zero. Hence, we can always find the value of a peak (or trough) by setting $\Delta y / \Delta x = 0$. Applying this to our calculation:

$$\frac{\Delta \Gamma}{\Delta x} = 0 \quad \text{hence} \quad c\frac{\Delta \Gamma}{\Delta x} = 0 = \frac{1}{\beta_1 \gamma_1} - \frac{1}{\beta_2 \gamma_2} \quad \text{so } \beta_1 = \beta_2.$$

The maximum proper time occurs when the object moves with a constant velocity, rather than over two sub-paths ending up at the same event. Consequently, the world line of the object is straight. This illustrates a fundamentally important principle – between any two events in (flat) space-time, the maximum proper time is experienced along the straight world line connecting the events (the geodesic). This obviously has relevance to the twin paradox, which we mentioned back in Section 4.5. In essence, the twin remaining on Earth is following a geodesic path and is hence moving inertially, while the twin on the spacecraft is following a world line that is not a geodesic.

9.3.1 Proper Time Intervals in the General Theory

Moving across to the general theory, we set up the proper time interval as follows:

$$\Delta \Gamma = \frac{1}{c} \int_{t_1}^{t_2} \sqrt{\sum_{u,v} g_{\overline{uv}} \Delta x_u \Delta x_v} \ .$$

If the world line of an object is parameterised, so that $x_u = x_u(\lambda)$, then we have the following relationship for a small change:

$$\Delta x_u = \left(\frac{\Delta x_u}{\Delta \lambda} \right) \Delta \lambda.$$

The metric is also a function of the co-ordinates, so it will be a function of λ as well: $g_{\overline{uv}} = g_{\overline{uv}}(\lambda)$. The integral then becomes:

$$\Delta \Gamma = \frac{1}{c} \int_{\lambda_1}^{\lambda_2} \sqrt{\sum_{u,v} g_{\overline{uv}}(\lambda) \frac{\Delta x_u(\lambda)}{\Delta \lambda} \frac{\Delta x_v(\lambda)}{\Delta \lambda}} \ d\lambda$$

which can be evaluated, either analytically or using numerical approximation techniques, if we know the functions $x_u(\lambda)$. Keep in mind the physical picture behind the mathematics: we are adding small proper time intervals along the world line, with the metric being responsible for transforming from the co-ordinate distances and times into the physical ones.

For time-like intervals, we can parameterise the world line by the proper time, which allows us to write our integral in the form:

$$\Delta \Gamma = \frac{1}{c} \int_{(x_0)_1}^{(x_0)_2} \sqrt{\sum_{u,v} g_{\overline{uv}} \frac{\Delta x_u}{\Delta x_0} \frac{\Delta x_v}{\Delta x_0}} \ dx_0 = \frac{1}{c} \int_{(x_0)_1}^{(x_0)_2} \sqrt{g_{\overline{00}} - \sum_{i,k} g_{\overline{ik}} \frac{\Delta x_i}{\Delta x_0} \frac{\Delta x_k}{\Delta x_0}} \ dx_0$$

$$= \frac{1}{c} \int_{(x_0)_1}^{(x_0)_2} \sqrt{1 - \sum_{i,k} \frac{g_{\overline{ik}}}{g_{\overline{00}}} \frac{\Delta x_i}{\Delta x_0} \frac{\Delta x_k}{\Delta x_0}} \sqrt{g_{\overline{00}}} \ dx_0.$$

If we evaluate this in the co-moving system, then Δx_{ij} is not changing with x_0 and we are reduced to:

$$\Delta\Gamma = \frac{1}{c}\int_{(x_0)_1}^{(x_0)_2}\sqrt{g_{\overline{00}}}\,dx_0.$$

For any other system, we take a closer look at:

$$\sum_{i,k}\frac{g_{\overline{ik}}}{g_{\overline{00}}}\frac{\Delta x_i}{\Delta x_0}\frac{\Delta x_k}{\Delta x_0}=\sum_i\frac{g_{\overline{ii}}}{g_{\overline{00}}}\frac{\Delta x_i}{\Delta x_0}\frac{\Delta x_i}{\Delta x_0}=\sum_i\left(\frac{\sqrt{g_{\overline{ii}}}\Delta x_i}{\sqrt{g_{\overline{00}}}\Delta x_0}\right)\left(\frac{\sqrt{g_{\overline{ii}}}\Delta x_i}{\sqrt{g_{\overline{00}}}\Delta x_0}\right).$$

In the first step I have assumed that we are using a co-ordinate system where the metric has no off-diagonal terms. If we interpret $\left(\dfrac{\sqrt{g_{\overline{ii}}}\Delta x_i}{\sqrt{g_{\overline{00}}}\Delta x_0}\right)$ as a component of the instantaneous relative β between the co-moving system and the system we are working in, then the integral becomes:

$$\Delta\Gamma = \frac{1}{c}\int_{(x_0)_1}^{(x_0)_2}\sqrt{1-\beta^2}\,\sqrt{g_{\overline{00}}}\,dx_0 = \frac{1}{c}\int_{(x_0)_1}^{(x_0)_2}\frac{\sqrt{g_{\overline{00}}}}{\gamma(x_0)}\,dx_0.$$

Clearly an adaptation of the same expression we obtained before in the context of the special theory. Given the similarity between the expressions in the general and special theories, we are motivated to extend the conclusion we came to earlier:

DEEP RULE 1

Along the time-like geodesic connecting two events, the proper time is a maximum compared with any other world lines between the events.

DEEP RULE 1A

The path of any object through space-time, if the object is not under some form of propulsion, always maximises the proper time for the object.

This rule can be turned around as an alternative definition of a geodesic: and the geodesic equation:

$$\frac{\Delta}{\Delta\tau}\left(\frac{\Delta x_k}{\Delta\tau}\right)=-\sum_i\sum_i\frac{\Delta x_i}{\Delta\tau}\frac{\Delta x_j}{\Delta\tau}\Gamma_{ij}^k$$

can be derived on that basis. Unfortunately, the proof requires an advanced form of mathematics called the *calculus of variations*, which is extremely beautiful but outside the scope of what we can achieve in this book.

It's important not to over-generalise these rules, which strictly speaking apply to local groups of world lines. If we take two events in space-time and consider all the world lines linking the two, then out of the sub-set of time-like world lines that are 'close' to each other, the geodesic will be the one with the largest proper time (the local maximum). In some instances, events can be linked by more than one time-like geodesic, in which case each will carry the maximum proper time out of the collection of close possibilities, but one could well have a bigger proper time than the other (the global maximum).

Space-like geodesics have the shortest proper length out of sub-sets of world lines that are the same in time span, but different (although close) in spatial distance; they have the longest proper time from those that have the same spatial distance, but different (close) times spans. Null geodesics can't be characterised in this way, as their intervals are always zero.

9.4 Gravitational Lenses

If we take a ray of light that is propagating radially away from a gravitating mass, then as for all rays of light, the space-time interval $\Delta s = 0$:

$$0 = \left(1 - \frac{2GM}{c^2 r}\right)c^2 (\Delta t)^2 - \frac{(\Delta r)^2}{\left(1 - \frac{2GM}{c^2 r}\right)},$$

which can be used to find $\Delta r / \Delta t$:

$$\frac{\Delta r}{\Delta t} = c\left(1 - \frac{2GM}{c^2 r}\right).$$

If this is interpreted as the *co-ordinate velocity of light*, it is evidently a function of position. Einstein used this idea to calculate the deflection of starlight passing close to the Sun, the prediction that was the motivation for Eddington's 1919 eclipse expedition. His approach was to treat the space-time in the vicinity of the Sun as a refractive 'material' reducing the co-ordinate velocity for light rays passing through. He was then able to apply standard techniques from optics to calculate the deflection angle. (Although of course he did not use the Schwarzschild metric to do this.)

Modern day relativists regard the co-ordinate velocity as non-physical and approach the problem by calculating geodesics for light rays passing close to the Sun. The result for the angle of deflection, $\Delta\theta$, is stated in terms of the impact parameter, b, which is the perpendicular distance from the initial path of the light ray to the centre of the gravitating mass:

$$\Delta\theta = \frac{4GM}{c^2 b}.$$

Using this formula, we can estimate the deflection angle for rays grazing the stellar surface, if we take b as the radius of the Sun:

$$\Delta\theta = \frac{4GM}{c^2 b} = \frac{4 \times 6.674 \times 10^{-11} \times 1.989 \times 10^{30}}{\left(2.998 \times 10^8\right)^2 \times 6.955 \times 10^8} = 8.49 \times 10^{-6}\,\text{rad.}$$

These days, this effect is commonly known as *gravitational lensing*, although strictly the term is not accurate. With a lens, the greatest refraction happens towards the edges, and less towards the centre, which is why the lens has a *focal point* and hence *focal length*. Gravitating masses bend light more the closer the ray's path is to the mass, the opposite to an optical lens.

Einstein first discussed the possibility that gravitating masses could produce images of more distant objects in an article published in 1936.[7] In principle, if a luminous object, a gravitating mass and an observer are in exact linear alignment, the observer will see a ring image of the object around the mass (Figure 9.14).

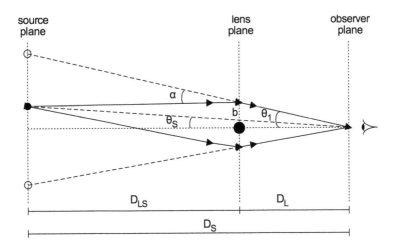

FIGURE 9.14 The geometry of a gravitational lens. (Image credit: Krishnavedala – Own work, CC0, https://commons.wikimedia.org/w/index.php?curid=20298731.)

Any misalignment will result in the image breaking into arc segments instead. At the time, Einstein was sceptical that the effect could be observed in practice, due to the necessary but highly unlikely co-incidental alignment of stars, as well as the small size of the image being beyond the resolution of instruments at the time. However, the astronomer Fritz Zwicky[8] became an enthusiastic proponent of the idea, refining it to suggest that the issues of foreground glare masking the effect could be overcome if the lensing matter was a galaxy or similar, rather than a star. Zwicky published three articles on the subject in 1937 and constantly urged astronomers to search for lensing effects for the rest of his life. Unfortunately, the first discovery was made 5 years after Zwicky's death.

There are now many recorded instances where at least partial ring images have been discovered; the gravitating masses being galaxies, clusters, black holes or regions containing concentrations of *dark matter*. In fact, this effect provides some of the most convincing evidence for the existence of dark matter – a mysterious component of the universe which does not interact with electromagnetic waves, yet can be detected by its gravitational effect (Section 12.5.3).

Beautiful pictures of gravitational lenses can easily be found on the internet. One of the most impressive is Figure 9.15 which shows an incredible double ring due to almost perfect alignment between two distant galaxies and a third (doing the imaging) nearer to us.

9.5 The Orbit of Mercury

Newtonian orbital theory shows that if one object is considerably more massive than the other (e.g. the Sun and any of the planets), then the less massive one executes an elliptical orbit around the other, which is at one focus of the ellipse. However, simple orbital theory seldom, if ever, applies. For one thing, the presence of other masses, such as planetary siblings, will disturb this pattern. As a result, the point of closest approach on the ellipse (known as the *perihelion* in our solar system) rotates around the Sun, somewhat as in Figure 9.16.

The biggest influence on the precession of the perihelion is the presence of other planets, although the fact that the Sun bulges out somewhat at its equator also has a minor effect.

In 1859, Urbain Le Verrier recognised that Mercury's orbit was a puzzle. Using timing data for transits of Mercury across the face of the Sun from 1697 to 1848 he was able to show that the precession of the orbit disagreed with the Newtonian prediction by 0.0106° per century. This was later upgraded to 0.0119° per century by Simon Newcomb in 1882. A variety of suggestions were made to explain this anomalous

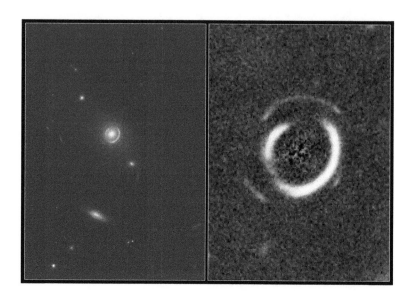

FIGURE 9.15 This is an image of gravitational lens system SDSSJ0946+1006 as photographed by Hubble Space Telescope's Advanced Camera for Surveys. The gravitational field of an elliptical galaxy warps the light of two galaxies exactly behind it. The massive foreground galaxy is almost perfectly aligned in the sky with two background galaxies at different distances. The foreground galaxy is 3 billion light-years away, the inner ring and outer ring are comprised of multiple images of two galaxies at a distance of 6 and approximately 11 billion light-years. The odds of seeing such a special alignment are estimated to be 1 in 10,000. The right panel is a zoom onto the lens showing two concentric partial ring-like structures after subtracting the glare of the central, foreground galaxy. (Image credit: NASA, ESA, and R. Gavazzi and T. Treu [University of California, Santa Barbara].)

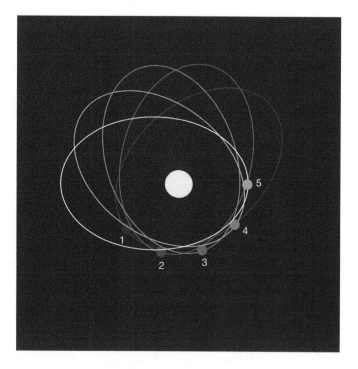

FIGURE 9.16 The precession of Mercury's orbit. As the planet loops around the Sun, its point of closest approach (the perihelion) moves from 1 to 5.

precession, but they did not come to anything convincing and the phenomenon remained a puzzle until Einstein's work on the general theory.

Much to Einstein's delight, working with an approximation scheme rather than a specific metric, he was able to calculate the orbit of Mercury and show that the precession came out exactly as measured. It's not easy to point to a specific difference between general relativity and Newtonian gravity that accounts for the precession, it is just a second level approximation beyond the Newtonian limit. Unfortunately, the calculation is a little too intricate for us.

Perihelion precession is not limited to Mercury. All planets in the solar system suffer the same effect, but as they are further out the angular rates were too small to be measured until modern times. As it stands, the precessions of Earth and Venus have been measured producing good agreement with the calculations. This has also been done for the binary pulsar system that Hulse and Taylor discovered (Section 1.5).

Notes

1. Letter from K Schwarzschild to A Einstein, 22 December 1915, *The Collected Papers of A Einstein*, Vol 8a doc No. 169.
2. Sort of a differential equation – I have not been at all punctilious about the distinction between $\Delta z / \Delta r$, $\delta z / \delta r$ and dz/dr...
3. In business terms, metrics are the statistics or figures used to measure results. They are also defined as a means of deriving a quantitative measurement or approximation for otherwise qualitative phenomena.
4. The parameter λ is then called an *affine parameter,* a term that often comes up in developments of the general theory.
5. There are various ways of doing this. One approach is to use proper time for a sub-light particle's path between two events and to move this into an abstract parameter as you take the limit of the object moving faster and faster. You will then need to re-scale to make sure that the null path is still finite in length. The parameter is simply a way of mapping out the geodesic, so need not in any way be physical. In the case of time-like geodesics, it obviously makes sense to use proper time as it is available for the job.
6. Gödel, K., 1949. An example of a new type of cosmological solutions of Einstein's field equations of gravitation. *Reviews of Modern Physics*, **21**, p. 447.
7. Einstein, A., 1936. Lens-like action of a star by the deviation of light in the gravitational field. *Science*, **84**, pp. 506–507. Bibcode:1936Sci....84..506E.
8. Fritz Zwicky, 1898–1974, a Swiss astronomer working at Caltech in the USA. Zwicky made many important contributions to theoretical and observational astronomy.

10

Black Holes

A big misconception is that a black hole is made of matter that has just been compacted to a very small size. That's not true. A black hole is made from warped space and time.

K Thorne[1]

10.1 Gravitational Collapse

In the early days of the general theory, it was assumed that objects would always be larger than their Schwarzschild radius. As the Schwarzschild metric is a solution of the vacuum equations, it does not apply within the confines of the distributed mass of an object, hence avoiding the problem of its odd behaviour at $r = R_s$. For example, the Sun's Schwarzschild radius is:

$$R_s = \frac{2GM_{Sun}}{c^2} = \frac{2 \times 6.67 \times 10^{-11} \times 2.00 \times 10^{30}}{\left(3 \times 10^8\right)^2} = 3,000\,m$$

evidently well inside the radius of the Sun. Table 10.1 confirms that astronomical objects that would have been familiar to physicists in the first half of the 20th century all have Schwarzschild radii deep inside the object, and so not physically relevant.

However, as theoretical analysis became more sophisticated, it became clear that dying stars collapse under their own gravity, possibly shrinking to within their own Schwarzschild radius.

10.1.1 Deaths of Stars

Mature stars exist in a delicate balance between expansion, driven by pressure resulting from high internal temperatures, and collapse due to the gravitational effect of their own mass. Deep in their cores, reactions take place between hydrogen nuclei, liberating vast amounts of energy which maintains the high temperatures (typically $\sim 2 \times 10^7$ °C for the most massive category of star). This continual energy production counterbalances the energy that a star radiates into space.

TABLE 10.1 The Schwarzschild Radius for Different Objects

Object	Schwarzschild Radius	Comment
Earth	8.9 mm	
Sun	3 km	
VY Canis Majoris	50 km	Red supergiant star
Sgr A*	1.2×10^{10} m	Supermassive black hole at the centre of our galaxy
NGC 1277	1.8×10^{13}–1.1×10^{13} m	Most massive known black hole, somewhere between 6 billion and 37 billion solar masses

Like all hot gases, the material in a star exerts a pressure to expand due to the rapid random thermal motion of the particles. However, in a ball of gas as massive as a star, there is significant gravity which tends to draw the material towards the centre. While the energy production continues, the balance can be maintained and the star is stable. Once the hydrogen fuel in the core is depleted, the reactions decline and the pressure drops, triggering a steady inward collapse. What happens next depends on the mass of the star, which is typically categorised as a fraction of the Sun's mass, M_s.

Low-Mass Stars, i.e. $< 0.8\,M_s$

Such stars have expected lifetimes longer than the universe has currently been in existence (~ 14 billion years) so we have no observational evidence of their ultimate fate. *Red dwarfs* ($\sim 0.1 M_s$) are modelled to burn[2] for between $6-12 \times 10^{12}$ years and then slowly fade down into a *white dwarf* (I'll explain what these are shortly). At the more massive end of this category, we expect the stars to expand outwards and form *red giants*. As the core collapses inwards, hydrogen that was outside the core reaches a higher temperature and that triggers refreshed nuclear reactions. The energy production in what was an outer layer causes the material to expand outwards while at the same time becoming colder and more diffuse. The result is a star with many times its initial radius (up to 200 times greater) glowing with a dull red colour (Figure 10.1). However, due to their vast surface area, red giants can be among the most luminous stars.[3]

Red giants formed from these lower mass stars have a gentle decline once the refreshed nuclear burning has ceased. They also end up as white dwarfs.

Medium Mass Stars, i.e. $0.8-10\,M_s$

Such stars evolve into red giants once their initial hydrogen burning in the core has declined. However, these more massive red giants can also ignite reactions involving helium, due to their higher temperatures, which gives them the ability to remain stable at this stage for significantly longer than their lower mass siblings. Once they end this phase of their lives, the core once again contracts and

FIGURE 10.1 The constellation Orion, showing the red giant star Betelgeuse (ringed) which is about 600 light years from us. The star is so large, that in our solar system it would nearly reach to Saturn's orbit. With the naked eye, the star can be seen with a red tinge on a clear night. (Image credit: TheStarmon, via Wikkimedia.)

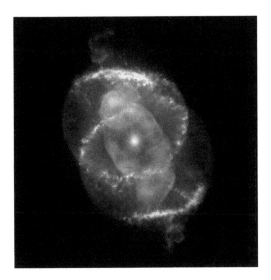

FIGURE 10.2 An X-ray/optical composite image of the planetary nebula NGC 6543, the Cat's Eye Nebula (Image credit: X-ray: NASA/UIUC/Y. Chu et al., Optical: NASA/HST.)

warms up further, but no new nuclear reactions can ignite. The outer layers of the star now become detached from the collapsing core and drift away to form a large shell of expanding and gently glowing gas surrounding the super-hot core. Such *planetary nebulae* are among the most attractive astrophysical objects that can be imaged (Figure 10.2).

The central core of the star is now a white dwarf (Figure 10.3) – a stellar remnant that is no longer generating energy, but is still radiating due to accumulated thermal energy that is not yet completely exhausted.

FIGURE 10.3 The white dwarf star Sirius B (ringed), a close orbital companion of the main star, Sirius A. The diagonal lines and the circular rings around the main star are artefacts of the imaging system. (Image credit: By NASA, ESA, H. Bond (STScI), and M. Barstow [University of Leicester], http://www.spacetelescope.org/images/heic0516a/, CC BY 3.0, https://commons.wikimedia.org/w/index.php?curid=477445.)

Typically, white dwarfs have masses comparable to that of the Sun ($\leq 1.4\,M_s$) but sizes similar to that of Earth. This makes their compositional material exceptionally dense.

White dwarfs are stabilised against gravitational collapse by a quantum mechanical effect known as *electron degeneracy pressure*. Briefly, the temperature in a white dwarf is sufficient to strip all electrons from their atoms. In concert, these free electrons act somewhat like a gas, exerting significant outward pressure. One of the basic quantum rules that governs a class of particles called *fermions*, of which the electron is a significant member, is that each distinct physical state can only be occupied by a single fermion. Consequently, the electrons in a white dwarf occupy higher energy states than we would classically expect (as they can't all sit in lower energy states). As a result, they have greater kinetic energy than the molecules in an equivalent gas at that temperature[4] and exert sufficient pressure to hold off the gravitational collapse.

We have observed many white dwarfs in our own galaxy, but the universe is still too young for such stars to have completely radiated away their thermal energy. So, we have not as yet seen a *black dwarf*, which is their expected end after something like 10^{15} years: a cold, dark, stellar remnant no longer radiating significant energy to its surroundings and detectable only by its gravitational effects.

High Mass Stars, i.e. $> 10\,M_s$

The quantum mechanical theory underpinning our understanding of white dwarf stability was worked out by Subrahmanyan Chandrasekhar[5] in the 1930s. As part of his calculations, Chandrasekhar demonstrated that electron degeneracy pressure could support a white dwarf of up to $1.4\,M_s$, but beyond that the gravitational collapse passes through the white dwarf stage and compresses the dying star to even smaller sizes.

In the case of high mass stars, once their red giant phase has completed, and having shed significant mass to the surroundings, the core is still above the Chandrasekhar limit. As the collapsing star shrinks, the converging electron and nuclear geodesics bring electrons inside the confines of the nucleus where they react with protons to form neutrons, releasing energy and particles called *neutrinos*. Once the inner core has been converted into neutrons, other infalling material bounces off causing an outwards propagating shock wave. The flux of weakly interacting extremely low-mass neutrinos is sufficient to cause the surrounding layers of material to absorb considerable energy, which is re-radiated and the star's luminosity vastly increases. This is a *supernovae*; a stellar explosion of such luminous violence, the exploding star's output can exceed that of all other stars in the same galaxy combined (Figure 10.4).

The resulting remnant has a complex structure, predominantly composed of neutrons. Short-range repulsive forces between the neutrons combine with neutron degeneracy pressure (neutrons are also fermions) to oppose the gravitational collapse and a *neutron star* has been formed (Figure 10.5).

During any form of stellar collapse, the contracting star tends to spin at an ever-faster rate. Simplistically, this comes about due to the same angular momentum effect that causes an ice skater performing a twirl to rotate more quickly when they draw their arms into their sides. In the case of neutron stars, this spin rate can be exceptional. The fastest known is that of PSR J1748–2446ad (such an evocative name...) which turns on its axis 716 times per second. Given that neutron stars are typically only ~20 km in diameter and in this case ~32 km, that puts the linear speed of the material on its surface at $0.24c$.

Neutron stars are hot. When freshly formed, their surface temperatures are ~10^{11} K, but due to the energy carried away by escaping neutrinos, this drops to ~10^6 K within a few years. They also generate extraordinarily strong magnetic fields, between 10^4 and 10^{11} Tesla (T) at their surface. In comparison, a stable magnetic field of 16 T has been generated in the laboratory, which was sufficient to magnetically levitate a frog.[6] More commonly, patients experiencing an MRI scan experience a field of $0.5-1.5$ T.

As yet, we do not completely understand how neutron stars generate these magnetic fields, but they are responsible for accelerating charged particles near the magnetic poles, which then emit intense beams of radiation. As the magnetic poles are not necessarily aligned with the rotation axis, these beams

FIGURE 10.4 This Hubble Space Telescope image shows Supernova 1994D (SN1994D) (ringed) in galaxy NGC 4526. The supernova is brighter than all the stars in his host galaxy combined. Supernova. (Image credit: By NASA/ESA, CC BY 3.0, https://commons.wikimedia.org/w/index.php?curid=407520.)

FIGURE 10.5 The Crab Nebula supernova remnant. This cloud of luminous gas is the result of a supernova explosion in our own galaxy. At the centre of the cloud is a rotating neutron star (pulsar) which was first detected by its radio emissions and was later seen pulsing on and off optically. (Image credit: NASA, ESA, J. Hester and A. Loll [Arizona State University].)

of radiation most often sweep out from the neutron star rather like that of a lighthouse. In some cases, the beam sweeps back and forth across the Earth's field of view, and a *pulsar* (pulsating star) is observed. You will recall that observations of a binary neutron star system that includes a pulsar have led to convincing evidence for gravitational waves (Section 1.5).

However, neutrons stars are not the end of the story. There is a limit to the maximum mass of a neutron star, just as the Chandrasekhar limit caps the mass of a white dwarf. The most massive neutron star that we are currently aware of is $2.01\,M_s$. LIGO has detected the gravitational waves emitted by a pair of neutron stars that spiralled together and merged. The gravitational wave profile indicated that the pair formed a black hole, and constrained the upper mass limit for a neutron star $\sim 2.17\,M_s$. Theoretical work, which is far from complete, suggests that the mass limit is between $1.5\,M_s$ and $3.0\,M_s$, although these estimates do not consider the spin of the neutron star. In terms of the mass of the progenitor star, this translates to stars between $15-20\,M_s$. Beyond these limits, the forces mustered by the neutrons are insufficient to resist the curvature of space-time and the star collapses through the neutron star phase just as it passed through being a white dwarf.

We now reach the situation where the star can and will shrink inside its own Schwarzschild radius, so we have to face the up to the strange distortions of space-time that appear to happen at that borderline.

10.2 Singularities

From a mathematical standpoint, the Schwarzschild radius is a *singularity* in the metric: a value of r that leads to an impossible result. In the case of simple functions, this is often caused by trying to input a value that the function is not designed to process. For example, the function $f(x)=1/(x-a)$ is not defined for the value $x=a$. A singularity of a different kind occurs when the rate of change 'blows up', rather than the function. For example, $f(x)=|(x-a)|$ has a well-defined value at $x=a$, but its rate of change at that point blows up (Figure 10.6) due to the sharp corner or cusp.

In the case of the Schwarzschild metric, there are two singularities:

$$(\Delta s)^2 = \left(1-\frac{2GM}{c^2 r}\right)c^2(\Delta t)^2 - \frac{(\Delta r)^2}{\left(1-\frac{2GM}{c^2 r}\right)} - r^2(\Delta\theta)^2 - r^2\sin^2(\theta)(\Delta\varphi)^2$$

$$r = R_s = \frac{2GM}{c^2} \text{ which causes the spatial term to blow up}$$

$$r = 0 \text{ which detonates everything}\ldots$$

The first of these is the less significant of the two, as it is a *co-ordinate singularity.*

10.2.1 Co-ordinate Singularities

Co-ordinate singularities come about when something goes wrong with your co-ordinate system. Across the surface of the Earth we identify locations via lines of latitude and longitude (Figure 10.7).

In this co-ordinate system, the distance between any two points on the surface is given by:

$$(\Delta s)^2 = R^2(\Delta\varphi)^2 + R^2\cos^2(\varphi)(\Delta\theta)^2,$$

where φ is the latitude angle with $\varphi=0$ at the equator, and θ is the longitude with $\theta=0$ at the prime meridian. As straightforward as this appears, there is something fishy: at the North and South Poles, there are *an infinite number of θ s which have the same φ.* Consequently, longitude is technically not defined at the poles and in that sense they are singular points in the co-ordinate system.

In the general theory, with its general covariance, we have to keep an eye out for singularities that are due to the co-ordinate system and not arising from the geometry of the space-time (which is the second type of singularity in the Schwarzschild metric). For many years the singularity at $r=R_s$ was thought to be a physics (space-time geometry) issue and hence people were reluctant to accept that a star could

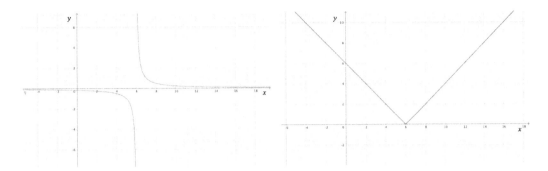

FIGURE 10.6 Two types of singularity. On the left, the function $y = 1/(x-6)$ is singular at the point $x = 6$. On the right, the function $y = |x-6|$ has a rate of change that is singular at the point $x = 6$.

contract through this radius. However, a co-ordinate transformation shows that the singularity is an artefact of the co-ordinates. One possibility is to shift the co-ordinate time so that:

$$ct = v - r - R_s \log\left(\left|\frac{r}{R_s} - 1\right|\right)$$

with the new co-ordinate system being known as *Eddington-Finkelstein*[7] *co-ordinates* (E–F co-ordinates). If we do this, then we have:

$$c\Delta t = \Delta v - \Delta r - \frac{R_s \Delta r}{r - R_s}$$

a step that relies on the result:

$$\frac{\Delta}{\Delta r}\left(R_s \log\left(\left|\frac{r}{R_s} - 1\right|\right)\right) = \frac{R_s}{r - R_s}$$

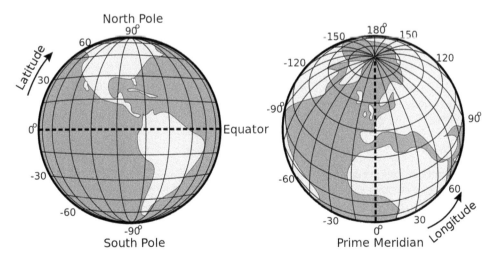

FIGURE 10.7 The lines of latitude and longitude used as a co-ordinate system across the surface of the Earth. (Image credit: Djexplo via Wikkimedia commons.)

but the proof would be a bit of a distraction at the moment. If you will just accept the result, then a bit of algebraic manipulation shows that the Schwarzschild interval converts into:

$$(\Delta s)^2 = \left(1 - \frac{2GM}{c^2 r}\right)(\Delta v)^2 - 2\Delta v \Delta r - r^2 (\Delta\theta)^2 - r^2 \sin^2(\theta)(\Delta\varphi)^2,$$

which does not have any singular behaviour at $r = \dfrac{2GM}{c^2} = R_s$. From the graph of the log function (Figure 10.8), we see that $R_s \log\left(\left|\dfrac{r}{R_s} - 1\right|\right) \to -\infty$ as $r \to R_s$, so that it cancels the tendency of the spatial interval to blow up in ordinary co-ordinates.

This is why this singularity does not occur in the E–F co-ordinate system.

However, the $r = 0$ singularity cannot be removed by a deft change in co-ordinates. It is a genuine physical singularity and so will always be present in any system. We will discuss such singularities in Section 10.2.3.

10.2.2 Geodesics in E–F Co-ordinates

From the interval in E–F co-ordinates, we can extract the geodesic paths followed by light rays by setting $(\Delta s)^2 = 0$, so that:

$$0 = \left(1 - \frac{R_s}{r}\right)(\Delta v)^2 - 2\Delta v \Delta r - r^2 (\Delta\theta)^2 - r^2 \sin^2(\theta)(\Delta\varphi)^2.$$

If we focus on radial geodesics i.e. $\Delta\theta = \Delta\varphi = 0$ then we are reduced to:

$$0 = \left(1 - \frac{R_s}{r}\right)(\Delta v)^2 - 2\Delta v \Delta r \quad \text{or} \quad \left(1 - \frac{R_s}{r}\right)(\Delta v)^2 = 2\Delta v \Delta r$$

which tells us that *either* $\Delta v = 0$ *or* $\left(1 - \dfrac{R_s}{r}\right)\Delta v = 2\Delta r$. The first possibility resolves into $v = $ constant while the second gives us another differential equation:

$$\frac{\Delta v}{\Delta r} = \frac{2}{\left(1 - \dfrac{R_s}{r}\right)}$$

which has solutions of the form:

$$v = 2R_s \log(r - R_s) + 2r.$$

These two types of null geodesic can be plotted using $y = v - r$ on a vertical axis, as in Figure 10.9. Three light cones are also shown, one for a distance $r > R_s$ and two inside R_s. The light cones are tipped and contracted by the space-time geometry inside the Schwarzschild radius. As all time-like geodesics starting from an event must lie within the span of the light cone originating at that event, this geometry ensures that all time-like world lines inside the Schwarzschild radius end up, eventually, at the singularity $r = 0$. Notice that this also implies that world lines from events with $r \leq R_s$ can't cross into $r > R_s$. The Schwarzschild radius marks an *event horizon*: no information from events taking place inside that radius can ever be communicated to observers on the outside.

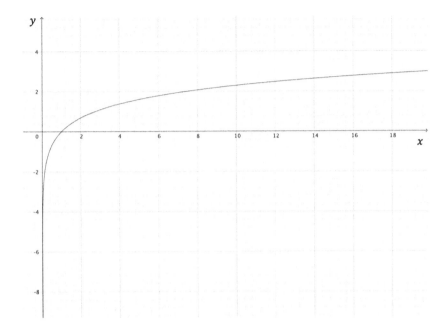

FIGURE 10.8 The function $y = \log(x) \to -\infty$ as $x \to 0$.

In the Newtonian force-based view of gravity, the minimum velocity needed to escape the gravitational pull of a mass, M, is calculated by setting the total energy equal to zero:

$$\text{Total energy E} = \text{kinetic energy} + \text{gravitational PE} = \frac{1}{2}mv_{escape}^2 - \frac{GMm}{r} = 0,$$

so that:

$$v_{escape}^2 = \frac{2GM}{r}.$$

If this escape velocity equals the speed of light, then the object's radius is:

$$r = \frac{2GM}{c^2} = R_s,$$

which is the same as the Schwarzschild radius. Given that the speed of light is the fastest possible speed, then a Newtonian gravitational view, seasoned by some special relativity, would say that the gravitational pull is so great that not even the mass of the star itself can resist, and it gets crushed to smaller and smaller sizes.

From the space-time geometry standpoint, we consider the geodesics lying inside the Schwarzschild radius and note that they all converge at the singularity. In this view, there is no more a force pulling us towards the singularity than there is a 'force' pulling us towards the future in a flat space-time. Meisner, Thorne and Wheeler[8] put it succinctly:

That unseen power of the world which drags everyone forward willy-nilly from age twenty to forty and from forty to eighty also drags {us towards the singularity}

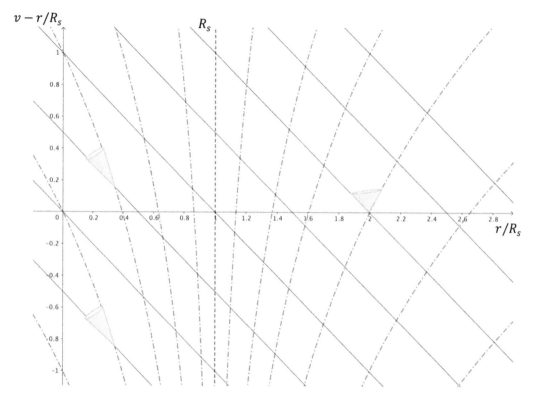

FIGURE 10.9 In going and outgoing null geodesics in ingoing Eddington-Finkelstein co-ordinates. The light cones at three events are also shown. Note how for $r < R_s$, all possible futures of an event end up at the singularity $r = 0$.

Of course, everything we have said about objects inside the Schwarzschild radius applies equally to the stellar material collapsing through the event horizon, if the star is too massive to stabilise at a neutron star. Once the star has reached its Schwarzschild radius, the collapse is irresistible and irreversible. All the star's matter is destined to converge on the singularity at $r = 0$. A *black hole* has formed.

10.2.3 Unavoidable Singularities

Black holes are very exotic objects. Technically, a black hole puts a huge amount of mass inside of zero volume. So our understanding of the centre of black holes doesn't make sense, which is a big clue to physicists that we don't have our physics quite right.

A Ghez[9]

The Schwarzschild singularity at $r = 0$ is a genuinely catastrophic distortion of space-time and cannot be removed by some cunning co-ordinate transformation. As it stands, no known form of matter can resist the geodesic deviation that comes about near to a singularity, but the general theory does not tell us what happens as a result. All the matter from the collapsing star, and anything that follows it over the event horizon at a later stage, will end up at the singularity and so crushed to infinite density. Consequently, the mass/energy density at the singularity is infinite.

The modern definition of a physical singularity relates to *terminating geodesics*: those that cannot be extended beyond a certain value of their parameter, i.e. at the singularity.

The conventional way of confirming that a singularity is physical and not a co-ordinate artefact is to check the behaviour of an invariant (and so co-ordinate independent) quantity such as $\sum_{i,j,k,l} R_{ijkl} R_{\overline{ijkl}}$.

If that blows up, then there is a physical singularity involved.

At the moment our theory is inadequate to deal with the nature of such points. The singularity is a radical curvature in our understanding, as well as space-time. The hope is that some well-developed quantum gravity will rescue us from the existence of singularities and replace them, presumably with some very exotic and highly dense objects. As we expect that quantum gravity effects will become significant at the *Planck Scale*, which corresponds to ~10^{-35} m (see Section 13.1.1), this may well represent the order of magnitude of the quantum object's size.

Due to established theorems, we know that singularities are an inevitable consequence of gravitational collapse, as well as being present for the 'start' of the Big Bang. So, the general theory predicts the existence of singularities, but gives us no mechanism to understand them. However, as we like to believe in the consistency of our theories (at least once they have been reasonably well established), Roger Penrose[10] has proposed the *cosmic censorship hypothesis*: all singularities (with the conspicuous exception of that involved in the Big Bang) are clothed within an event horizon, which prevents their being seen by observers outside the Schwarzschild radius. In that way, the theory is clothed as well. Singularities may exist and the theory not be able to make any predictions about them, but the predictions could not be checked anyway, so it does not matter!

The term *black hole* applies to a singularity with accompanying event horizon; otherwise, we have a (hypothetical) *naked singularity*. Calculations show that geodesics also terminate at a naked singularity, so in practical terms the distinction is moot.

As yet the cosmic censorship hypothesis lacks a precise formal definition and it is certainly not established as a proven truth of general relativity. Indeed, in 1992 Stuart Shapiro and Saul Teukolsky[11] created a computer simulation of a rotating plane of dust that generated a naked singularity. However, several simplifying assumptions (such as spherical symmetry) went into the simulation, and it could be that in a realistic situation, an event horizon would be present.

The *Kerr metric* represents a rotating neutral black hole (Section 10.5), and its mathematical properties show that above a certain angular momentum, there can be no event horizon. Again, however, the physical realism of this situation is uncertain.

So, the cosmic censorship hypothesis is still undecided, but physically attractive, at least until some complete and correct quantum gravity establishes the situation one way or another.

10.3 Falling Towards a Black Hole

When you fall into a black hole, everything that falls in after you over millions of years, as seen by you inside the black hole, comes pounding down on you in a fraction of a second, because of the enormous differences of time flow.

K Thorne[12]

To investigate what happens when an object crosses an event horizon, we imagine an observer in free fall towards a black hole. Assuming that they start a distance r_0 away from the singularity, where they are initially at rest, we can calculate their proper time to reach a distance $r < r_0$ in the following way.

First we take the geodesic equation, parameterised by proper time:

$$\frac{\Delta}{\Delta\tau}\left(\frac{\Delta x_k}{\Delta\tau}\right) + \sum_i \sum_j \frac{\Delta x_i}{\Delta\tau}\frac{\Delta x_j}{\Delta\tau}\Gamma_{ij}^k = 0$$

and find the time equation, which has $k=0$. The only relevant non-zero Christoffel symbols for the Schwarzschild metric are:

$$\Gamma^0_{01} = \Gamma^0_{10} = \frac{GM}{r^2 c^2 \left(1 - \dfrac{2GM}{c^2 r}\right)} = \frac{R_s}{2r^2 \left(1 - \dfrac{R_s}{r}\right)},$$

so:

$$\frac{\Delta}{\Delta \tau}\left(\frac{\Delta x_0}{\Delta \tau}\right) + \frac{\Delta x_0}{\Delta \tau}\frac{\Delta x_1}{\Delta \tau}\Gamma^0_{01} + \frac{\Delta x_1}{\Delta \tau}\frac{\Delta x_0}{\Delta \tau}\Gamma^0_{10} = 0$$

which, using $x_0 = ct, x_1 = r$, becomes:

$$\frac{\Delta}{\Delta \tau}\left(\frac{\Delta t}{\Delta \tau}\right) + 2\frac{\Delta r}{\Delta \tau}\frac{\Delta t}{\Delta \tau}\left(\frac{R_s}{2r^2\left(1 - \dfrac{R_s}{r}\right)}\right) = 0.$$

Doing some re-arrangement gives:

$$\left(1 - \frac{R_s}{r}\right)\frac{\Delta}{\Delta \tau}\left(\frac{\Delta t}{\Delta \tau}\right) + \frac{\Delta r}{\Delta \tau}\frac{\Delta t}{\Delta \tau}\left(\frac{R_s}{r^2}\right) = 0.$$

To help solve this, we 'spot' the interesting relationship:

$$\frac{\Delta}{\Delta \tau}\left(\left(1 - \frac{R_s}{r}\right)\left(\frac{\Delta t}{\Delta \tau}\right)\right) = \frac{R_s}{r^2}\frac{\Delta r}{\Delta \tau}\frac{\Delta t}{\Delta \tau} + \left(1 - \frac{R_s}{r}\right)\frac{\Delta}{\Delta \tau}\left(\frac{\Delta t}{\Delta \tau}\right) = 0$$

coming from the rule:

$$\frac{\Delta}{\Delta \tau}(u \times v) = \frac{\Delta u}{\Delta \tau} \times v + u \times \frac{\Delta v}{\Delta \tau}.$$

Writing $\left(1 - \dfrac{R_s}{r}\right)\left(\dfrac{\Delta t}{\Delta \tau}\right) = F$, we see that $\dfrac{\Delta}{\Delta \tau}(F) = 0$ and if the rate of change of a quantity is zero, that implies that the quantity is constant. Hence we have solved this equation in a rather elegant way:

$$\left(1 - \frac{R_s}{r}\right)\left(\frac{\Delta t}{\Delta \tau}\right) = K \quad \text{or} \quad \left(\frac{\Delta t}{\Delta \tau}\right) = \frac{K}{\left(1 - \dfrac{R_s}{r}\right)},$$

where K is some constant that we need to figure out. A nice way to do that is to see what happens as we let $r \to \infty$, in other words we move well outside of the influence of the gravitating mass. In that case, $\left.\dfrac{\Delta t}{\Delta \tau}\right|_{r=\infty} = K$ and we can expect the metric to be flat, so that the special theory is appropriate. Thinking back to the temporal component of the 4-Momentum:

$$P = \begin{pmatrix} E/c \\ p \end{pmatrix} = m \frac{\Delta}{\Delta \tau} \begin{pmatrix} ct \\ x \end{pmatrix}$$

we can extract:

$$\left. \frac{\Delta t}{\Delta \tau} \right|_{r=\infty} = \frac{E}{mc^2} = K \quad \text{hence} \quad \left(\frac{\Delta t}{\Delta \tau} \right) = \frac{E}{mc^2 \left(1 - \dfrac{R_s}{r}\right)}.$$

From chapter 8, we have the relationship:

$$\frac{\Delta \tau}{\Delta t} = \sqrt{1 - R_s/r} \quad \text{or} \quad \frac{\Delta t}{\Delta \tau} = 1/\sqrt{1 - R_s/r}$$

for a stationary object. Hence at r_0 where the fall starts:

$$\frac{E}{mc^2 \left(1 - \dfrac{R_s}{r_0}\right)} = \frac{1}{\left(1 - \dfrac{R_s}{r_0}\right)^{1/2}}$$

or:

$$E = mc^2 \left(1 - \frac{R_s}{r_0}\right)^{1/2} \approx mc^2 - \frac{GMm}{r_0}$$

in the Newtonian limit.

Next we take a look at the Schwarzschild interval for radial motion, $(\Delta \theta = \Delta \varphi = 0)$:

$$(\Delta s)^2 = \left(1 - \frac{R_s}{r}\right) c^2 (\Delta t)^2 - \frac{(\Delta r)^2}{\left(1 - \dfrac{R_s}{r}\right)}.$$

We can divide through by $(\Delta s)^2$ and use $(\Delta s)^2 = c^2 (\Delta \tau)^2$ to give:

$$1 = \left(1 - \frac{R_s}{r}\right)\left(\frac{\Delta t}{\Delta \tau}\right)^2 - \frac{1}{c^2 \left(1 - \dfrac{R_s}{r}\right)} \left(\frac{\Delta r}{\Delta \tau}\right)^2.$$

Substituting the result $\left(\dfrac{\Delta t}{\Delta \tau}\right) = \dfrac{K}{\left(1 - \dfrac{R_s}{r}\right)}$ produces:

$$1 = \frac{K^2}{\left(1 - \dfrac{R_s}{r}\right)} - \frac{1}{c^2 \left(1 - \dfrac{R_s}{r}\right)} \left(\frac{\Delta r}{\Delta \tau}\right)^2,$$

which we can re-arrange to give:

$$\left(\frac{\Delta r}{\Delta \tau}\right)^2 = K^2 c^2 - c^2 \left(1 - \frac{R_s}{r}\right) \quad \text{or} \quad \frac{\Delta r}{\Delta \tau} = -c \sqrt{K^2 - \left(1 - \frac{R_s}{r}\right)}$$

and we have elected to take the negative square root to ensure that r is *reducing* as τ *increases*.

This differential equation takes a bit more work to solve than the previous one:

$$c\Gamma = c\int_0^\Gamma d\tau = -\int_{r_0}^r \frac{dr}{\sqrt{K^2 - \left(1 - \dfrac{R_s}{r}\right)}},$$

where Γ is the finite proper time taken to fall from the starting point, r_0, to a distance r from the singularity. If we suggest that our intrepid observer falls from rest at infinity, then $E = mc^2 - \dfrac{GMm}{r_0} \approx mc^2$ and so $K = 1$, which transforms the integral into:

$$c\Gamma = c\int_0^\Gamma d\tau = -\int_{r_0}^r \frac{dr}{\sqrt{1^2 - \left(1 - \dfrac{R_s}{r}\right)}} = -\int_{r_0}^r \frac{dr}{\sqrt{\dfrac{R_s}{r}}} = -\frac{1}{\sqrt{R_s}}\int_{r_0}^r \sqrt{r}\, dr.$$

This is one of the standard integrals, giving the result:

$$c\Gamma = -\frac{1}{\sqrt{R_s}}\left[\frac{2}{3} r^{3/2}\right]_{r_0}^r.$$

Finally, after a little more massaging, we get:

$$\Gamma = \frac{2}{3c\sqrt{R_s}}\left\{r_0^{3/2} - r^{3/2}\right\}.$$

Note that this is a finite answer, even for $r = R_s$ and indeed for $r = 0$.

10.3.1 Time to Fall

Rather than resting on our laurels, having found this result, we press on to find the *co-ordinate time taken* for the same fall. To do this, we start from:

$$\left(\frac{\Delta t}{\Delta \tau}\right) = \frac{K}{\left(1 - \dfrac{R_s}{r}\right)} = \frac{1}{\left(1 - \dfrac{R_s}{r}\right)},$$

which we manipulate to get:

$$T = \int_0^T dt = \int_0^\Gamma \frac{d\tau}{\left(1 - \dfrac{R_s}{r}\right)}.$$

We can substitute in from the previous equation:

$$\frac{\Delta r}{\Delta \tau} = -c\sqrt{K^2 - \left(1 - \frac{R_s}{r}\right)} = -c\sqrt{1 - \left(1 - \frac{R_s}{r}\right)} = -c\sqrt{\frac{R_s}{r}}$$

to turn the integral over $d\tau$ into one over dr

$$T = \int_0^T dt = -\int_{r_0}^r \frac{\sqrt{r}\,dr}{c\sqrt{R_s}\left(1 - \frac{R_s}{r}\right)}.$$

On-line integrators allow us to find a solution, but the result is unpleasant (those of a slightly nervous mathematical disposition should look away now):

$$T = \frac{1}{c}\left\{\frac{-2}{3\sqrt{R_s}}\left((r)^{3/2} - r_0^{3/2} + 3R_s(r)^{1/2} - 3R_s(r_0)^{1/2}\right)\right\}$$

$$+ \frac{R_s}{c}\log\left\{\frac{\left[\left((r)^{1/2} + (R_s)^{1/2}\right)\left((r_0)^{1/2} - (R_s)^{1/2}\right)\right]}{\left[\left((r_0)^{1/2} + (R_s)^{1/2}\right)\left((r)^{1/2} - (R_s)^{1/2}\right)\right]}\right\}.$$

This expression diverges when $r = R_s$, as in Figure 10.10, which shows the proper time and the co-ordinate time for an observer falling towards Sgr A* from an initial distance of 5 times the Schwarzschild radius.

In order to make physical sense of this, we need to remember that the co-ordinate time corresponds to the proper time as seen from a point outside the influence of the gravitating mass, $(r > R_z)$. Such an observer would watch our falling volunteer approaching the Schwarzschild radius, but taking an infinite time to get there.

Our redshift calculations from Chapter 8 suggest that the light from the falling observer would get shifted to longer and longer wavelengths as they approached the event horizon. This shift would be greater than that calculated in Chapter 8, which imagined the source of light to be stationary. In this case, the falling observer is moving away from the distant one, so the relativistic Doppler shift has to be taken into account as well.

If the instruments observing the fall could keep up with this combined frequency shift, then the image of the falling observer would effectively freeze on the brink of the event horizon.

Not only will this image be redshifted, but it will also be dimmer. At first, the light from the falling observer gets dimmer simply due to the distance from the stationary one, as all light sources get dimmer with distance. However, as the event horizon gets nearer, then other general relativistic effects chime in and the brightness decreases exponentially.

Of course, all the effects that we have described apply equally well to the stellar material crossing the event horizon during the initial gravitational collapse that forms the black hole. There would be a frozen last image of the star, at massive redshift and very little luminosity imprinted just at the event horizon.

From the perspective of the falling observer or the stellar material, they would cross over the event horizon without noticing anything different and continue on to their eventual meeting with the singularity, in a finite time (a little under 300s for the case illustrated in Figure 10.10).

10.3.2 Geodesic Deviation Destruction

In truth, there wouldn't be an imprint of our observer emblazoned onto the event horizon, or at least not an image of an intact observer. Regrettably, they would get lethally stretched out in the radial direction, and compressed transversely: the inevitable consequence of geodesic deviation in the space-time. With a stellar mass black hole, this unfortunate fate comes about long before the event horizon is reached. Consequently, the last image of an observer in that case is likely to be rather gruesome. For supermassive black holes, the same fate comes to the observer eventually, but within the Schwarzschild radius, rather than before it. With the vastly greater mass of such black holes, the event horizon is much further out than for stellar varieties and the geodesic effect much gentler at these distances. Whenever it may

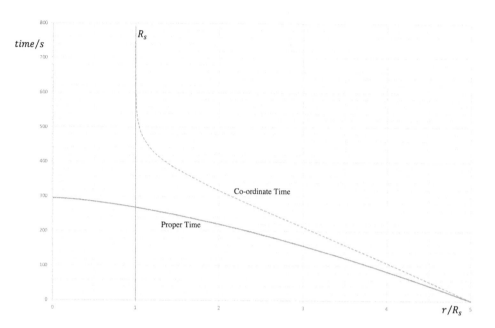

FIGURE 10.10 The proper time and the co-ordinate time for an object falling into Sgr A* from a distance of 5 times the Schwarzschild radius.

happen, eventually all forms of matter will encounter sufficient stresses to pull their structure apart, before the singularity is reached.

10.3.3 Inside the Schwarzschild Radius

While we do not understand the singular end point of gravitational collapse, we have some measure of confidence about the surrounding space-time inside the Schwarzschild radius. You may have already noticed that when $r < R_s$, $R_s/r > 1$, in which case:

$$\left(\Delta s\right)^2 = -\left|1 - \frac{R_s}{r}\right| c^2 \left(\Delta t\right)^2 + \frac{\left(\Delta r\right)^2}{\left|1 - \frac{R_s}{r}\right|} - r^2 \left(\Delta \theta\right)^2 - r^2 \sin^2 \left(\theta\right) \left(\Delta \varphi\right)^2$$

making it clear that the signs of the 00 and 11 components are reversed compared to $r > R_s$. Restricting ourselves to events at the same location, so that $\Delta r = \Delta \theta = \Delta \varphi = 0$, results in an interval which is space-like:

$$\left(\Delta s\right)^2 = -\left|1 - \frac{R_s}{r}\right| c^2 \left(\Delta t\right)^2 < 0,$$

which tells us that *two temporally separated events at the same location cannot be causally connected.* So, it is impossible for an object to remain stationary in this space-time; otherwise, it breaks causal connection with its own future! On the other hand, the interval between two spatially separated events with $\Delta t = 0$ is time-like.

$$\left(\Delta s\right)^2 = \frac{\left(\Delta r\right)^2}{\left|1 - \frac{R_s}{r}\right|} > 0.$$

There is important physics going on here, but it would be an exaggeration to say, as sometimes appears in popular accounts, that space and time have '*changed places*' inside a black hole, or that '*space and time are reversed*'. For one thing, co-ordinate time is calibrated to the proper time of an observer sitting outside of the influence of the gravitating mass $(r > R_z)$. As the event horizon prevents any communication with an observer placed far beyond that boundary, it is difficult to see how this practical programme could be carried through once you have fallen inside the Schwarzschild radius.

Actually, this flipping of space-like and time-like behaviour is another aspect of this specific co-ordinate system. There are other systems appropriate to this metric where this effect does not happen. However, what we can say is this: outside the event horizon an observer, using whatever means of propulsion lies to hand, can manoeuvre towards or away from the Schwarzschild radius. Once inside the event horizon, any event A inside the future light cone of an earlier event B *cannot be further from the singularity than A*. The singularity lies in the future, no matter what. In this sense, Δr takes on a 'time-like' aspect. There is now only one direction, towards the singularity, in which r can change.

10.4 Accretion Discs

Given that space is black, and that a black hole is... well, black, it would seem that scientists have their work cut out in obtaining observational evidence for their existence.

Two astronomers are credited with the discovery of Sgr A*, Bruce Balick and Robert L Brown after their 1974 paper[13] about a compact but powerful radio source at the centre of the galaxy. Subsequent orbital analysis of stars in the vicinity showed the influence of a gravitating object of about 4 million solar masses, but confined to a region much smaller than our solar system. With that sort of density, Sgr A* really has to be a supermassive black hole (Figure 10.11).

FIGURE 10.11   The region surrounding Sgr A* in X-ray and Infra-Red light. (Image credit: X-ray: NASA/UMass/D. Wang et al.; IR: NASA/STScI.)

The gravitational imprint of a black hole is not the only way it gives away its presence. Nature has provided another mechanism: the *accretion disc* (Figure 10.12).

There are many different situations where discs of diffuse material circulate around a central massive object. For example, when stars form, they are typically orbited by discs of material that can eventually coalesce into clumps that go on to form planets. In the case of black holes, however, gravitational effects are more marked and make the accretion discs highly efficient converters of mass into radiation.

In Newtonian physics, when an object's orbit is disturbed, it falls inwards towards the gravitating mass (or flies off into the distance). As its radius reduces, so does its gravitational potential energy which is converted into kinetic energy. In the case of an accretion disc, collisions between the particles in the diffuse material produce frictional and viscous effects which transform some of this kinetic energy into thermal energy. So, the inner radii of the disc are at much higher temperatures than the outer edges, and hence radiate electromagnetic radiation. Around a black hole, we need to take relativistic effects into account and as a result, material near to the event horizon becomes hot enough to emit X-rays. Overall between 10% and 40% of the mass of the material falling into the black hole can be converted into radiant energy. This contrasts markedly with the 0.7% conversion that takes place in typical fission reactions that feed nuclear power stations. In some cases, the accretion discs are accompanied by jets of material along their rotation axes. The mechanisms that form these jets are not understood, but they provide useful observational markers for astronomers.

Strictly speaking, in the presence of an accretion disc, the Schwarzschild metric is not appropriate as the material in the disc is a source of gravitation, so vacuum solutions are not valid. However, ignoring this factor is an approximation that we can live with.

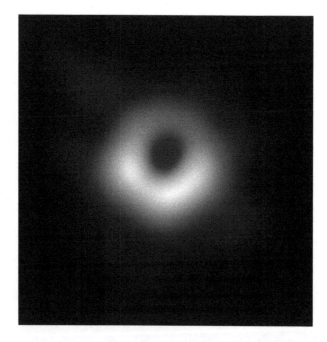

FIGURE 10.12 In this historic millimetre-wavelength image, the supermassive black hole in galaxy M87 is imaged for the first time. The bright ring is created by emissions from hot material circulating near the black hole. That the image is brighter towards the bottom is down to relativistic effects and shows that the disk is at a slight angle to our line of sight (compare with Figure 10.13 remembering that the jets emerge 90° to the plane of the disk), so that material towards the bottom of the image is moving towards us. The black hole is rotating counter-clockwise as viewed from Earth. The central shadow is determined by the radius of the closest stable photon orbit and is two to three times bigger than R_s. (Image credit: the Event Horizon Telescope collaboration.)

10.4.1 Active Galaxies

We now think that most galaxies have a supermassive black hole at their centre, although the mechanism that forms them is unclear. There certainly seems to be a link between the mass of a galaxy and the mass of the black hole at its core, suggesting that both are fed by the same mechanism.

These black holes are generally accompanied by accretion discs that are highly active and powerfully luminous, at least across some sections of the electromagnetic spectrum. Such *active galactic nuclei* (AGNs) can be distinguished as their characteristic emissions indicate that their luminosity is not attributable to stellar processes. *Active galaxies* with excess non-stellar emissions have been observed in every part of the spectrum from radio through to X-rays. Oftentimes, they are accompanied by jets that can be imaged from Earth (Figure 10.13).

As we observe more and more distant objects, so we see further back in time and as a result uncover a higher proportion of active galaxies in the earlier history of the universe. Presumably many of them have died down by now, as their accretion discs have been starved of material.

Some of these very early active galaxies are so distant they can't be imaged as galaxies at all. They appear as star-like dots of light. Such *quasars* (quasi-stellar objects) were a profound mystery when they were first observed, as astronomers in the 1960s took them to be stars in our own galaxy with extremely peculiar properties. As more detailed measurements of their spectra came in, so it became evident that these were exceptionally distant (highly redshifted) and profoundly old objects that had to be incredibly powerful luminous sources to be visible, even as dots of light, at these distances.

10.4.2 Cygnus X-1

As of 2018, there are fewer than 10 confirmed or suspected black hole candidates in our own galaxy. Of these, probably the most famous is Cygnus X-1. As you might suspect from its designation, Cygnus X-1 is a powerful X-ray source in the constellation of Cygnus, the swan. Cosmic X-rays do not penetrate the atmosphere, so this object was first observed in a series of instrument-carrying rocket launches in the 1960s. These were followed up by satellite measurements in the 1970s that revealed fluctuations in

FIGURE 10.13 A jet composed of electrons and other subatomic particles issuing from the active galactic nucleus of galaxy M87. (Image credit: J. A. Biretta et al., Hubble Heritage Team [STScI /AURA], NASA.)

the X-ray intensity over a timescale of a few fractions of a second.[14] Given such data, you can estimate the size of the emitting object.

Consider assembling a team of observers and spreading them across a large circular area on a parade ground or similar. You have equipped each one of them with a powerful torch that they point towards the sky. The aim is to get all the observers to switch their torches on and off at the same time, thus signalling to the stars. In order to synchronise the torches, an audible command is issued via speakers or similar at the centre of the circle. This all sounds terribly easy, until the speed of sound is taken into account. With a large enough area of observers, and a fast enough switching rate, those at the edge just get the command 'ON' as the command 'OFF' is being issued at the centre. Then the synchronisation breaks down and rather than a coherent signal being emitted, you get a 'grey' average.

So, seeing fluctuations ~ 0.4 s in Cygnus X-1 suggests that the source has to be ~ $c \times 0.4$ s in size, which is ~ 1.2×10^8 m, as this represents the largest region that can have synchronised emissions. To give a sense of scale, the Sun is ~ 1.4×10^9 m across.

Further refined analysis of the observational data has revealed that the X-ray source is comprised of a binary system with a compact massive object (the black hole) and a blue supergiant star in orbit about their common centre of mass with a period of ~ 5.6 days. The mass of the compact object has been estimated at ~ 15 solar masses with a Schwarzschild radius of about 44 km.

Astronomers believe that the accretion disc surrounding the black hole is being fed by material that is drawn from the supergiant, somewhat as represented in Figure 10.14.

The infalling material is being frictionally heated to a temperature that prompts it to emit X-rays: something like 3×10^5 K at least. In comparison, the Sun's surface temperature is ~ 6,000 K.

FIGURE 10.14 An artist's diagram showing material being gravitationally drawn from a blue supergiant variable star designated HDE 226868 onto a black hole known as Cygnus X-1. Note that gravitational lensing has not been taken into account in the rendering of the accretion disc and black hole. (Image credit: By ESA - European Homepage for the NASA/ESA Hubble Space Telescope: Copyrighted free use, https://commons.wikimedia.org/w/index.php?curid=6707568.)

10.5 Rotating Black Holes

One of the doyens of black hole physics, John Wheeler[15] coined the odd motto *black holes have no hair* to remind us that a black hole retains no history of its formation. The only measurable properties of a black hole are it mass, electrical charge and spin rate (angular momentum). There is nothing else that can tell us the nature of the matter that formed the black hole, nor anything about the varieties of stuff that have fallen in since. As all black holes must have mass, there are only four distinct types that can exist (Table 10.2).

We will not dwell on charged black holes. Given that matter tends to be neutral to high precision, we would expect most black holes to be neutral as well. In any case, if a black hole were to form with an electrical change, it is likely that it would attract nearby opposite charges and quickly become neutral. It is worth saying, however, that the charge would have an impact on the metric via the energy density that the accompanying electrical field would represent.

There is a further informal rough classification,[16] depending on the black hole's mass, which uses four categories (Table 10.3).

Some argue that the mini black hole category should be further sub-divided. There is as yet no observational evidence for their existence, but it is possible that the pressures involved in the dense turbulent early universe may have crushed material into these mini black holes. Intermediate black holes heavier than stellar masses are most likely developed from their lighter siblings by absorbing material or by merging with other black holes. The supermassive category seem linked to the cores of galaxies of sufficient mass, and may have their origins in the process of galaxy formation.

It may have crossed your mind to ponder what is meant by the mass of a black hole. After all, if all the matter has been destroyed by its encounter with the singularity, what material carries this mass? The mass of the black hole is the total mass of matter that has fallen into the event horizon and which generated the space-time curvature in the first place. The event horizon cuts off all forms of signal from inside its boundary, so even if the mass/energy within were to magically vanish, the outside would have no indication of that fact. Whatever the fate of matter at the singularity, its gravitational imprint remains. We will take up the issue of mass with respect to relativistic gravitational sources again in Section 11.4.1.

TABLE 10.2 Types of Black Hole with the Appropriate Metric

Mass	Spin	Charge	Metric
Yes	No	No	Schwarzschild
Yes	Yes	No	Kerr
Yes	No	Yes	Reissner–Nordström
Yes	Yes	Yes	Kerr–Newman

TABLE 10.3 Black Hole Categories by Mass

Mass	Category
$< 0.1\,M_s$	Mini black holes
$0.1 - 300\,M_s$	Stellar mass black holes
$300 - 10^5\,M_s$	Intermediate mass black holes
$10^5 - 10^{10}\,M_s$	Supermassive black holes

10.5.1 The Kerr Metric

The Kerr metric,[17] for a rotating neutral black hole is another vacuum solution of the field equations and generates the space-time interval:

$$(\Delta s)^2 = \left(1 - \frac{R_s r}{\rho^2}\right)c^2(\Delta t)^2 + \frac{2R_s r\alpha \sin^2 \vartheta}{\rho^2}c\Delta t\Delta\phi - \frac{\rho^2(\Delta r)^2}{D} - \rho^2(\Delta\vartheta)^2$$

$$- \left\{(r^2 + \alpha^2)\sin^2 \vartheta + \frac{R_s r\alpha^2 \sin^4 \vartheta}{\rho^2}\right\}(\Delta\phi)^2$$

in which ϕ is the standard azimuthal co-ordinate from the spherical-polar system, but ϑ and r are related to x, y, z by the following equations:

$$x = \sqrt{(r^2 + \alpha^2)}\sin \vartheta \cos \phi \quad y = \sqrt{(r^2 + \alpha^2)}\sin \vartheta \sin \phi \quad z = r\cos\vartheta,$$

which makes them slightly different to the spherical-polar system that we are familiar with. In particular, $r = 0$ does not mean $x = y = 0$, unless $\alpha = 0$.

In this metric, R_s is the Schwarzschild radius as before, and $\alpha = L/Mc$, where L is the angular momentum of the black hole.[18] The co-ordinates are set up so that the rotation brings about a change in ϕ only. Note that none of the metric's components depend explicitly on ϕ, which gives it symmetry over changes in that angle, which is one of the assumptions made in the metric's derivation. Unlike the Minkowski and Schwarzschild metrics, this one is not symmetrical in time: the metric is different if we change $\Delta t \to -\Delta t$.

The two functions $D = r^2 - R_s r + \alpha^2$ and $\rho^2 = r^2 + \alpha^2 \cos^2 \vartheta$ are introduced to tidy up the algebra.

Thanks to Birkoff's theorem (Section 8.1.4), we know that the Schwarzschild metric applies to the vacuum exterior of all spherically symmetrical mass distributions, even if they pulsate. There are also non-vacuum solutions for the interior of mass distributions that fold nicely into the Schwarzschild metric, which re-assures us about its applicability. However, there is no equivalent theorem for rotating structures and, so far, no non-vacuum solutions for a rotating mass distribution that merges with the Kerr metric. The assumption of symmetry about the rotation axis (axial symmetry) is quite a strong condition for real masses to adhere to, so it is conceivable that the Kerr metric might not be generally applicable. However, in the last moments of formation for a rotating black hole, we expect gravitational radiation to smooth things out, so the suspicion is that the Kerr metric will in practice work for all spinning neutral black holes (this is an aspect of the 'no-hair' conjecture).

10.5.2 Consistency of the Kerr Metric

If we take a snap-shot of Kerr space-time at fixed t (i.e. $\Delta t = 0$), then a surface of fixed r (i.e. $\Delta r = 0$) is an ellipsoid, the reduced interval being different to that on the surface of a sphere:

$$(\Delta s)^2 = -\rho^2(\Delta\vartheta)^2 - \left\{(r^2 + \alpha^2)\sin^2 \vartheta + \frac{R_s r\alpha^2 \sin^4 \vartheta}{\rho^2}\right\}(\Delta\phi)^2.$$

So, r is somewhat like a radial co-ordinate.

If we let $r \to \infty$ and in particular have $r \gg \alpha$, then $\rho \to r$ and $D \to r(r - R_s)$. The interval then becomes:

$$(\Delta s)^2 = \left(1 - \frac{R_s r}{r^2}\right)c^2(\Delta t)^2 + \frac{2R_s r\alpha \sin^2 \vartheta}{r^2}c\Delta t\Delta\phi - \frac{r^2(\Delta r)^2}{r(r - R_s)} - r^2(\Delta\vartheta)^2$$

$$- \left\{r^2 \sin^2 \vartheta + \frac{R_s r\alpha^2 \sin^4 \vartheta}{r^2}\right\}(\Delta\phi)^2.$$

A little cancelling down gives us:

$$(\Delta s)^2 = \left(1 - \frac{R_s}{r}\right)c^2(\Delta t)^2 + \frac{2R_s\alpha \sin^2 \vartheta}{r}c\Delta t\Delta\phi - \frac{(\Delta r)^2}{\left(1 - \dfrac{R_s}{r}\right)} - r^2(\Delta\vartheta)^2$$

$$- \left\{r^2 \sin^2 \vartheta + \frac{R_s\alpha^2 \sin^4 \vartheta}{r}\right\}(\Delta\phi)^2$$

revealing that as r continues to grow, the term $\dfrac{2R_s\alpha \sin^2 \vartheta}{r} \to 0$ and $\dfrac{R_s\alpha^2 \sin^4 \vartheta}{r} \to 0$. Equally, $\left(1 - \dfrac{R_s}{r}\right) \to 1$, reducing the Kerr metric to the Minkowski metric in spherical-polar co-ordinates, as we would expect a long way from the gravitating mass.

If α reduces to zero, i.e. the mass is not spinning, then the Kerr metric becomes the Schwarzschild metric and the co-ordinates also become the spherical-polar system that we are familiar with.

10.5.3 Singularities in the Kerr Metric

A casual inspection shows that the metric has singularities at $D = 0$ and $\rho = 0$. The first is a co-ordinate singularity, but the second is physical.

The physical singularity is a very curious animal (even more so than ordinary singularities) as it exists at $r = 0$, but only with $\vartheta = \pi/2$. Translating back to Cartesian co-ordinates for a moment, via:

$$x = \sqrt{(r^2 + \alpha^2)}\sin\vartheta\cos\phi \quad y = \sqrt{(r^2 + \alpha^2)}\sin\vartheta\sin\phi \quad z = r\cos\vartheta$$

shows us that the singularity lies at:

$$x = \alpha\sin\left(\frac{\pi}{2}\right)\cos\phi = \alpha\cos\phi \quad y = \alpha\sin\left(\frac{\pi}{2}\right)\sin\phi = \alpha\sin\phi \quad z = r\cos\left(\frac{\pi}{2}\right) = 0,$$

which is a ring of radius α about the Cartesian origin in the $\vartheta = \pi/2$ plane. Of course if $\alpha = 0$ this reduces to the Schwarzschild singularity.

The co-ordinate singularity at $D = 0$ equates to $r^2 - R_s r + \alpha^2 = 0$, creating a quadratic equation in r. If we write a quadratic in the form:

$$ax^2 + bx + c = 0 \quad \text{then } x = \frac{-b \pm \sqrt{b^2 - 4ac}}{2a}.$$

Applying this to our case gives:

$$r = \frac{R_s \pm \sqrt{R_s^2 - 4\alpha^2}}{2}.$$

Clearly, if $\alpha = 0$ this reduces to either $r = 0$ or $r = R_s$, which is the clue we need to see that the outer surface where:

$$r_1 = \frac{R_s + \sqrt{R_s^2 - 4\alpha^2}}{2} \to R_s \quad \text{if} \quad \alpha \to 0$$

is the event horizon for this form of black hole. It is interesting to see that this is smaller than the event horizon for a non-rotating ($\alpha = 0$) black hole of the same mass.

The inner surface:

$$r_2 = \frac{R_s - \sqrt{R_s^2 - 4\alpha^2}}{2} \to 0 \quad \text{if} \quad \alpha \to 0$$

is variously called the *Cauchy horizon* or the *inner event horizon*. The space-time within that region of the black hole behaves in a very odd manner, but more of that shortly.

In either case, there is a further condition:

$$R_s^2 - 4\alpha^2 \geq 0 \quad \text{giving} \quad R_s \geq 2\alpha$$

in order to ensure that the square root is a physically sensible value. This equates to a limit on the angular momentum of:

$$\alpha_{max} = \frac{R_s}{2} = \frac{GM}{c^2} \quad \text{or} \quad L \leq Mc\alpha_{max} = Mc\frac{R_s}{2} = \frac{GM^2}{c}.$$

Above this limit, there is no event horizon, rendering the singularity naked. However, it is not at all clear if exceeding this angular momentum limit is a physical possibility. Stellar collapse does tend to spin up the material, but various theoretical investigations starting from black holes with L below the limit have failed to produce a convincing mechanism that would increase L above the limit, via infalling matter. The issue is still an open one, as is the status of Penrose's cosmic censorship hypothesis.

Measurements of spin have been made for supermassive black holes at the centres of galaxies, which suggest that they are very close to the maximum possible (from 80% to over 90% with some variation between measurement techniques). These measurements either rely on the spin rate dependent broadening of spectral lines from ionised gas just above or below the accretion disc material close to the event horizon, or alternatively temperature dependent lower energy X-ray emissions from the material itself. It is possible that these black holes get their high spin rates during formation, or perhaps due to mergers with other black holes.

10.5.4 System Dragging

Looking back at the full Kerr metric:

$$(\Delta s)^2 = \left(1 - \frac{R_s r}{\rho^2}\right) c^2 (\Delta t)^2 + \frac{2R_s r\alpha \sin^2 \vartheta}{\rho^2} c\Delta t\Delta\phi - \frac{\rho^2 (\Delta r)^2}{D} - \rho^2 (\Delta\vartheta)^2$$

$$- \left\{ \left(r^2 + \alpha^2\right) \sin^2 \vartheta + \frac{R_s r\alpha^2 \sin^4 \vartheta}{\rho^2} \right\} (\Delta\phi)^2$$

and in particular the temporal component, $\left(1 - \frac{R_s r}{\rho^2}\right)$ we can see that if:

$$\frac{R_s r}{\rho^2} > 1 \quad \text{or} \quad \frac{\rho^2}{r} < R_s$$

then the component becomes negative. If we then 'sit' at a constant value of r, $\Delta r = 0$, in the $\vartheta = \pi/2$ $\Delta\vartheta = 0$, plane, we have:

$$(\Delta s)^2 = -\left|1 - \frac{R_s r}{\rho^2}\right| c^2 (\Delta t)^2 + \frac{2 R_s r \alpha}{\rho^2} c\Delta t \Delta \phi - \left\{ \left(r^2 + \alpha^2\right) + \frac{R_s r \alpha^2}{\rho^2} \right\} (\Delta \phi)^2.$$

In order for the overall interval to remain time-like, the unusual 'off-diagonal' term $\dfrac{2 R_s r \alpha}{\rho^2} c\Delta t \Delta \phi$ has to be positive, and large enough. In other words, $\Delta \phi > 0$ for $\Delta t > 0$. *Any object within that region of space-time has to be rotating along with the black hole.* This is known as *frame dragging*, or in our parlance that would be *system dragging*.

The boundary of this region can be found from the condition:

$$r = \frac{\rho^2}{R_s} \quad \text{which translates to} \quad r^2 - r R_s + \alpha^2 \cos^2 \vartheta = 0$$

generating the solutions:

$$r = \frac{R_s \pm \sqrt{R_s^2 - 4\alpha^2 \cos^2 \vartheta}}{2}.$$

The first marks what is called the *static limit*:

$$r_{SL} = \frac{R_s + \sqrt{R_s^2 - 4\alpha^2 \cos^2 \vartheta}}{2} \geq r_1$$

which extends to $2R_s$ at the equator and touches the outer event horizon at the poles (Figure 10.15). We will come back to the second solution shortly.

The region between the static limit (r_{SL}) and the event horizon (r_1) is called the *ergosphere*. Although it is not spherical in shape, it has the same basic geometry as a sphere. Again it is worth remembering that no forces operate to push any object in this region into rotating with the black hole, it is down to the nature of the space-time itself.

System dragging also occurs outside of the ergosphere, as $c\Delta t \Delta \phi$ does not vanish. As a result, any object falling freely towards the black hole on an initially radial path would collect non-radial components of motion due to the system dragging. The key point is that the effect can be resisted with appropriate propulsion outside the ergosphere, but not within.

From the point of view of an observer outside the gravitational influence of the black hole, a radially falling object curves around near to the event horizon and continues to orbit there for ever, this motion being tracked in co-ordinate distances and times. For an observer travelling with the object, the fall proceeds through the horizon just as with a Schwarzschild black hole.

10.5.5 System Dragging of Light

Consider a light ray orbiting a Kerr black hole at constant r, ϑ, i.e. $\Delta r = \Delta \vartheta = 0, \Delta \phi \neq 0$. The Kerr metric gives the null interval:

$$0 = \left(1 - \frac{R_s r}{\rho^2}\right) c^2 (\Delta t)^2 + \frac{2 R_s r \alpha \sin^2 \vartheta}{\rho^2} c\Delta t \Delta \phi - \left\{ \left(r^2 + \alpha^2\right) \sin^2 \vartheta + \frac{R_s r \alpha^2 \sin^4 \vartheta}{\rho^2} \right\} (\Delta \phi)^2.$$

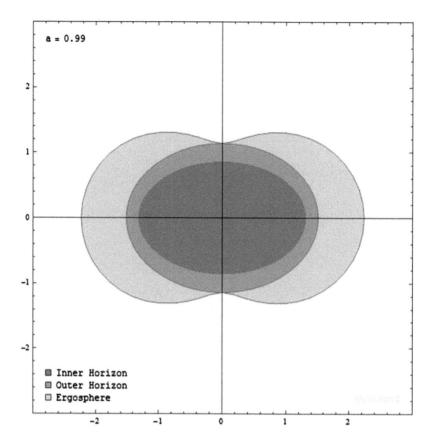

FIGURE 10.15 The ergosphere and horizons for a Kerr black hole rotating at $\alpha = 0.99$. The axes are in units of GM/c^2. (Image credit: Yukterez [Simon Tyran, Vienna].)

Dividing throughout by $c^2 (\Delta t)^2$ and inverting signs produces:

$$\left\{ \left(r^2 + \alpha^2\right)\sin^2 \vartheta + \frac{R_s r \alpha^2 \sin^4 \vartheta}{\rho^2} \right\} \frac{1}{c^2} \left(\frac{\Delta\phi}{\Delta t}\right)^2 - \frac{2R_s r \alpha \sin^2 \vartheta}{c\rho^2} \left(\frac{\Delta\phi}{\Delta t}\right) - \left(1 - \frac{R_s r}{\rho^2}\right) = 0$$

which is a fairly nasty quadratic equation in $\Delta\phi/\Delta t$. Writing:

$$A = \frac{1}{c^2} \left\{ \left(r^2 + \alpha^2\right)\sin^2 \vartheta + \frac{R_s r \alpha^2 \sin^4 \vartheta}{\rho^2} \right\} \quad B = -\frac{2R_s r \alpha \sin^2 \vartheta}{c\rho^2} \quad C = -\left(1 - \frac{R_s r}{\rho^2}\right)$$

sets up the quadratic solutions:

$$\frac{\Delta\phi}{\Delta t} = \frac{-B \pm \sqrt{B^2 - 4AC}}{2A}.$$

Now if we impose $C = 0$, which was the condition for the static limit, we see that:

$$\frac{\Delta\phi}{\Delta t} = \frac{-B \pm B}{2A} \quad \text{giving} \quad \frac{\Delta\phi}{\Delta t} = 0 \quad \text{or} \quad \frac{\Delta\phi}{\Delta t} = \frac{-B}{A}.$$

So, when viewed from the perspective of a distant observer, a light ray orbiting the black hole at the static limit but in the opposite sense to the black hole's rotation, would have a stationary ϕ co-ordinate (the $\Delta\phi/\Delta t = 0$ solution). At the static limit, the space-time is being dragged around the black hole's rotation axis at the speed of light, so any contra-rotating light is running against the tide and appears stationary to a distant observer.

The other solution is for a light ray orbiting in the same sense as the black hole, in which case:

$$\frac{\Delta\phi}{\Delta t} = \frac{-B}{A} = \frac{2cR_s r \alpha \sin^2 \vartheta}{\rho^2 \left\{ \left(r^2 + \alpha^2 \right) \sin^2 \vartheta + \dfrac{R_s r \alpha^2 \sin^4 \vartheta}{\rho^2} \right\}}.$$

As the ergosphere is further out than the event horizon, objects that cross into that region can escape the black hole again, if they have sufficient velocity. This leads to a possibility, called the *Penrose Process*, by which energy can be extracted from the rotation of the black hole. If an object falls into the ergosphere and is programmed to split into two (like the stages of a rocket), then provided that the 4-momenta of the two components are artfully arranged, one piece can escape with greater energy than whole object had originally. The other falls into the black hole slowing the rotation rate. We will discuss this further in Section 13.3.4.

System dragging is a remarkable feature of general relativity that shows how space-time becomes an active participant in physics, rather than the stage on which the universe plays out. Although its effects are most clearly and dramatically posited for rotating black holes, there have been various attempts to verify its existence via orbiting satellites. The Gravity Probe B mission flew between 2004 and 2005 and aimed to check system dragging via its effect on the direction of spin for four on-board gyroscopes. Unfortunately, non-uniform coatings on the gyroscopes produced unmodelled noise which complicated data analysis, but by 2008 NASA[19] confirmed that the system dragging effect had been observed to within 0.5% of the predicted value.

Another opportunity will lie in further observations of the orbital precession (as per Mercury in our own solar system) of the stars orbiting Sgr A*.

10.5.6 Left-Over Solutions

We have yet to consider two solutions:

the inner event horizon: $r_2 = \dfrac{R_s - \sqrt{R_s^2 - 4\alpha^2}}{2}$

the inner ergosphere: $r_{SL2} = \dfrac{R_s - \sqrt{R_s^2 - 4\alpha^2 \cos^2 \vartheta}}{2}$.

In the case of a black hole on the limit of rotation rate, $\alpha_{max} = R_s/2$, then $r_2 = r_1$ and the two event horizons merge into one.

With slower rotation rates, the space-time within the inner horizon has many intriguing features. Closed time-like curves are possible near to the singularity, and extensions to the co-ordinate systems suggest that it might be possible to avoid the ring singularity by flying through the centre into a new region of space-time. This has led to some theoretical speculations regarding links to other universes. However, analysis strongly suggests that the inner event horizon and the inner ergosphere are highly unstable – matter in their vicinity causes the structures to be destroyed. As a result, it is unlikely that these space-time regions would survive during the formation of a real black hole.

10.6 Gargantua

The plot of the 2014 movie *Interstellar* called for a fictional planetary system orbiting a rotating supermassive black hole, named *Gargantua*, in another galaxy. From the outset, the producers wanted the movie to be as scientifically accurate as possible, consistent with the needs of the central plot and the intelligibility of the script to an audience. Consequently, the visual effects team worked closely with astrophysicist Kip Thorne who produced detailed general relativistic calculations for bundles of light rays in the vicinity of a black hole. These equations were incorporated into new image rendering software that was specially written for the movie. The result was not only visually stunning imagery, but also scientific insight resulting in peer reviewed journal publications.

Gargantua's accretion disc was modelled to be rather benevolent, not fed by infalling material and stabilised to a temperature similar to the surface of the Sun. This served the needs of the plot in that heat from the disc made some of the planets habitable, although with extreme environments. It also ensured that there were no lethal levels of X-rays being produced by material cascading into the black hole, which would have fried the explorers.

The first surprising, but in hindsight obvious, aspect of the Gargantua images is the effect of gravitational lensing.[20] The result is a glorious arch over the top of the black hole showing the part of the accretion disc that is actually behind the black hole from the observer's perspective. A similar arch is present below the black hole. Many other artistic renditions portray black holes as eclipsing their accretion discs (e.g. Figure 10.14), which is not taking gravitational lensing into account. Figure 10.16 is one of the test images produced using Kip Thorne's calculations and the imaging software developed for the movie, even though it is not typical of the final images used.

The structure of this image can be broken down into several sections, with different physical causes for the image segments.

A: this image is formed from light emitted by the upper surface of the accretion disc in the region behind the black hole. The light is deflected over the top of the black hole to the observer's position.

B: the two parts of section B are both generated by light emitted from the underside of the disc from the observer's perspective. The wide bottom portion of the image is formed from light shining down from the underside of the disc behind the black hole and gravitationally deflected around as per A. The thin top image is due to light emitted from the underside of the disc at the front of the black hole. These rays have looped around the back of the black hole, so they are initially moving away from the observer, and are then deflected over the top and back towards the observation point.

C: this is light emitted from the upper surface of the disc that has travelled around the black hole once. A similar image should appear at the top for light that has made 1.5 loops around the black hole, but that image is unresolved in this render (Figure 10.17).

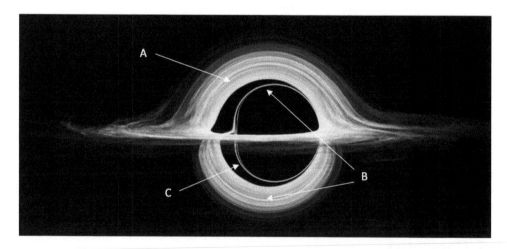

FIGURE 10.16   The fictional black hole Gargantua with its accretion disc viewed from a perspective that is slightly above the plane of the disc. The optical geometry leading to sections A, B, C of the image is explained in the text. (Image credit: Oliver James et al. 2015 Class. Quantum Grav. 32 065001.)

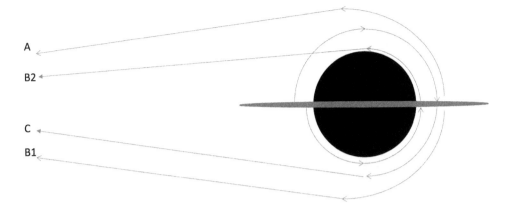

FIGURE 10.17 Ray paths for the accretion disc images around Gargantua. In reality the rays would pass around beyond the edge of the disc.

The calculations used for this image assumed a spin close to the maximum for a Kerr black hole with $\alpha = 0.999\alpha_{max}$, which is responsible for the distinct flattening of the thin inner rings on the left hand side (the black hole is rotating so that the left hand edge of the disc is moving towards us) and the evident off-centre location of the whole central portion. Images used in the movie were calculated for $\alpha = 0.6\alpha_{max}$, which gave much less flattening and off-centre results, as it was felt that the audience would be distracted by this puzzling aspect of the imagery. (Certain plot points, e.g. extreme time dilation experienced by some of the astronauts, were leveraged from the faster spin rate.) This image also does not include the effect of gravitational and relativistic Doppler shifting of light from the disc, which would produce a blue colouration on the left and red on the right. This same effect would also make the left hand side of the image much brighter than the right. The movie images also lacked this combined frequency shift. However, simulated lens flare, such as would be experienced by an IMAX camera imaging the black hole and accretion disc, was included in the movie renders.

Notes

1. 'Kip Thorne Quotes'. BrainyQuote.com. Brainy Media Inc., 2018. 18 October 2018. https://www.brainyquote.com/quotes/kip_thorne_682517.
2. Astrophysicists typically refer to the nuclear reactions in stars as 'burning'. This is picturesque but unhelpful in some ways, as burning is a chemical reaction with oxygen in the atmosphere, and that is certainly not what is happening in stars…
3. Luminosity relates to the energy output of the star. Brightness is how it appears in the sky, which depends on how far the star is from us. Very luminous stars can appear to be dim if they are a long way away.
4. The classical relationship between average kinetic energy and temperature is broken in such quantum systems.
5. Subrahmanyan Chandrasekhar (1910–1955), University of Chicago and Nobel Prize 1983. Acceptance of this work was significantly opposed by Eddington, who had a considerable sway over the community at the time.
6. It is not known if the frog enjoyed the experience.
7. David Ritz Finkelstein (1929–2016), Georgia Institute of Technology.
8. *Gravitation*, Misner, Thorne, Wheeler, W.H. Freeman & Co, (1973) § 31.3.
9. 'Galactic Explorer Andrea Ghez'. NOVA Interview, http://www.pbs.org. 31 October 2006.

10. Gravitational collapse: the role of general relativity, Penrose, R., 1969. *Noovo Ci-mento*, **1**, pp. 252–276.

11. Shapiro, S. L. and Teukolsky, S. A., 1992. Black holes, star clusters, and naked singularities: Numerical solution of Einstein's equations. *Philosophical Transactions of the Royal Society of London. Series A: Physical and Engineering Sciences*, **340**, 365–390. DOI:10.1098/rsta.1992.0073. Published 15 September 1992.

12. 'Kip Thorne Quotes'. BrainyQuote.com. Brainy Media Inc., 2018. 18 October 2018. https://www.brainyquote.com/quotes/kip_thorne_874789.

13. Balick, B. and Brown, R. L., 1 December 1974. Intense sub-arcsecond structure in the galactic center. *Astrophysical Journal*, **194**(1), pp. 265–270, bibcode: 1974ApJ...194..265B.

14. Oda, M., et al. 1999. X-Ray Pulsations from Cygnus X-1 Observed from UHURU. *The Astrophysical Journal*, 166, pp. L1–L7, Bibcode:1971ApJ...166L...1O, DOI:10.1086/180726.

15. John Archibald Wheeler, 1911–2008, Professor of Physics at Princeton.

16. Rough in the sense that the mass limits are not rigorously defined and universally accepted.

17. Discovered in 1963 by Roy Kerr (1934–), Mathematician at the University of New Zealand.

18. Angular momentum is mass multiplied by angular velocity for a point mass, for extended objects it is moment of inertia multiplied by angular velocity.

19. Everitt, C.W.F. and Parkinson, B.W., 2009. Gravity Probe B Science Results—NASA Final Report.

20. The Gargantua images are not the first renderings of black holes to correctly incorporate this aspect, but they are the first in popular culture.

11

Gravitational Waves

It's a spectacular signal. It's a signal many of us have wanted to observe since the time LIGO was proposed. It shows the dynamics of objects in the strongest gravitational fields imaginable, a domain where Newton's gravity doesn't work at all, and one needs the fully non-linear Einstein field equations to explain the phenomena. The triumph is that the waveform we measure is very well-represented by solutions of these equations. Einstein is right in a regime where his theory has never been tested before.

R Weiss[1]

Just as electromagnetic astronomy was begun in essence, at least modern astronomy, by Galileo pointing his telescope in the sky and discovering Jupiter's moons. This is the same thing but for gravitational waves. Gravitational waves are the only other kind of wave, besides electromagnetic that propagate across the universe, bringing us information about the universe, so initially we will see not just binary black holes. We will see neutron stars collide, tear each other apart, we will see black holes tearing neutron stars apart, we will see spinning neutron stars, pulsars, when the space-based LISA mission is operating hopefully by about 2030, we'll be exploring basically the birth of the universe, the earliest moments of the universe. And there will ever so much more I'm sure, including huge surprises, as the years wear on.

K Thorne, by permission[2]

11.1 The Hulse–Taylor Pulsar System

The pulsar system PSR B1913+16, which is also known as the Hulse–Taylor system, was first discovered from radio signals detected at the *Arecibo* telescope in 1974. This 305 m radio dish in Puerto Rico (Figure 11.1) was completed in 1963 and was the largest single dish telescope in the world until 2016, when it was eclipsed by the Five hundred metre Aperture Spherical Telescope (FAST) in China.

As we noted in Chapter 1, the discovery was made by Russell Alan Hulse and Joseph Hooton Taylor, Jr., of the University of Massachusetts Amherst, who were awarded the 1993 Nobel Prize in Physics for[3] *for the discovery of a new type of pulsar, a discovery that has opened up new possibilities for the study of gravitation.*

Timing of the Hulse–Taylor pulsar's signals shows that it has a period of 59 ms, meaning that it rotates on its axis 17 times every second. As more data about the pulsar came in, a systematic variation in the arrival time of the pulses became evident: sometimes they arrive 3 s earlier than at other times. This suggests that the object is in an orbit that sometimes brings it nearer to the Earth and sometimes further away. In fact, the orbit must be 3 light-seconds (~ 2/3 of the Sun's diameter) across to account for the variation.

FIGURE 11.1 The 305-m Arecibo radio telescope in Puerto Rico, used to discover the Hulse-Taylor pulsar system. (Image credit: Author H. Schweiker/WIYN and NOAO/AURA/NSF.)

We now know that the system comprises two neutron stars, both of ~ 1.4 solar masses, in mutual orbit with a period of about 7.75 h. The second neutron star is not a pulsar, presumably as its emissions are aimed at an angle that does not sweep across our line of sight.

It situations like this, both bodies orbit about their common centre of mass, as illustrated in Figure 11.2. At their closest (periastron) the neutron stars are separated by about ~ 1.1 times the radius of the Sun and at their furthest (apastron) ~ 4.8 solar radii.

This remarkable system has proven to be a valuable experimental test-bed for the general theory. First, the orientation of periastron rotates by about 4.2 degrees per year in accordance with the precession of orbits predicted by general relativity.[4] Second, as the pulsar orbits, so its speed changes along with its distance to the other neutron star. As a result, shifts take place in the pulse rate due to gravitational and Doppler effects, which have been confirmed to be in line with the relativistic predictions.

Most importantly, over the 40 odd years that the system has been observed, the two neutron stars have spiralled in towards each other. It would appear that energy is being lost from the system at a rate of 7.35×10^{24} W, which compares with the 3.85×10^{26} W that the Sun radiates.[5] As the orbits contract by 3.5 m per year, the orbital period reduces by 76.5 μs per year (Figure 11.3). Consequently, the predicted time to the final collision and merging of the neutron stars is 300 million years.

The ratio between the predicted rate of orbital decay, based on gravitational wave emission, and the measured rate[6] is 0.997 ± 0.002, with the biggest contributions to the uncertainty coming from various measurements, including the distance of the Sun from the galactic centre, the distance to the pulsar and its motion through the galaxy. It's unlikely that these figures will be improved in the near future, but this still represents impressive confirmation of the theory and most physicists see it as a persuasive indication that gravitational waves exist.

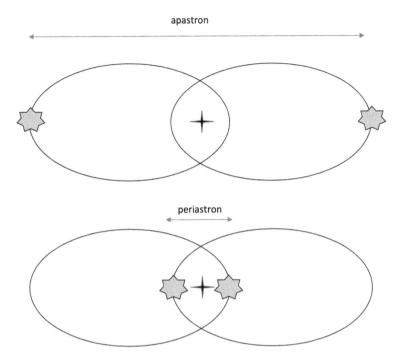

FIGURE 11.2 Two masses in orbit about their common centre of mass, indicated by the cross. Periastron is when the two objects are at their closest, and apastron when they are furthest apart.

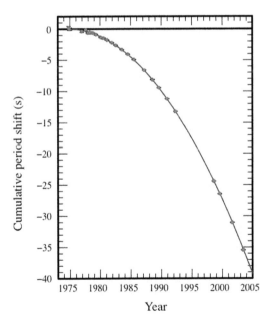

FIGURE 11.3 Cumulative shift in the periastron period in seconds for the binary star system PSR B1913+16 as the system loses energy by gravitational wave emission. Points are experimental data, and the line is the theoretical values predicted by general relativity. The data come from J. M. Weisberg and J. H. Taylor, Relativistic Binary Pulsar B1913+16: Thirty Years of Observations and Analysis, July 2004. (Image credit: *Inductiveload – Own work.*)

11.2 Linear Gravity

The theory of gravitational radiation is made much harder by the non-linearity of Einstein's field equations. If we presume that a gravitational wave carries energy (Section 11.4.3), then it must itself act as a source of gravitational effects. A wave propagating through otherwise empty space-time, would curve its own space-time curvature.

Of course, there is value in seeking solutions (exact, approximate, numerical or otherwise) of the full field equations, but we are a long way from being equipped for such a challenge. Instead, we will confine ourselves to gravitational waves that are so weak the contribution of their own energy to curvature can be neglected, which is acceptable as the gravitational waves detected on Earth are certainly weak.

This puts us squarely in the weak field limit, as discussed in Chapter 8. However, the results that we obtained there will not do, as one of the criteria we applied in approaching the Newtonian limit was that the fields were stationary (not changing with time). That is not the case with a gravitational wave! So, we need to obtain linear versions of the field equations that are not restricted to stationary situations. This is called *linear gravity theory*.

In Chapter 8, we expressed the metric as a small adjustment to the Minkowski flat-space version by writing $g_{\overline{ab}} = \eta_{\overline{ab}} + h_{\overline{ab}}$ with $h_{\overline{ab}} \ll 1$. Consequently, in linear gravity we can switch between contravariant and covariant indices using η rather than the full metric, g. So, for example[7]:

$$h_{a\overline{b}} = \sum_n \eta_{bn} h_{an} \quad \text{and} \quad h_{\overline{ab}} = \sum_{n,m} \eta_{am} \eta_{bn} h_{mn}$$

$$h_{ab} = \sum_n \left(\eta^{-1}\right)_{bn} h_{a\overline{n}} \quad \text{and} \quad h_{ab} = \sum_{n,m} \left(\eta^{-1}\right)_{am} \left(\eta^{-1}\right)_{bn} h_{\overline{mn}}.$$

The linearised Christoffel symbols are as follows:

$$\Gamma^j_{ik} = \frac{1}{2} \sum_m \eta_{jm} \left\{ \frac{\Delta h_{\overline{mk}}}{\Delta x_i} + \frac{\Delta h_{\overline{im}}}{\Delta x_k} - \frac{\Delta h_{\overline{ik}}}{\Delta x_m} \right\}$$

and as products in Γ can be dropped (weak field limit), the linearised Riemann tensor takes the form:

$$R_{\overline{iabc}} = \frac{\Delta\left(\Gamma^i_{ac}\right)}{\Delta x_b} - \frac{\Delta\left(\Gamma^i_{ab}\right)}{\Delta x_c} + \sum_n \left(\Gamma^n_{ac}\Gamma^i_{nb} - \Gamma^n_{ab}\Gamma^i_{nc}\right) \approx \frac{\Delta\left(\Gamma^i_{ac}\right)}{\Delta x_b} - \frac{\Delta\left(\Gamma^i_{ab}\right)}{\Delta x_c}.$$

From this we get to the linearised covariant Ricci tensor:

$$R_{\overline{ab}} \approx \sum_i R_{\overline{iabi}} = \sum_i \left\{ \frac{\Delta\left(\Gamma^i_{ai}\right)}{\Delta x_b} - \frac{\Delta\left(\Gamma^i_{ab}\right)}{\Delta x_i} \right\}.$$

At this point, we need to insert the Christoffel symbols and process a fair degree of algebra. The interested reader can pick up the argument in the online Appendix, while here we will skip to the intermediate result. However, some points from the proof are relevant to our future discussion:

- the notation

$$\Box^2(\phi) = \sum_{u,v} g_{uv} \frac{\Delta}{\Delta x_u}\left(\frac{\Delta\phi}{\Delta x_v}\right) = \frac{\Delta}{\Delta x_o}\left(\frac{\Delta\phi}{\Delta x_o}\right) - \nabla^2\phi$$

is commonly used to shorten appropriate mathematical expressions[8];

- it is helpful to define the contraction $h = \sum\limits_{i,m} \eta_{im} h_{\overline{mi}}$;

- the proof relies on our ability to choose co-ordinate systems without losing generality, as we discussed in Section 9.1. In particular, two selections are crucial. First a new metric related to $h_{\overline{ab}}$ is defined using $H_{\overline{ab}} = h_{\overline{ab}} - \frac{1}{2} \eta_{\overline{ab}} h$ and then the condition $\sum\limits_{a} \dfrac{\Delta H_{\overline{ab}}}{\Delta x_a} = 0$ is applied to further constrain that choice. This does not exhaust our ability to adjust co-ordinates, and a further step like this is taken as we develop the argument into gravitational waves.

Once we get to the end of the algebraic fiddling, the Ricci tensor has become:

$$R_{\overline{ab}} = \frac{1}{2} \Box^2 \left(h_{\overline{ab}} \right)$$

making the Ricci scalar:

$$\mathcal{R} = \sum_{a,b} \eta_{ab} R_{\overline{ab}} = \frac{1}{2} \sum_{a,b} \eta_{ab} \Box^2 \left(h_{\overline{ab}} \right) = \frac{1}{2} \sum_{a,b} \Box^2 \left(\eta_{ab} h_{\overline{ab}} \right) = \frac{1}{2} \Box^2 (h),$$

which may seem slightly odd, as neither of them involve H, but our final result will.

11.2.1 Linear Field Equations

The field equations we obtained in Chapter 8 using the linear approximations are:

$$\mathbb{R}_{\overline{ab}} - \frac{1}{2} \mathcal{R} \eta_{\overline{ab}} = -\frac{8\pi G}{c^4} \mathbb{T}_{\overline{ab}}.$$

Essentially, we used the fact that $h_{ab} \ll 1$ to use η rather than g in front of the Ricci scalar.
   Substituting our linearised Ricci tensor and scalar:

$$\frac{1}{2} \Box^2 \left(h_{\overline{ab}} \right) - \frac{1}{2} \left(\frac{1}{2} \Box^2 (h) \right) \eta_{\overline{ab}} = \frac{1}{2} \Box^2 \left(h_{\overline{ab}} - \frac{1}{2} \eta_{\overline{ab}} h \right) - \frac{8\pi G}{c^4} \mathbb{T}_{\overline{ab}}.$$

Remembering that we have defined $H_{\overline{ab}} = h_{\overline{ab}} - \frac{1}{2} \eta_{\overline{ab}} h$ gives us a final version:

$$\frac{1}{2} \Box^2 \left(H_{\overline{ab}} \right) = -\frac{8\pi G}{c^4} \mathbb{T}_{\overline{ab}} \quad \text{or} \quad \Box^2 \left(H_{\overline{ab}} \right) = -\frac{16\pi G}{c^4} \mathbb{T}_{\overline{ab}}, \tag{11.1}$$

which is an amazingly compact and elegant result.
   The vacuum equations come from setting $\mathbb{T}_{\overline{ab}} = 0$:

$$\Box^2 \left(H_{\overline{ab}} \right) = 0,$$

which you may recognise as a type of wave equation, as we will discuss shortly.
   This set of linear field equations for the components of $H_{\overline{ab}}$ are still tricky to solve, but a lot better than the complete equations were. They suffice for nearly all the classic experimental tests of general relativity, including starlight deflection, the shift in Mercury's orbit, and tests performed by bouncing radar pulses off deep space satellites. They can also be used to discuss the working of gravitational lenses, and gravitational wave motion and detection. However, gravitational wave *emission* requires the full theory,

e.g. for the analysis of the Hulse–Taylor binary pulsar system and the modelling of black hole and neutron star merger for the LIGO experiment. Cosmology is also a fertile ground for the full theory.

11.2.2 Weak(ish) Gravitation

Our linear gravity theory has to be consistent with our previous explorations of the Newtonian limit. Imposing the extra conditions (static field, slow moving objects) restricts $\mathbb{T}_{ab}$ so that the dominant term is $\mathbb{T}_{\overline{00}} = \rho c^2$ (as we discussed in Chapter 8). Hence, the linear equations (Equation 11.1) become:

$$\frac{1}{2}\Box^2\left(H_{\overline{00}}\right) = -\frac{8\pi G}{c^4}\rho c^2.$$

The expression $\Box^2\left(H_{\overline{00}}\right)$ reduces to $-\nabla^2\left(H_{\overline{00}}\right)$ as the field is static (there is no temporal rate of change). So, now we have:

$$-\frac{1}{2}\nabla^2\left(H_{\overline{00}}\right) = -\frac{8\pi G}{c^4}\rho c^2.$$

Comparing this to Poisson's equation $\nabla^2\left(\phi\right) = 4\pi G\rho$ tells us that $H_{\overline{00}} = 4\phi/c^2$, which is not quite what we had in Chapter 8. However, our previous result was for h not H and the two are connected by $H_{\overline{ub}} = h_{\overline{ub}} - \frac{1}{2}\eta_{\overline{ub}}h$ where $h = \sum_i h_{\overline{ii}}$. Contracting $H_{\overline{ub}}$ into H we obtain:

$$H = \sum_{u,b}\eta_{ub}H_{\overline{ub}} = \sum_{u,b}\eta_{ub}h_{\overline{ub}} - \frac{1}{2}\sum_{u,b}\eta_{ub}\eta_{ub}h = h - \frac{1}{2}\times 4\times h = -h.$$

As all terms in $H_{\overline{ab}}$ are small compared with $H_{\overline{00}}$, in the Newtonian limit, the sum $\sum_i H_{\overline{ii}}$ is approximately equal to the value of the $H_{\overline{00}}$ term, making $H = -h = 4\phi/c^2$. Now we can calculate $h_{\overline{ab}}$:

$$h_{\overline{ab}} = H_{\overline{ab}} + \frac{1}{2}\eta_{\overline{ab}}h = H_{\overline{ab}} - \frac{1}{2}\eta_{\overline{ab}}\frac{4\phi}{c^2}$$

allowing us to extract the components $h_{\overline{00}}$ and $h_{\overline{ii}}$:

$$h_{\overline{00}} = H_{\overline{00}} + \frac{1}{2}\eta_{\overline{00}}h = \frac{4\phi}{c^2} - \eta_{00}\frac{2\phi}{c^2} = \frac{2\phi}{c^2}\quad\text{as before in Chapter 8}$$

$$h_{\overline{ii}} = H_{\overline{ii}} + \frac{1}{2}\eta_{\overline{ii}}h = 0 - \eta_{\overline{ii}}\frac{2\phi}{c^2} = \frac{2\phi}{c^2}\quad\text{where}\quad i=1,2,3.$$

This means that we can construct a full (linear, approximate) metric, $g_{\overline{ab}} = \eta_{\overline{ab}} + h_{\overline{ab}}$, with the equivalent interval:

$$(\Delta s)^2 = \left(1 + \frac{2\phi}{c^2}\right)c^2(\Delta t)^2 - \left(1 - \frac{2\phi}{c^2}\right)\left((\Delta x)^2 + (\Delta y)^2 + (\Delta z)^2\right).$$

Using the Newtonian definition of the potential $\phi = -GM/r$, we can expose this metric in a more revealing form:

$$(\Delta s)^2 = \left(1 - \frac{2GM}{c^2 r}\right)c^2 (\Delta t)^2 - \left(1 + \frac{2GM}{c^2 r}\right)\left((\Delta x)^2 + (\Delta y)^2 + (\Delta z)^2\right).$$

The Schwarzschild metric is:

$$(\Delta s)^2 = \left(1 - \frac{2GM}{c^2 r}\right)c^2 (\Delta t)^2 - \frac{(\Delta r)^2}{\left(1 - \frac{2GM}{c^2 r}\right)} - r^2 (\Delta \theta)^2 - r^2 \sin^2 (\theta)(\Delta \varphi)^2$$

and if $\frac{2GM}{c^2 r} < 1$, we can use the good old $(1+x)^n \approx 1+nx$ approximation to change the term in Δr into:

$$\frac{1}{\left(1 - \frac{2GM}{c^2 r}\right)} \approx \left(1 + \frac{2GM}{c^2 r}\right)$$

giving:

$$(\Delta s)^2 \approx \left(1 - \frac{2GM}{c^2 r}\right)c^2 (\Delta t)^2 - \left(1 + \frac{2GM}{c^2 r}\right)(\Delta r)^2 - r^2 (\Delta \theta)^2 - r^2 \sin^2 (\theta)(\Delta \varphi)^2,$$

which is clearly:

$$(\Delta s)^2 \approx \left(1 + \frac{2\phi}{c^2}\right)c^2 (\Delta t)^2 - \left(1 - \frac{2\phi}{c^2}\right)(\Delta r)^2 - r^2 (\Delta \theta)^2 - r^2 \sin^2 (\theta)(\Delta \varphi)^2$$

similar to our new metric, once we have sorted out the difference between Cartesian and spherical-polar co-ordinates.

What we have here is a second stage approximation midway between the Newtonian limit of temporal curvature only and full-on general relativity. If we need to take spatial curvature into account, this metric will do provided the masses involved are not too large. If we want full accuracy, or the masses are large (so that we can't say $\frac{2GM}{c^2 r} < 1$) then we need to deploy the Schwarzschild metric, or whichever is determined by the situation. You might say that our new metric is appropriate for weak(ish) gravitation…

One way to classify the strength of gravitational effects is to look at the ratio between the (Newtonian) gravitational potential energy for a particle and its (Einsteinian) mass energy at rest:

$$\varepsilon = \frac{GMm/r}{mc^2} = \frac{GM}{c^2 r}.$$

Earlier in Chapter 8, we calculated values for the Sun and Sgr A*. Now, for completeness we add in some more possibilities in Table 11.1.

TABLE 11.1 Strength of Gravitational Effects in Different Situations

Situation	ε
The Sun	2.2×10^{-6}
Sgr A*	0.22
Earth	7.0×10^{-10}
Jupiter	2.0×10^{-8}

The strong field situation is taken as $\varepsilon \to 1$ with the weak field being characterised by $\varepsilon \ll 1$, leaving the middle levels for weak(ish) gravitation.

11.3 Gravitational Wave Theory

The simplest form of wave motion can be described by a sine function of the type:

$$y = A \sin kx,$$

where y represents the displacement caused by the wave, A is the maximum displacement, otherwise known as the *amplitude*, and k is a constant that we will discuss further in a moment. As it stands, this formula represents a stationary form which looks like Figure 11.4.

Now consider shifting the y-axis to the right by two units, as per Figure 11.5, so that there is a new x' scale related to the old by $x = x' + 2$. The function now becomes:

$$y = A \sin kx = A \sin k(x' + 2).$$

To tweak this further, imagine that the y-axis was sliding along the x-axis at speed v, so that $x = x' + vt$, now the wave is:

$$y = A \sin k(x' + vt) = A \sin(kx' + kvt).$$

Back in Section 5.2 we introduced the following variables for waves:

$$k = \frac{2\pi}{\lambda} \quad \text{and} \quad \omega = 2\pi f$$

with the wave speed being determined by

$$v = f\lambda \quad \text{or} \quad v = \frac{\omega}{2\pi} \times \frac{2\pi}{k} = \frac{\omega}{k}.$$

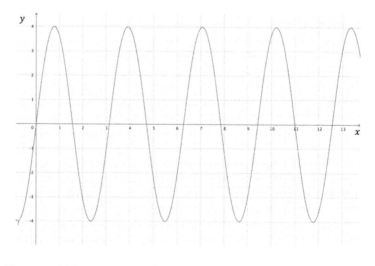

FIGURE 11.4   The sinusoidal function $y = A \sin kx$, here drawn for $k = 2$, $A = 4$.

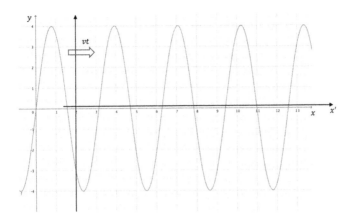

FIGURE 11.5 Displacing the *y*-axis to the right generates a new *x'*-axis, here drawn slightly above the *x*-axis for clarity.

Using these, we now have the standard formula for a moving (progressive) wave:

$$y = A\sin(kx' + \omega t).$$

As we moved the *axis to the right*, this is the equation for a *wave moving to the left*. By convention we normally consider waves moving to the right, so the orthodox equation is:

$$y = A\sin(kx - \omega t + \varphi)$$

switching back to *x* co-ordinates, simply for convenience, and adding the *phase factor*, φ, which tells us the initial displacement in the wave at $t = x = 0$. Generally the axes are chosen so that $\varphi = 0$.

11.3.1 The Wave Equation

There are two standard results from the theory of rates of change that we need for the next step:

$$\frac{\Delta}{\Delta x}(A\sin kx) = Ak\cos kx \qquad \frac{\Delta}{\Delta x}(A\cos kx) = -Ak\sin kx.$$

Applying these, twice, to our wave formula gives us:

$$\frac{\Delta}{\Delta x}\left(\frac{\Delta}{\Delta x}(A\sin(kx - \omega t))\right) = -Ak^2\sin(kx - \omega t) = -k^2 y$$

$$\frac{\Delta}{\Delta t}\left(\frac{\Delta}{\Delta t}(A\sin(kx - \omega t))\right) = -A\omega^2\sin(kx - \omega t) = -\omega^2 y,$$

so that

$$\frac{1}{\omega^2}\frac{\Delta}{\Delta t}\left(\frac{\Delta}{\Delta t}(A\sin(kx - \omega t))\right) = \frac{1}{k^2}\frac{\Delta}{\Delta x}\left(\frac{\Delta}{\Delta x}(A\sin(kx - \omega t))\right)$$

which is the same as:

$$\frac{k^2}{\omega^2}\frac{\Delta}{\Delta t}\left(\frac{\Delta}{\Delta t}\left(A\sin\left(kx-\omega t\right)\right)\right)=\frac{\Delta}{\Delta x}\left(\frac{\Delta}{\Delta x}\left(A\sin\left(kx-\omega t\right)\right)\right).$$

Rearranging into a more often quoted form we have:

$$\frac{1}{v^2}\frac{\Delta}{\Delta t}\left(\frac{\Delta y}{\Delta t}\right)-\frac{\Delta}{\Delta x}\left(\frac{\Delta y}{\Delta x}\right)=0$$

which is the standard equation for waves. Although we have obtained this by manipulating one specific solution, there are a wide range of different solutions to this equation. A powerful and beautiful theorem tells us that all possible solutions of the wave equation can be built by artful combinations of $y(k)=A\sin\left(kx-\omega t+\varphi\right)$, with appropriate choices of k and the other various constants. Hence, we have not lost any generality by starting from this one simple solution.

Next we need to have an ah ha! moment, and recognise that:

$$\frac{1}{v^2}\frac{\Delta}{\Delta t}\left(\frac{\Delta y}{\Delta t}\right)-\frac{\Delta}{\Delta x}\left(\frac{\Delta y}{\Delta x}\right)=0$$

looks very similar to the vacuum field equations:

$$\Box^2\left(H_{\overline{ab}}\right)=\frac{\Delta}{\Delta x_o}\left(\frac{\Delta H_{\overline{ab}}}{\Delta x_o}\right)-\nabla^2 H_{\overline{ab}}=0$$

making sure that we are in Cartesian co-ordinates:

$$\Box^2\left(H_{\overline{ab}}\right)=\frac{1}{c^2}\frac{\Delta}{\Delta t}\left(\frac{\Delta H_{\overline{ab}}}{\Delta t}\right)-\nabla^2 H_{\overline{ab}}=0,$$

emphasises that this is a (3D) wave equation and its solutions have velocity c.

11.3.2 Gravitational Wave Solutions

Given the similarity between the vacuum versions of the linear field equations and the wave equation, we are motivated to look for a solution like $y=A\sin\left(kx-\omega t\right)$.

The first step is to recall the frequency 4-vector that we introduced in Section 5.2:

$$F=\begin{pmatrix}\omega/c\\k\end{pmatrix},$$

so that our solution takes the form:

$$H_{\overline{ab}}=A_{\overline{ab}}\sin\left(-\sum_i F_iX_{\overline{i}}\right)=A_{\overline{ab}}\sin\left(\sum_{i=1,2,3}k_ix_i-\omega t\right)$$

and the amplitude is represented by the tensor $A_{\overline{ab}}$, which has constant components.

As we set up the linear field equations, we picked the condition:

$$\sum_a \frac{\Delta H_{a\bar{b}}}{\Delta x_a} = 0.$$

Applying that condition:

$$\sum_a \frac{\Delta H_{a\bar{b}}}{\Delta x_a} = \sum_a \frac{\Delta}{\Delta x_a} \left(\sum_j \eta_{aj} A_{\bar{j}\bar{b}} \sin\left(-\sum_i F_i x_{\bar{i}} \right) \right) = 0$$

and as the amplitude tensor has constant components, this becomes:

$$-\sum_{j,a} \eta_{aj} F_a A_{\bar{j}\bar{b}} \cos\left(-\sum_i F_i x_{\bar{i}} \right) = 0$$

which tells us that:

$$\sum_a F_a A_{a\bar{b}} = 0.$$

Normally, the product of a vector into a tensor gives the components of a new vector. In this case, we see that all the components are zero. Putting a geometrical interpretation on this, the product of the vector into the tensor is telling us the extent to which the tensor's components are 'pointing' in the same direction as the vector. Our vector is F which is spatially oriented along the wave's direction of travel. The tensor is $A_{a\bar{b}}$, the amplitude, so as all product's components are zero, the amplitude is at right angles to the direction of travel. In other words, this is a *transverse wave*.[9]

11.3.3 Polarisation of Gravitational Waves

To simplify things, we assume that we have a wave that is propagating along the z-axis. This makes the covariant 4-frequency:

$$\bar{F} = \begin{pmatrix} \omega/c \\ 0 \\ 0 \\ -k \end{pmatrix}.$$

The amplitude tensor can be expressed in matrix form by writing:

$$A = \begin{pmatrix} A_{00} & A_{01} & A_{02} & A_{03} \\ A_{10} & A_{11} & A_{12} & A_{13} \\ A_{20} & A_{21} & A_{22} & A_{23} \\ A_{30} & A_{31} & A_{32} & A_{33} \end{pmatrix}.$$

So multiplying the amplitude tensor by the frequency 4-vector gives the result:

$$\begin{pmatrix} A_{00} & A_{01} & A_{02} & A_{03} \\ A_{10} & A_{11} & A_{12} & A_{13} \\ A_{20} & A_{21} & A_{22} & A_{23} \\ A_{30} & A_{31} & A_{32} & A_{33} \end{pmatrix} \begin{pmatrix} \omega/c \\ 0 \\ 0 \\ -k \end{pmatrix} = \begin{pmatrix} A_{00}\omega/c - kA_{03} \\ A_{10}\omega/c - kA_{13} \\ A_{20}\omega/c - kA_{23} \\ A_{30}\omega/c - kA_{33} \end{pmatrix}.$$

We already know that this product is zero, for all the terms, and hence $A_{i0} \omega/c = kA_{i3}$ and as $\omega/c = k$, $A_{i0} = A_{i3}$. The co-ordinate freedom that we used earlier has not been exhausted; there is sufficient flexibility left for us to impose a further condition which is $A_{0i} = 0$, so the tensor becomes:

$$\begin{pmatrix} 0 & 0 & 0 & 0 \\ A_{10} & A_{11} & A_{12} & A_{13} \\ A_{20} & A_{21} & A_{22} & A_{23} \\ A_{30} & A_{31} & A_{32} & A_{33} \end{pmatrix}.$$

Proposing that the tensor is symmetrical (like other metrics) brings us:

$$\begin{pmatrix} 0 & 0 & 0 & 0 \\ 0 & A_{11} & A_{12} & 0 \\ 0 & A_{12} & A_{22} & 0 \\ 0 & 0 & 0 & 0 \end{pmatrix}$$

as $A_{i0} = A_{i3} = 0$.

Now we exhaust our co-ordinate freedom (this is the 4th constraint) with another requirement, which is $A = \sum_i A_{\bar{i}\bar{i}} = 0$, making the final amplitude tensor:

$$A = \begin{pmatrix} 0 & 0 & 0 & 0 \\ 0 & A_+ & A_\times & 0 \\ 0 & A_\times & -A_+ & 0 \\ 0 & 0 & 0 & 0 \end{pmatrix}.$$

In this form, A_+ and $A_\times$ are the two *polarisation modes* of the wave. The metric is:

$$H_{\bar{a}\bar{b}} = A_{\bar{a}\bar{b}} \sin\left(\sum_{i=1,2,3} k_i x_i - \omega t \right) = \begin{pmatrix} 0 & 0 & 0 & 0 \\ 0 & A_+ & A_\times & 0 \\ 0 & A_\times & -A_+ & 0 \\ 0 & 0 & 0 & 0 \end{pmatrix} \sin\left(\sum_{i=1,2,3} k_i x_i - \omega t \right),$$

which we will write as:

$$H_{\bar{a}\bar{b}} = \begin{pmatrix} 0 & 0 & 0 & 0 \\ 0 & H_+ & H_\times & 0 \\ 0 & H_\times & -H_+ & 0 \\ 0 & 0 & 0 & 0 \end{pmatrix}$$

for compactness, but we need to remember that the oscillation of the wave is inside the H components in this notation.

11.3.4 Particle Motion Under the Influence of Gravitational Waves

Now it's sensible to ask what effect these waves have on particles in space. To gain some traction on that, we go back to the geodesic equation:

$$\frac{\Delta}{\Delta\tau}\left(\frac{\Delta x_k}{\Delta\tau}\right) = -\sum_i\sum_j U_i U_j \Gamma_{ij}^k \tag{7.2}$$

and consider a particle at rest, so that $X = \begin{pmatrix} x_0 \\ 0 \\ 0 \\ 0 \end{pmatrix}, U = \begin{pmatrix} 1 \\ 0 \\ 0 \\ 0 \end{pmatrix}$ this turns the geodesic equation into:

$$\frac{\Delta}{\Delta\tau}\left(\frac{\Delta x_k}{\Delta\tau}\right) = -\Gamma_{00}^k.$$

The Christoffel symbols are:

$$\Gamma_{00}^k = \frac{1}{2}\sum_m \eta_{km}\left\{\frac{\Delta H_{\overline{m0}}}{\Delta x_0} + \frac{\Delta H_{\overline{0m}}}{\Delta x_0} - \frac{\Delta H_{\overline{00}}}{\Delta x_m}\right\},$$

and if we inspect one of the terms:

$$\frac{\Delta H_{\overline{m0}}}{\Delta x_0} = \left(\frac{\Delta A_{\overline{m0}}}{\Delta x_0}\right)\sin\left(\sum_{i=1,2,3} k_i x_i - \omega t\right) + A_{\overline{m0}}\frac{\Delta}{\Delta x_0}\left(\sin\left(\sum_{i=1,2,3} k_i x_i - \omega t\right)\right)$$

the part $\dfrac{\Delta A_{\overline{m0}}}{\Delta x_0} = 0$, as the metric terms are constant, and the second part is zero as $A_{\overline{m0}} = 0$. The same argument applies to each of the terms in the Christoffel symbols, so we end up with $\Gamma_{00}^k = 0$. As a result, $\dfrac{\Delta}{\Delta\tau}\left(\dfrac{\Delta x_k}{\Delta\tau}\right) = 0$, so the 4-velocity of the particle is constant, and given that we started with a particle that is not moving, we conclude that *the co-ordinate location of the particle does not change as the gravitational wave passes*. However, that's just the co-ordinate location, to find the *proper location* we need to factor in the metric.

Remember that we set up the whole linear theory with $g_{\overline{ab}} = \eta_{\overline{ab}} + h_{\overline{ab}}$ which we shifted to $g_{\overline{ab}} = \eta_{\overline{ab}} + H_{\overline{ab}}$. Hence, the space-time interval is:

$$(\Delta s)^2 = \left(\eta_{\overline{00}} + H_{\overline{00}}\right)c^2\left(\Delta t\right)^2 + \left(\eta_{\overline{11}} + H_{\overline{11}}\right)\left(\Delta x\right)^2 + \left(\eta_{\overline{22}} + H_{\overline{22}}\right)\left(\Delta y\right)^2$$

$$+ \left(\eta_{\overline{33}} + H_{\overline{33}}\right)\left(\Delta z\right)^2 + \left(\eta_{\overline{12}} + H_{\overline{12}}\right)\Delta x \Delta y + \left(\eta_{\overline{21}} + H_{\overline{21}}\right)\Delta y \Delta x,$$

where the cross terms come from the off-diagonal parts of H. Putting in the values of η:

$$(\Delta s)^2 = (1+0)c^2(\Delta t)^2 + (-1+H_+)(\Delta x)^2 + (-1-H_+)(\Delta y)^2$$
$$+ (-1+0)(\Delta z)^2 + (0+H_\times)\Delta x\Delta y + (0+H_\times)\Delta y\Delta x$$

or:

$$(\Delta s)^2 = c^2(\Delta t)^2 + (H_+ - 1)(\Delta x)^2 - (1+H_+)(\Delta y)^2 - (\Delta z)^2 + 2H_\times\Delta y\Delta x.$$

For the moment, we consider the case when $H_\times = 0, H_+ \neq 0$ and start with a pair of particles with co-ordinates $(d, 0, 0)$ and $(-d, 0, 0)$ at the same co-ordinate time (so $\Delta t = 0$). Then $\Delta x = 2d$ and the interval between the particles is:

$$(\Delta s)^2 = (H_+ - 1)(2d)^2 = -(1-H_+)(2d)^2 \quad \text{or} \quad (\Delta\sigma_x)^2 = (1-H_+)(2d)^2.$$

Another pair of particles located $(0, d, 0)$ and $(0, -d, 0)$ will have the interval:

$$(\Delta s)^2 = -(1+H_+)(2d)^2 \quad \text{or} \quad (\Delta\sigma_y)^2 = (1+H_+)(2d)^2.$$

These equations give us, in turn:

$$\Delta\sigma_x = 2\sqrt{1-H_+}\, d \quad \text{and} \quad \Delta\sigma_y = 2\sqrt{1+H_+}\, d.$$

Now we use $H \ll 1$ and our $(1+x)^n \approx 1+nx$ approximation to give:

$$\Delta\sigma_x = 2\left(1 - \frac{1}{2}H_+\right)d \quad \text{and} \quad \Delta\sigma_y = 2\left(1 + \frac{1}{2}H_+\right)d.$$

Referring back to Figure 11.4, as the sine function contained in H_+ is growing towards its maximum value, we can see that $\Delta\sigma_x$ is getting *smaller* and $\Delta\sigma_y$ is getting *bigger*. Indeed, if we trace the effect of the wave's variation on the particles it will look like Figure 11.6.

To carry out a similar analysis for the case $H_\times \neq 0, H_+ = 0$, we start with a pair of particles located at $\frac{1}{\sqrt{2}}(d, d, 0)$ and $\frac{1}{\sqrt{2}}(-d, -d, 0)$ alongside another pair at $\frac{1}{\sqrt{2}}(d, -d, 0)$ and $\frac{1}{\sqrt{2}}(-d, d, 0)$ (the four will then mark the corners of a square). The interval is:

$$(\Delta s)^2 = -(\Delta x)^2 - (\Delta y)^2 + 2H_\times\Delta y\Delta x,$$

and in the case of our first pair of particles:

$$\Delta x = \frac{2d}{\sqrt{2}} = \sqrt{2}d \quad \Delta y = \frac{2d}{\sqrt{2}} = \sqrt{2}d,$$

so that:

$$(\Delta s)^2 = -\left(\sqrt{2}d\right)^2 - \left(\sqrt{2}d\right)^2 + 2H_\times\sqrt{2}d\sqrt{2}d = -4d^2(1-H_\times)$$

leading to:

$$\Delta\sigma_1 = 2d\left(1 - \frac{1}{2}H_\times\right).$$

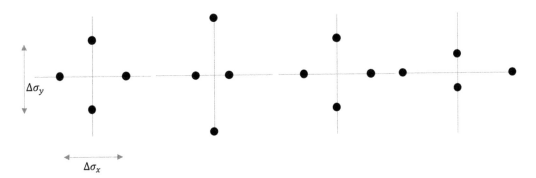

FIGURE 11.6 The effect of an H_+ gravitational wave on an array of particles. The left-most diagram is the starting arrangement, then the next three show the progressive distortion to the proper distances as the wave passes vertically through the page.

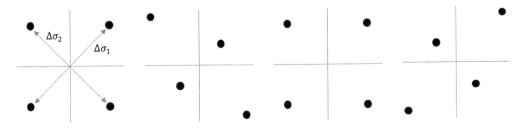

FIGURE 11.7 The effect of an $H_\times$ gravitational wave on an array of particles. The left-most diagram is the starting arrangement, then the next three show the progressive distortion to the proper distances as the wave passes vertically through the page.

For the second pair of particles, we have:

$$\Delta x = \sqrt{2}d \quad \Delta y = -\sqrt{2}d$$

making the interval:

$$(\Delta s)^2 = -\left(\sqrt{2}d\right)^2 - \left(\sqrt{2}d\right)^2 - 2H_\times \sqrt{2}d\sqrt{2}d = -4d^2\left(1 + H_\times\right)$$

and so:

$$\Delta \sigma_2 = 2d\left(1 + \frac{1}{2}H_\times\right).$$

The effect of this polarisation mode on the array of particles is shown in Figure 11.7.

As we know from Chapter 1, gravitational waves have been detected at the LIGO observatory, via their effect on the proper distance covered by light rays bouncing back and forth along two perpendicular arms.

11.4 Energy in General Relativity

At various times in previous chapters, I have mentioned that the whole issue of gravitational energy is complicated and unclear in the general theory. No less an expert than Roger Penrose has written[10]:

> I believe that it is fair to say that we do not yet have a complete understanding of gravitational mass/ energy

The subject of gravitational energy is not always explored in first level text books on the general theory, so it would be overly ambitious of us to delve too deeply here. Nevertheless, I feel bound to make some comments, if only because my own physical intuition needs something to work with… We need to start by discussing the mass of a relativistic gravitational source.

11.4.1 The Mass of a Relativistic Source

For any gravitating object, no matter how relativistic it might be, we can always move far enough away that linear gravity theory applies. The only constraint is that the material making up the object must be localised in a region of space, not spread throughout the universe. It would then be tempting to suppose that the metric:

$$(\Delta s)^2 = \left(1 + \frac{2\phi}{c^2}\right)c^2(\Delta t)^2 - \left(1 - \frac{2\phi}{c^2}\right)\left((\Delta x)^2 + (\Delta y)^2 + (\Delta z)^2\right)$$

would apply to the space-time at this distance from the source. In turns out that this metric is appropriate, but it is not immediately obvious why that should be. When we obtained this metric, we were assuming that the gravitation was weak *everywhere*, including at the source itself. After all we wrote:

$$\frac{1}{2}\Box^2(H_{\overline{00}}) = -\frac{8\pi G}{c^4}\rho c^2,$$

which is applying our weak field Einstein tensor (the left hand side) to the interior region of the source. With a relativistic source, we can't assume that the field will be weak everywhere and certainly not in the interior of the object. We also made the identification $\phi = -GM/r$ as a result of the weak field applying to the region of the source in the Newtonian limit. At the moment, we can't be sure what ϕ is for a relativistic object, so we have no way to apply the metric without some expression for ϕ. To solve this problem, we assume for the moment that the source of the field is static ($\mathbb{T}_{\overline{ab}} \neq \mathbb{T}_{\overline{ab}}(x_0)$), so that when we go far enough away that the linear field equations apply (Equation 11.1), not only do we have $\Box^2(H_{\overline{ab}}) = 0$ but also $H_{\overline{ab}} \neq H_{\overline{ab}}(x_0)$ so that $\nabla^2(H_{\overline{ab}}) = 0$ which has a solution:

$$H_{\overline{ab}} \approx \frac{C_{\overline{ab}}}{r}$$

with the $C_{\overline{ab}}$ being constants (see the on-line Appendix for a justification of this).

Applying the condition $\displaystyle\sum_a \frac{\Delta H_{a\overline{b}}}{\Delta x_a} = 0$ we obtain:

$$\sum_a \frac{\Delta}{\Delta x_a}\left(\sum_i \eta_{ai}\frac{C_{\overline{ib}}}{r}\right) = \sum_a\left(\sum_i \eta_{ai}\frac{\Delta}{\Delta x_a}\left(\frac{C_{\overline{ib}}}{r}\right)\right) = -\sum_{a,i}\eta_{ai}\frac{C_{\overline{ib}}}{r^2}\frac{\Delta r}{\Delta x_a} = 0,$$

and the only way that this can be true under all circumstances is if $C_{\overline{ib}} = 0$. As the field is static, the index a in the summation runs over the *spatial components only*. So, we have not found out the value of $C_{\overline{00}}$ as yet, only that the spatial components are zero. Hence, the only surviving component of $H_{\overline{ab}}$ is $H_{\overline{00}} \approx \dfrac{C_{\overline{00}}}{r}$. This is just as we found earlier for the static weak field, but now we have established its plausibility for a *relativistic source*, provided we are a long way from its boundary region. It is then very tempting to write:

$$H_{\overline{00}} = \frac{4\phi}{c^2} \approx \frac{C_{\overline{00}}}{r}$$

making $\phi \approx c^2 C_{\overline{00}}/4r$. If we then make the further association $C_{\overline{00}} = -4M/c^2$ we can *define* the mass, M, of a relativistic source in this way. In Section 11.1.2, we obtained the metric

$$(\Delta s)^2 = \left(1 + \frac{2\phi}{c^2}\right) c^2 (\Delta t)^2 - \left(1 - \frac{2\phi}{c^2}\right)\left((\Delta x)^2 + (\Delta y)^2 + (\Delta z)^2\right)$$

from $H_{\overline{00}} = 4\phi/c^2$ and we can now follow the same argument again with impunity, justifying the use of this metric at an appropriate distance from a relativistic source. Any space-time which reduces to this metric at an appropriate distance from the source we term *asymptotically flat*.

Far from our relativistic source, small masses (so they do not themselves curve the space-time) will follow the geodesics of this metric, which we could calculate now that we have an expression for ϕ. In any practical situation, matching the movement or orbit of an object to the calculated geodesic would enable us to determine the mass, M, that goes into the metric, effectively measuring the relativistic object's mass (as we determine the mass of Sgr A* by the orbits of stars around it). This is how we can make a sensible definition and determination of a black hole's mass (Section 10.5). Crucially, this is a *direct determination* and definition of the *total* mass (and hence energy) of the relativistic object. We have not obtained the mass via any summation over the components that make up the object.

In the case of non-static sources, defining the mass is somewhat trickier. Time varying sources can generate gravitational waves, which do not reduce to Newtonian fields as their spatial components are as large as the temporal ones. However, there are still some circumstances where this definition of total mass can be appropriate, for example if the waves are weak. Also, if the source was stationary in the past before it started to radiate, then we can always step far enough away that the radiation has not had time to reach that part of space-time and apply the definition there. Outside of these cases a much more advanced treatment than we can provide is needed.

11.4.2 Energy Issues

When we were deriving the field equations back in Section 7.6.3, we mentioned that $\sum_a \dfrac{\Delta \mathbb{T}_{ab}}{\Delta x_a} = 0$ was an expression of local energy-momentum conservation. However, this does not generalise to a global conservation law. Over a larger slice of space-time we can expect particles to gain or lose gravitational energy, but we have specifically excluded any energy associated with gravitation from $\mathbb{T}_{ab}$, so $\sum_a \dfrac{\Delta \mathbb{T}_{ab}}{\Delta x_a} = 0$ is going to fail globally. Constructing an appropriate energy density tensor for gravitation is fraught with difficulties. The metric does not distinguish between the dynamic aspects (e.g. propagating gravitational waves) and the background space-time; it's all rolled into one and there is no clear way of consistently distinguishing them in order to allocate energy to the dynamic part. However, under certain conditions it is possible to define a total energy for an isolated system.

11.4.3 Energy for Isolated Systems

In the Newtonian picture, if we assemble an object out of constituent masses there will be an internal gravitational energy associated with the resulting system. The isolated components have a larger (less negative) gravitational potential energy at their (assumed) distant starting points. When we move them together, their gravitational potential energy *reduces*, i.e. it becomes *more negative*. As a result, the total energy of the assembly must be less than the sum of the energies of the isolated constituents. This energy difference is not associated with the particles, but resides in the negative contribution of the internal gravitational field. As the mass of the composite object is related to its total energy, the total mass must be *less* when the particles are bound together, as $E = mc^2$.

If you did not know how to calculate the energy in the internal gravitational field at least you could write:

$$\frac{\text{Energy in internal gravitational field}}{c^2} = \text{total mass of assembly} - \text{sum of indidual isolated masses}$$

and determine it that way: perhaps this is a way of obtaining gravitational energy.

However, transferring this picture to the general theory is not straightforward. We defined the total mass of a relativistic object in the previous section (but *not* by adding constituent masses), so we have a way of finding the total mass of the assembly. However, we will struggle to get a unique sum of the isolated masses (4-momenta) to subtract.

In traditional branches of physics, the total energy and momentum of a collection of particles is the sum of their separate energies and momenta. However, in the general theory this can't be done as adding a vector at A to a vector at B is not defined. You can only add vectors if they are at the same point. If not, then the best that you can do is parallel transport the vector at B back to A and add them there. That's fine (and actually what we really do) if the space-time is flat. In curved space-time, the value of the vector once it has been parallel transported *depends on the path taken*. So, *there is no unique way to define the sum of the vectors*. Without that sum, we can't find the internal gravitational energy of the composite object, at least not uniquely.

We can define the total energy of a system (as per the previous section), but there is no well-defined way of decomposing it into gravitational and particle contributions.

Another troubling aspect is the difficulty with conservation laws. We are used to the notion that energy is conserved: the total energy in the universe now is the same as what is was a second, a minute, or a million years ago. Equally, that total will extend into the future. Unfortunately, as simultaneity is destroyed in relativity this becomes problematical. We can define 3D space-like hypersurfaces and suggest that any conserved quantity summed across a surface would be the same no matter what surface we chose. However, for the reasons we have just explained, we can't uniquely total up a vector quantity across such a surface. Even if we restrict ourselves to something that is scalar (invariant) we still get a headache. Counting up scalar quantities (such as charge) normally involves studying the flux of some vector through the surface bounding the charge (we look at the electrical field emerging through a sphere around the charge for example), but then it becomes difficult to be sure which side of the surface the charge actually lies!

Let's not mention the problems that emerge if the metric itself varies with time...

Given the difficulty in defining exactly what is meant by energy in the general theory, perhaps it doesn't make any sense to think of gravitational waves as carrying energy after all. If we did take that view though, there would be two points that would give us difficulty:

1. Systems like Hulse-Taylor are clearly losing energy via some mechanism.
2. A simple 'thought experiment' shows that it is likely that gravitational waves carry energy. This suggestion was first put forward by Richard Feynman,[11] but was developed and popularised by Hermann Bondi.[12] Consider a thin rod, which is not friction free, on which is mounted two beads that are free to slide back and forth. If the rod is oriented at 90° to a gravitational wave, changes in the metric due to the wave will distort the space-time in the vicinity of the rod and the beads. The atomic forces within the rod will be able to resist the metrical change, but the proper distance between the beads will alter. As a result, the beads slide back and forth along the rod. Friction between the beads and the rod will dissipate the kinetic energy as heat. The rod warms up. The energy for this has to come from somewhere, so the gravitational wave must be carrying energy with it.

In truth, these arguments are not completely water-tight. All that we have established is that gravitational wave *production* can *extract* energy and gravitational wave *absorption* can *donate* energy. This is not the same as saying that gravitational waves *carry energy*. In principle, the energy could just vanish from Hulse-Taylor and (some of it) appear in the rod. We have to consider that possibility due to the problems of establishing global conservation laws applicable for all space-times. However, it makes sense to at least try and develop some notion of energy carrying gravitational waves, even given the rather bleak prognosis we have developed for energy in the general theory. It is possible to make some progress with gravitational wave energy, if we apply some appropriate simplifications.

11.4.4 Energy in Gravitational Waves

When we carried out the metrical shift $g_{\overline{ab}} = \eta_{\overline{ab}} + h_{\overline{ab}}$ we effectively split the metric so that there was a background space-time, and a term that can 'ripple' propagating as gravitational wave. This opens up the possibility of being able to define an energy density tensor for these approximate conditions and hence get a global energy conservation.

We have been making excellent use of the approximation $(1+x)^n \approx 1+nx$, but this is only the first term of an *approximation series* that looks like this:

$$(1+x)^n \approx 1+nx+\frac{n(n-1)}{1\times 2}x^2+\frac{n(n-1)(n-2)}{1\times 2\times 3}x^3+\frac{n(n-1)(n-2)(n-3)}{1\times 2\times 3\times 4}x^4+\dots$$

Each term representing a *higher order* in the series; nx being first order, $\frac{n(n-1)}{1\times 2}x^2$ second order, etc. With this terminology in mind, we turn to how we might approximate things in linear gravity theory to extract some understanding of gravitational wave energy.

Starting from our standard metric written as $g_{\overline{ab}} = \eta_{\overline{ab}} + h_{\overline{ab}}$, with $h_{\overline{ab}} \ll 1$, we would write the vacuum field equations as $\mathbb{G}^{(1)}_{\overline{ab}}(\eta+h)=0$ where $\mathbb{G}^{(1)}_{\overline{ab}}$ are the components of the Einstein tensor, as a function of the metric, to first order. If we wanted a slightly better approximation, we would add to the metric:

$$g_{\overline{ab}} = \eta_{\overline{ab}} + h_{\overline{ab}} + h^{(2)}_{\overline{ab}}$$

with $h^{(2)}_{\overline{ab}}$ being an additional disturbance to the flat metric, to second order. Think of g being expanded out like $(1+x)^n$ with η playing the part of the '1' in the expansion, h is doing the role of nx and so h^2 is acting like $\frac{n(n-1)}{1\times 2}x^2$. Following from this, we would have a second order version of the Einstein tensor as well, making the field equations:

$$\mathbb{G}^{(1)}_{\overline{ab}}\left(\eta+h^{(2)}\right)+\mathbb{G}^{(2)}_{\overline{ab}}\left(\eta+h\right)=0$$

eliminating any terms in $hh^{(2)}$, etc., which would be third order. Now we do something creative and move the terms around:

$$\mathbb{G}^{(1)}_{\overline{ab}}\left(\eta+h^{(2)}\right)=-\mathbb{G}^{(2)}_{\overline{ab}}\left(\eta+h\right)$$

write:

$$\mathbb{G}^{(2)}_{\overline{ab}} = \frac{8\pi G}{c^4}t_{\overline{ab}}$$

and interpret $t_{\overline{vb}}$ as being *the energy density term for the gravitational wave in a vacuum*. The field equations are now:

$$\mathbb{G}^{(1)}_{\overline{ab}}\left(\eta+h^{(2)}\right)=-\frac{8\pi G}{c^4}t_{\overline{ab}}$$

as we are used to seeing them.

It might appear that we have miraculously solved our problems, albeit that we have not written down what the $t_{\overline{ab}}$ terms are as functions of the metric. This celebration, however, would be premature. In turns out that $t_{\overline{ab}}$ is co-ordinate system dependent (which should not be too much of a surprise given our earlier discussion). Indeed it is not really a tensor at all. It has approximate tensor-like behaviour in the linear theory, but this all breaks down if we try and use it with the full equations. However, this has not been a total waste of time, as it is possible to use the co-ordinate system dependent $t_{\overline{ab}}$ to calculate quantities that are not co-ordinate system dependent, in a restricted sense. First we have:

$$\mathbb{E} = \int_{\mathbb{S}_1} t_{\overline{00}}\, dx\, dy\, dz$$

with $\mathbb{E}$ being the total energy across a space-like hypersurface $\mathbb{S}_1$ (i.e. a surface formed by connecting together events with space-like separations at the same co-ordinate time). This is a sensible measure of the total energy if the space-time is asymptotically flat, and it is co-ordinate system independent if we restrict ourselves to transforming between sets of co-ordinates that do not ruin the asymptotic flatness of the metric. $\mathbb{E}$ is invariant under such transforms.

Second is the quantity

$$\Delta\mathbb{E} = \int_{\mathbb{S}_2} \sum t_{\overline{0v}}\eta_{vu}\epsilon_{\overline{u}}\,dxdydt,$$

which is the energy passing through a time-like surface, $\mathbb{S}_2$, formed from a 2-sphere a great distance from the source of radiation (in essence at infinity) with a 'thickness' given by a period of time and $\epsilon_{\overline{u}}$ is a unit-vector pointing at 90° out of the surface. This calculation is co-ordinate system independent if the space-time is asymptotically flat and starts off static (not changing with time) passes through a period where it is changing with time and then becomes static again. In essence, the gravitational source starts off static, has a finite period when it is radiating gravitational waves, and then settles down again.

Using calculations like this, physicists work out the amount of energy radiated away from the Hulse-Taylor system, or indeed coming from a black hole or neutron star merger.

Unfortunately, the theory we would need to study gravitational wave *production* is too advanced for us at this level. In summary, gravitational waves are generated when massive objects accelerate, provided the acceleration is not spherically symmetrical (i.e. a spherical mass is expanding or contracting isotropically). A structure that was shaped like a dumbbell would not radiate if it simply spins about some axis. It will radiate if it starts to tumble. In a hand-wavy sort of way, you can picture two black holes as being the masses on the ends of the 'dumbbell' and their orbital decay and inspiralling as being tumbling, especially if the two are rather different in mass.

11.5 The Weyl Tensor

In mass/energy free region of space-time, the Ricci tensor is zero. However, as I have stressed before this does not mean that the region is free of gravitational effects. The Riemann tensor need not be zero even if the Ricci tensor, which is built from some of its components, happens to be. The remains of the Riemann tensor can be assembled into the Weyl tensor, $C_{\overline{ijkl}}$, which has ten independent components:

$$C_{\overline{ijkl}} = R_{\overline{ijkl}} - \left(g_{\overline{ik}}R_{\overline{lj}} - g_{\overline{il}}R_{\overline{kj}} - g_{\overline{jk}}R_{\overline{li}} + g_{\overline{jl}}R_{\overline{ki}}\right) + \frac{1}{3}\left(g_{\overline{ik}}g_{\overline{lk}} - g_{\overline{il}}g_{\overline{kj}}\right)\mathcal{R}.$$

As the Ricci tensor is the only contraction of the Riemann tensor, it follows that all contractions of the Weyl tensor, which is what you get from the Riemann tensor when you subtract the Ricci bits, are zero.

Furthermore, if the Ricci tensor is zero (and consequently the Ricci scalar is zero as well), then the Weyl tensor and the Riemann tensor are the same.

We already know that the elements of the Ricci tensor are determined by the energy density tensor via the Einstein field equations. When you factor in the Weyl tensor as well, it is evident that the space-time curvature in a region is not completely determined by the energy density in that region. As an extreme example, recall that the vacuum equations are:

$$R_{\overline{ab}} - \frac{1}{2} \mathcal{R} g_{\overline{ab}} = 0$$

which have many possibly solutions: the Minkowski metric, the Schwarzschild metric and the Kerr metric are all examples that we have come across. In all of these cases, the Ricci tensor is zero, but the Weyl tensor is not: it is encoding the curvature of the space-time in the vacuum.

What picks out the solution relevant to a given situation? This is the role of the boundary conditions, which we mentioned in Chapter 6, and any other initial conditions that apply to the situation. However, it is important to realise that even though the components of the Weyl tensor are not directly determined by the energy density, they are connected *indirectly*. Back in Chapter 7, we mentioned that the Riemann tensor has many symmetries, one of which is known as the *second Bianchi identity*:

$$\frac{D}{\Delta x_m}\left(R_{\overline{ijkl}}\right) + \frac{D}{\Delta x_k}\left(R_{\overline{ijlm}}\right) + \frac{D}{\Delta x_l}\left(R_{\overline{ijmk}}\right) = 0.$$

Putting the mathematics to one side for a moment, this identity is showing us how the Riemann tensor components at a point are related to those in a nearby region via the rates of change applicable. Again, this is an expression of how space-time curvature causes space-time curvature. It is also characteristic of what happens when waves are generated.

11.6 LIGO's Detection of Neutron Star Merger

The clinching discovery at the LIGO stations, as discussed in Section 1.5, has opened up a new chapter in the development of the general theory. Gravitational waves have been firmly established as a confirmed prediction of the theory, and attention now moves to studying their properties and what they can tell us about the universe and extreme regions of space-time where events take place with sufficient energy to generate detectable waves.

Since the initial discovery, there have been five further detections, the most significant of which is labelled GW170817. This observation was also recorded by a completely independent team at the VIRGO detector[13] (two of the five were also picked up by VIRGO).

GW170817 involved two neutron stars with total mass ~3 solar masses (the individual masses are less certain) merging to form a black hole. The gravitational wave signal lasted about 100s and rose from an initial frequency of ~ 24 Hz to a few hundred Hz. Independently, a 2 second γ-ray burst was detected by orbiting satellites, starting 1.7 s after the gravitational wave signal was first picked up. An automatic computer alert was triggered within 16 seconds of the γ-ray detection and was already 'on the books' when the software that monitors the LIGO data stream produced its warning, 6 minutes after the gravitational waves had been detected. With the γ-ray burst arriving within such a short period of time after the gravitational waves, it was assumed to be due to the same event, prompting astronomers to search for other signals across the electromagnetic spectrum.

Comparing the arrival times at the two LIGO and the VIRGO detector, allowed a rough direction to be established, giving other astronomers a region to focus attention on. Eleven hours into this intense search, an optical signal was picked up from the galaxy NGC 4993 which is ~130 million light years away in the direction of the constellation of Hydra[14] (Figure 11.8). As further data accumulated, it became

FIGURE 11.8 The galaxy NGC 4993 in the constellation of Hydra as imaged by the Hubble space telescope. The small square is the optical signal form the neutron star merger and that region is reproduced in the larger square top right. (Image credit: By Hubble Space Telescope, NASA and ESA, https://www.nasa.gov/press-release/nasa-missions-catch-first-light-from-a-gravitational-wave-event, Public Domain, https://commons.wikimedia.org/w/index.php?curid=63409545.)

clear that the electromagnetic signals were coming from a fast moving cloud of neutron rich material that was rapidly cooling. This is just what we would expect to see as a consequence of a neutron star merger. Black hole mergers are not thought to produce electromagnetic energy.

Eventually, over a period of some weeks, observations amassed from 70 different observatories across seven continents covering a wide range of the electromagnetic spectrum.

Given the very precise picture formed of this event, scientists have looked back at a γ-ray burst detected back in 2015, and from its notable similarities to GW170817 in the γ-ray, optical and X-ray data, now conclude that it was also most likely caused by a neutron star merger.

Some delightful additional benefits have sprung from GW170817. For example, by assuming that the first photons were produced less than 10 s after the peak of the gravitational wave emission, physicists have been able to place a limit on the difference between the speed of light and the speed of gravitational waves. Clearly, from our earlier analysis, we expect gravitational waves to propagate at exactly the speed of light, so it is an interesting test of the theory to confirm their speed. GW170817 has enabled the previous constraints on the difference, $v_{grav} - c$, to be improved by 14 orders of magnitude. We now have $-9 \times 10^{-7} \leq v_{grav} - c \leq 2.1 \times 10^{-7}$. Additionally, some rival gravitational theories have been disproven by the information coming from this event and we now have evidence that neutron star mergers may be important contributors to the synthesis of elements heavier than iron.

11.7 Next Steps in Gravitational Wave Research

Detectable gravitational waves are emitted from regions of space-time where the masses are large and the speeds are close to that of light. In many cases, such regions are either surrounded by dense clouds of matter that absorb electromagnetic radiation, or do not themselves emit such radiation to any detectable extent (e.g. black holes). Consequently, gravitational wave astronomy opens up a new window on the universe.

Figure 11.9 shows the spectrum of gravitational waves and the different sources that would generate waves in the different bands.

The possibility that gravitational waves from the early history of the universe could be detected is especially exciting as there is a limit to how far back we go with electromagnetic information (to be discussed in Chapter 12).

As of Autumn 2018, there are four operational gravitational wave observatories, the two LIGO detectors and GEO600 in Germany, which is another laser interference experiment with 600 m arms and is capable of detecting gravitational waves in the frequency range between 50 Hz and 1.5 kHz. LIGO is sensitive to the range between 100 Hz and ~10 kHz. GE0600 has been taking data simultaneously with LIGO since September 2015; however, it was being tested when the LIGO initial discovery was made. In any case, it is not sensitive enough to have detected that particular signal. The VIRGO detector, with 3 km arms, near Pisa in Italy has undergone upgrades, and has been operational in this form since late 2017. KAGRA is in the tunnels of the Kamioka mine in Japan. It is also a laser interference detector with 3 km arms and will, like VIRGO, be sensitive from a few hertz up to 10 kHz. Hopefully, KAGRA will become active in 2019. Agreement has been reached to build another LIGO device in India and other ground-based detectors are in planning.

With such an array of detectors across the globe, physicists will be able to pin down the direction to the origin of a gravitational wave and hence help to identify triggering events.

Off the surface, the European Space Agency (ESA) has plans to launch the Evolved Laser Interferometer Space Antenna (eLISA) in 2034. This will be a constellation of three satellites in an orbit which trails behind that of Earth. The satellites will be arranged in an equilateral triangle with 1 million km along each side. They will bounce laser beams along the edges of the triangle with two of the satellites being set up as reflectors and the third as a detector. This observatory should bring the 10^{-3} Hz frequency range into play. There are even more ambitious satellite detectors being planned.

The International Pulsar Timing Array is a collaboration between three consortia trying to use the precise clocks provided by pulsar[15] to detect gravitational waves. Each group times around twenty pulsars with millisecond periods per month, out of a combined set thirty pulsars in total. Any gravitational waves should trigger a flutter in the timing of a pulsar, in a characteristic form. The array is sensitive to the 10^{-6} Hz frequency range of gravitational waves, which offers the prospect of detecting gravitational waves from quantum mechanical fluctuations in the space-time metric of the early universe.

The Gravitational Wave Spectrum

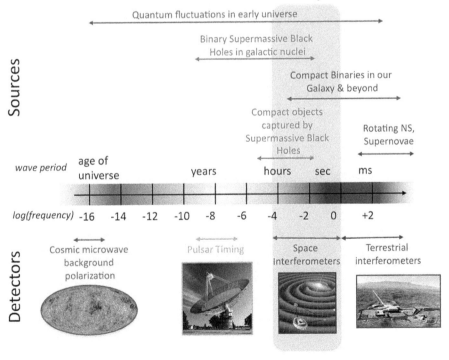

FIGURE 11.9 Sources of gravitational waves and their characteristic frequencies alongside the detectors capable of their study. The range open to the eLISA satellite based system is highlighted in the grey box. (Image credit: NASA Goddard Space Flight Center.)

Notes

1. Interview, MIT News. http://news.mit.edu/2016/ligo-first-detection-gravitational-waves-0211.
2. Nobel Prize interview. https://www.nobelprize.org/prizes/physics/2017/thorne/interview/.
3. https://www.nobelprize.org/prizes/physics/1993/press-release/.
4. This pulsar's periastron moves as far per day as Mercury's moves per century!
5. The Sun is radiating electromagnetically, whereas the Hulse-Taylor system is losing its energy via gravitational waves.
6. Weisberg, J.M., Nice, D.J. and Taylor, J.H., 2010. Timing measurements of the relativistic binary pulsar PSR B1913+16. *Astrophysical Journal*, **722**(2), 1030–1034. Bibcode:2010ApJ...722.1030W.
7. Remember that $\eta^2 = 1$ so $\eta = \eta^{-1}$ and hence $(\eta^{-1})_{ab} = \eta_{\overline{ab}} = \eta_{ab}$. Hence they can be used interchangeably to taste and for consistency.
8. Sometimes the symbol $\square$ is used, but I prefer $\square^2$ as it emphasises the scalar nature of the combination.
9. The definition of a transverse wave is that the vibration is at 90° to the energy flow. An example would be the type of wave that you can set up on a skipping rope or a guitar string. The ripples that propagate across the surface of a pound are also transverse.
10. *The Road to Reality*, R. Penrose, Vintage; New Ed edition (2005), ISBN-10: 0099440687 p. 486.
11. Richard Feynman (1918–1988), Professor at Caltech and Nobel Prize winner (1965).
12. Hermann Bondi (1919–2005), Professor of Physics at Kings College London.
13. To be discussed further in the next section.
14. For superhero fans, I will resist the temptation to say "hail Hydra" at this point…
15. This rotation rate is specifically selected as pulsars in this range suffer the least natural wobble in their pulse rate.

12

Cosmology

12.1 The Big Bang

> The standard Big Bang theory says nothing about what banged, why it banged, or what happened before it banged.
>
> **A Guth**[1]

The science of cosmology has a fully functioning and widely accepted paradigm: *the Big Bang theory*. That in itself is a remarkable statement: we have a detailed understanding of universe right back to within fractions of a second into history.

The Big Bang is often portrayed as a theory explaining the *origin* of the universe. I prefer to think of it as a theory of the universe's *evolution*. The Big Bang is not something that happened once in the past to kick start the universe; it is taking place right now.

We know that the early universe was very different to the dwelling we currently occupy. It started as a hot, dense 'soup' of fundamental particles (including dark matter, whatever that is) and electromagnetic radiation. General relativity melded with our best theories of particle physics shows us how this uniform soup, filling the expanding space-time, steadily evolved structure until the stars, galaxies and planets were formed. That evolution has not stopped: black holes are forming, stars are being born and dying in various ways and the universe's expansion appears to be accelerating, not slowing as first suspected. The Big Bang is still going on.

Arguably, this model (supplemented by the notion of *inflation* to be discussed in Chapter 13) works exceptionally well right back to within 10^{-43} s of the start of history. As to the moment, or cause, of creation, that is still a subject of speculation. For one thing, language does not help us in this arena. Due to the intimate fusion of space and time in the general theory, creation resulted in the origin of time as well as space. How do we then speak of a time 'before' the Big Bang[2]? Such speculation must rely on quantum gravity, which is at best a threadbare patchwork at the moment. To the extent that we can, we will touch on this in Chapter 13. For now, the rest of this chapter explores how the general theory can be applied to the universe as a whole.

12.2 The Foundations of Cosmology

> The real reason, though, for our adherence here to the Cosmological Principle is not that it is surely correct, but rather, that it allows us to make use of the extremely limited data provided to cosmology by observational astronomy. If we make any weaker assumptions …. Then the metric would contain so many undefined functions… that the data would be hopelessly inadequate to determine the metric.
>
> **S Weinberg**[3] **(although the quality of data has increased markedly since this was written in 1972)**

The science of cosmology rests on three interconnected assumptions:

1. that general relativity is the correct theory of gravitation and is applicable to the universe as a whole;
2. that a set of *fundamental observers* can be defined within the universe, allowing a universal time to be established (Weyl's hypothesis);
3. that the universe is *homogeneous* and *isotropic* for all fundamental observers (the *cosmological principle*).

The second and third assumptions make specific claims about the distribution of matter across the cosmos, claims that are not true in detail, but can be seen as good approximations when we take the broad picture. The extent to which these claims hold true (especially Weyl's hypothesis) in the very early universe, when quantum effects are likely to come into play, is still a matter of active research and conjecture (see Chapter 13).

12.2.1 Weyl's Hypothesis

Accounts of the universe's evolution talk about different epochs of history and the conditions present in the universe, generally in terms of density, temperature and matter content, at that time. All of this would be impossible without a *universal time* that is applicable throughout the universe. From what we know of the general theory, this is clearly not an obvious possibility and the fact that such a universal time can be defined is a significant constraint on the space-time and the matter distribution in the universe.

Weyl's hypothesis asserts the existence of fundamental observers who are in free fall and co-moving with respect to the local matter distribution. Consequently, their world lines are geodesics. We then define the universal time as being the proper time experienced by these fundamental observers. As they are in free fall, they are subject to a local Minkowski metric and can be treated as being effectively at rest with respect to the local matter distribution. It is then possible to define a space-like hypersurface comprised of all events with the same value of universal time. The world lines of all fundamental observers cross these hypersurfaces at 90°. If that were not the case, they would have a local velocity parallel to the hypersurface.

The universe can then be divided up into a collection of space-like hypersurfaces that are locally perpendicular to observers' world lines, and labelled by the proper time of the observer as they pass through the surface (Figure 12.1).

As an observer is locally at rest, their 4-displacement is $X_{\text{obs}} = \begin{pmatrix} x_0(\alpha) & x_1 & x_2 & x_3 \end{pmatrix}$, indicating that they are at a *fixed spatial location*, with α parameterising their temporal co-ordinate. So, as they move through time from one hypersurface to the next, their co-ordinate location is the same in each hypersurface. This will resonate in Section 12.3.1 when we discuss co-moving co-ordinates and the scale of the universe.

Across the space-like hypersurface, any event has a 4-displacement $X_S = \begin{pmatrix} x_0 & x_1(\beta) & x_2(\beta) & x_3(\beta) \end{pmatrix}$ i.e. the spatial co-ordinates are free (within the surface) but the temporal one is fixed as the 'marker' of that hypersurface. Following on from our discussion in Section 9.2.1, we can write the tangents to these two 4-displacements as:

$$U_{\text{obs}} = \begin{pmatrix} \dfrac{\Delta x_0(\alpha)}{\Delta \alpha} & 0 & 0 & 0 \end{pmatrix}$$

$$U_S = \begin{pmatrix} 0 & \dfrac{\Delta x_1(\beta)}{\Delta \beta} & \dfrac{\Delta x_2(\beta)}{\Delta \beta} & \dfrac{\Delta x_3(\beta)}{\Delta \beta} \end{pmatrix}.$$

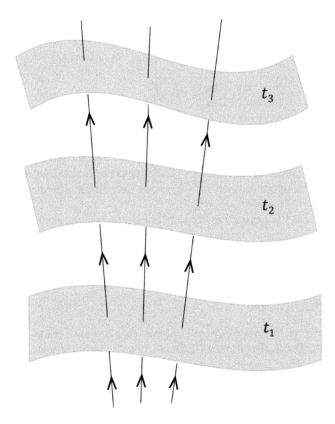

FIGURE 12.1 A collection of hypersurfaces (here rendered in 2D) with the world lines of fundamental observers. Each hypersurface is labelled by the proper time of a fundamental observer as they cross the hypersurface.

Given that the observers' world lines cross the hypersurfaces at 90°, we must have (product of orthogonal tangent vectors is zero):

$$\begin{pmatrix} 0 & \dfrac{\Delta x_1(\beta)}{\Delta\beta} & \dfrac{\Delta x_2(\beta)}{\Delta\beta} & \dfrac{\Delta x_3(\beta)}{\Delta\beta} \end{pmatrix} \begin{pmatrix} g_{\overline{00}} & g_{\overline{01}} & g_{\overline{02}} & g_{\overline{03}} \\ g_{\overline{10}} & g_{\overline{11}} & g_{\overline{12}} & g_{\overline{13}} \\ g_{\overline{20}} & g_{\overline{21}} & g_{\overline{22}} & g_{\overline{23}} \\ g_{\overline{30}} & g_{\overline{31}} & g_{\overline{32}} & g_{\overline{33}} \end{pmatrix} \begin{pmatrix} \dfrac{\Delta x_0(\alpha)}{\Delta\alpha} \\ 0 \\ 0 \\ 0 \end{pmatrix} = \begin{pmatrix} 0 \\ 0 \\ 0 \\ 0 \end{pmatrix}.$$

Once we have done the matrix multiplication, we get:

$$g_{\overline{10}}\left(\frac{\Delta x_1(\beta)}{\Delta\beta}\right)\left(\frac{\Delta x_0(\alpha)}{\Delta\alpha}\right) + g_{\overline{20}}\left(\frac{\Delta x_2(\beta)}{\Delta\beta}\right)\left(\frac{\Delta x_0(\alpha)}{\Delta\alpha}\right) + g_{\overline{30}}\left(\frac{\Delta x_3(\beta)}{\Delta\beta}\right)\left(\frac{\Delta x_0(\alpha)}{\Delta\alpha}\right) = 0.$$

The only way that this can be true for any observer and any location on the hypersurface is if $g_{\overline{10}} = g_{\overline{20}} = g_{\overline{30}} = 0$. Then, by assuming that the metric is symmetrical, we have reduced the universal metric to:

$$g_{ab} = \begin{pmatrix} g_{\overline{00}} & 0 & 0 & 0 \\ 0 & g_{\overline{11}} & g_{\overline{12}} & g_{\overline{13}} \\ 0 & g_{\overline{21}} & g_{\overline{22}} & g_{\overline{23}} \\ 0 & g_{\overline{31}} & g_{\overline{32}} & g_{\overline{33}} \end{pmatrix} \quad \text{i.e.} \quad (\Delta s)^2 = c^2 (\Delta t)^2 - (\Delta \ell)^2,$$

where I am using $(\Delta \ell)^2$ to indicate the spatial part of the interval: $(\Delta \ell)^2 = \sum_{i,j} g_{\overline{ij}} \Delta x_i \Delta x_j$. The co-ordinate time, t, is the universal time, i.e. the proper time of the fundamental observers. Time synchronisation can be maintained as we can agree to set clocks by the value of the local density or by the temperature of the cosmic microwave background, as both of these scale with time in a predictable manner, as we will discuss at length later in this chapter.

Another condition, which is an aspect of Weyl's hypothesis, is that the world lines of the fundamental observers never cross each other. If that were the case, there would be an ambiguity in the co-ordinate system. However, that does leave open the possibility that they converge, if we run back in time, at the initial singularity of the Big Bang.[4]

12.2.2 A Smooth Universe

In a *homogeneous* universe, the overall space-time looks the same at all places at the same moment in time. Using Weyl's hypothesis, the phrase 'at the same moment in time' translates into picking out one of our hypersurfaces, $\mathbb{S}_t$. Then, if we take any two events on this surface, e_1 & e_2, homogeneity implies the existence of a co-ordinate transformation that will take us from e_1 to e_2 (or vice versa) and which leaves the metric $g_{\overline{uv}}$ unchanged.

An *Isotropic* universe looks the same in any direction. In broad terms, this is not true for our universe, as it is easy to find observers who don't see it in this fashion. At the very least, someone moving through our galaxy at a good fraction of the speed of light would see much brighter light ahead of them than behind. However, as we defined our fundamental observers to be in free fall co-moving with the local matter distribution, then the notion of isotropy can be consistently applied to all such observers. Using a set of co-ordinates centred on a fundamental observer, isotropy implies that the form of the metric does not change if we rotate this system into a new orientation.

A universe can be homogeneous without being isotropic (Gödel's universe is like this), but the opposite is not true.[5] Consider an observer sitting inside a cloud of matter that is isotropically distributed about them. This means that the density of the material can only be a function of radius, as measured from their location. Also, as there can be no preferred direction, any matter moving away or towards them, must be doing so equally in all directions. Now, if the universe is isotropic about one observer, it seems plausible that it should be isotropic about all similar observers, at least for any reasonable universe which does not contain 'special locations'. Hence, we if adopt part of the cosmological principle and say that the universe is isotropic for all fundamental observers, it follows that such a universe must be homogeneous, justifying the full cosmological principle. The argument for this can be found in the online Appendix.

Of course, casual observation suggests that the universe is anything but homogeneous. The density of matter in my vicinity, for example, is far from uniform. However, this is taking an overly parochial view of things. We need to step back onto a larger scale.

Figure 12.2 shows the data from the *Sloan Digital Sky Survey* (SDSS), with each dot representing a galaxy. The large-scale clustering of galaxies into *filaments* with *voids* (~30 – 300 million light years across) between them is a challenge to scientists wishing to explain the processes behind galaxy formation. However, the impression is still that of a broadly uniform distribution of matter across the universe.

FIGURE 12.2 This image represents a slice of the universe 6 billion light years wide, 4.5 billion light years high and 500 million light years thick. Each dot represents a galaxy and the lighter the colour the nearer that galaxy is to Earth. There are 48,741 galaxies in this picture, which covers roughly 1/20th of the sky. Clearly visible is the clustering of galaxies together into filaments with large empty voids between them. However, the overall impression is of a uniform broad distribution of matter across the universe. (Image credit: Daniel Eisenstein and the SDSS-III collaboration.)

In Section 12.6, we will discuss the cosmic microwave background. Detailed maps of this radiation field have been made (see Figure 12.8) revealing tiny variations in the equivalent temperature in different directions into the cosmos. However, these variations are generally $\sim 10^{-5}$ K in an average temperature ~ 2.7 K, which is an extraordinary level of isotropy. As this map accurately reflects the density of matter in the universe 378,000 years into history, it provides the best evidence yet for the validity of the cosmological principle. Indeed, there is an argument that this map is too isotropic, which we will discuss in Chapter 13.

12.3 Universal Metrics and Symmetry

Without going into a detailed justification, we will assume that a universe which obeys the cosmological principle must have the same *curvature* at every point. That being the case, there are only three possibilities, with their associated *spatial* metrics:

1. Positive curvature – (top of Figure 12.3) this is a 3D sphere curved in 4D. The metric takes the form $(\Delta \ell)^2 = \sum_{i,j} \left\{ \delta_{ij} \Delta x_i \Delta x_j \right\} + (\Delta u)^2$ with u being the fourth spatial dimension in which the sphere is embedded, subject to $\sum_{i,j} \left\{ \delta_{ij} x_i x_j \right\} + u^2 = a^2$ where a is the radius of the sphere.

2. Negative curvature – (middle of Figure 12.3) the equivalent 2D surface is a 'saddle' shape (a Pringle!). The metric is $(\Delta \ell)^2 = \sum_{i,j} \left\{ \delta_{ij} \Delta x_i \Delta x_j \right\} - (\Delta u)^2$ subject to $\sum_{i,j} \left\{ \delta_{ij} x_i x_j \right\} - u^2 = -a^2$ where a is a constant that can't be interpreted as a radius.

3. Flat space – (bottom of Figure 12.3) with a Minkowskian metric: $(\Delta \ell)^2 = \sum_i \left\{ \delta_{ij} \Delta x_i \Delta x_j \right\}$.

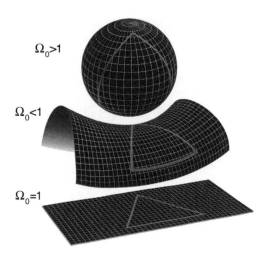

FIGURE 12.3   The three possible geometries of a 2D space with constant curvature. The Ω parameter on the diagram is related to the density of the universe and will be explained in a later section. (Image credit: NASA/ WMAP Science Team.)

Options 2 and 3 are infinite universes that have no boundaries. Option 1 is a closed universe, with no boundary but a finite size. If you consider walking across the surface of a sphere, it's easy to see that you can do a complete lap and end up where you started from – a finite but unbounded space.

Rescaling the co-ordinates by a factor of a, so that $x_i = ax_i'$ and $u = au'$, we can group cases 1 and 2 together using the metric:

$$(\Delta \ell)^2 = a^2 \left[\sum_{i,j} \left\{ \delta_{ij} \Delta x_i' \Delta x_j' \right\} \pm (\Delta u')^2 \right] \text{ subject to } \sum_{i,j} \left\{ \delta_{ij} x_i' x_j' \right\} \pm (u')^2 = \pm 1.$$

Applying our rate of change formula, $\dfrac{\Delta(UV)}{\Delta x_i} = U \dfrac{\Delta(V)}{\Delta x_i} + \dfrac{\Delta(U)}{\Delta x_i} V$, to the case $\sum_{i,j} \left\{ \delta_{ij} x_i' x_j' \right\} \pm (u')^2 = \pm 1$, it is easy to show that $2 \sum_{i,j} \delta_{ij} \left(\Delta x_i' \right) x_j' = \mp 2u \Delta u$ giving:

$$\Delta u = \mp \frac{\displaystyle\sum_{i,j} \delta_{ij} \Delta x_i' x_j'}{u} = \mp \frac{\displaystyle\sum_{i,j} \delta_{ij} \Delta x_i' x_j'}{\sqrt{1 \mp \delta_{ij} x_i' x_j'}}.$$

Putting this into the metric produces:

$$(\Delta \ell)^2 = a^2 \left[\sum_{i,j} \left\{ \delta_{ij} \Delta x_i' \Delta x_j' \right\} \pm \left(\frac{\displaystyle\sum_{i,j} \delta_{ij} \Delta x_i' x_j'}{\sqrt{1 \mp \displaystyle\sum_{i,j} \delta_{ij} x_i' x_j'}} \right)^2 \right] = a^2 \left[\sum_{i,j} \left\{ \delta_{ij} \Delta x_i' \Delta x_j' \right\} \pm \frac{\left(\displaystyle\sum_{i,j} \delta_{ij} \Delta x_i' x_j' \right)^2}{1 \mp \displaystyle\sum_{i,j} \delta_{ij} x_i' x_j'} \right].$$

If we now introduce a parameter $k = -1, 0, +1$, we can unify this metric with the Minkowskian one, which would be the $k = 0$ case:

$$(\Delta\ell)^2 = a^2 \left[\sum_{i,j} \left\{ \delta_{ij}\Delta x_i'\Delta x_j' \right\} + \frac{k\left(\sum_{i,j} \delta_{ij}\Delta x_i'x_j' \right)^2}{1 - k\sum_{i,j} \delta_{ij}x_i'x_j'} \right].$$

To tidy things up somewhat, it is convenient to apply spherical-polar co-ordinates (effectively stepping into intrinsic co-ordinates) by using the relationships:

$$\sum_{i,j} \delta_{ij}\Delta x_i'\Delta x_j' = (\Delta r)^2 + r^2(\Delta\theta)^2 + r^2\sin^2(\theta)(\Delta\phi)^2$$

$$\sum_i \delta_{ij}\Delta x_i'x_j' = r\Delta r \qquad \sum_i \delta_{ij}x_i'x_j' = r^2$$

making the metric:

$$(\Delta\ell)^2 = a^2 \left[\sum_i \left\{ \delta_{ij}\Delta x_i'x_j' \right\} + \frac{k\left(\sum_{i,j} \delta_{ij}\Delta x_i'x_j' \right)^2}{1 - k\sum_{i,j} \delta_{ij}x_i'x_j'} \right] = a^2 \left[(\Delta r)^2 + r^2(\Delta\theta)^2 + r^2\sin^2(\theta)(\Delta\phi)^2 + \frac{k(r\Delta r)^2}{1 - kr^2} \right],$$

which simplifies to:

$$(\Delta\ell)^2 = a^2 \left[\frac{(\Delta r)^2}{1 - kr^2} + r^2(\Delta\theta)^2 + r^2\sin^2(\theta)(\Delta\phi)^2 \right].$$

The final step is to take this spatial metric and absorb it into the space-time interval:

$$(\Delta s)^2 = c^2(\Delta t)^2 - (\Delta\ell)^2 = c^2(\Delta t)^2 - a^2(t) \left[\frac{(\Delta r)^2}{1 - kr^2} + r^2(\Delta\theta)^2 + r^2\sin^2(\theta)(\Delta\phi)^2 \right],$$

where the part in [] brackets applies to the space-like hypersurfaces that we discussed earlier. Inserting $a(t)$ as the scale parameter, rather than just a, takes account of any expansion or contraction of the universe, but we will need to use the field equations to find the function $a(t)$.

Our final result is the *Friedmann–Lemaître–Robertson–Walker* (FLRW) metric[6]:

$$(\Delta s)^2 = c^2(\Delta t)^2 - a^2(t) \left[\frac{(\Delta r)^2}{1 - kr^2} + r^2(\Delta\theta)^2 + r^2\sin^2(\theta)(\Delta\phi)^2 \right],$$

which will be the basis for our cosmological investigations.

It is interesting to note that we have used symmetry arguments to reduce a general metric down to just a function $a(t)$ and a parameter k that have to be fixed by the physics. This meshes with our discussion at the start of Chapter 9.

12.3.1 Proper Distances and Expansion

As it stands, the FLRW metric provides limited information, as we need to apply the field equations in order to determine the appropriate $a(t)$ and k. Having said that, there are a couple of interesting conclusions that we can draw from the metric on its own.

Firstly, the apparent singularity at $kr^2 = 1$ is a co-ordinate singularity, as can be demonstrated by switching to one of a range of alternative co-ordinate systems.

Secondly and undoubtedly confusingly, there are *three* different distance measurements wrapped up in this metric. To see how this works, we are going to calculate the (radial) distance between two objects. This is quite neatly done as, due to the homogeneity of the universe, we are entitled to pick any point as the origin of co-ordinates. So, if we wish to calculate the distance between any two objects, we define the origin to be at one of them and point in a radial direction to the other. The radial condition imposes $\Delta\theta = \Delta\varphi = 0$, and we will include $\Delta t = 0$. Drawing on the discussion of Chapter 9, we can jump to writing:

$$\Sigma = a(t) \int_0^r \frac{dr}{\sqrt{1-kr^2}},$$

which gives three different results depending on the value of k:

$$k = -1 \rightarrow \Sigma = a(t) \int_0^r \frac{dr}{\sqrt{1+r^2}} = a(t)\log\left(r + \sqrt{1+r^2}\right)$$

$$k = 0 \rightarrow \Sigma = a(t) \int_0^r dr = a(t)r$$

$$k = +1 \rightarrow \Sigma = a(t) \int_0^r \frac{dr}{\sqrt{1-r^2}} = a(t)\sin^{-1}(r).$$

The *co-ordinate distance* between two events, r, is unchanging over time. The *co-moving distance* is $\chi = \chi(r)$ where:

$$k = -1 \rightarrow \chi(r) = \log\left(r + \sqrt{1+r^2}\right)$$

$$k = 0 \rightarrow \chi(r) = r$$

$$k = +1 \rightarrow \chi(r) = \sin^{-1}(r)$$

is also unchanging with time, but takes account of the metric. This would be the distance measured by a team of fundamental observers as they are all co-moving. Finally, the *proper distance* is $\Sigma = a(t)\chi(r)$. This is the physical distance between events, accounting for the metric and the changing scale due to expansion (or indeed, hypothetically, contraction).

Armed with the proper distance, it is amusing to contemplate the *proper velocity*, v: the rate of change of proper distance with universal time. As we have defined the universal time as being the proper time measured by fundamental observers, it is the co-ordinate time used in the metric. So we have:

$$k = -1 \rightarrow v = \frac{\Delta\left(a(t)\chi(r)\right)}{\Delta t} = \frac{\Delta\left(a(t)\right)}{\Delta t}\chi(r) = \left(\frac{\Delta a}{\Delta t}\right)\log\left(r + \sqrt{1 + r^2}\right)$$

$$k = 0 \rightarrow v = \frac{\Delta\left(a(t)\chi(r)\right)}{\Delta t} = \frac{\Delta\left(a(t)\right)}{\Delta t}\chi(r) = \left(\frac{\Delta a}{\Delta t}\right)r$$

$$k = +1 \rightarrow v = \frac{\Delta\left(a(t)\chi(r)\right)}{\Delta t} = \frac{\Delta\left(a(t)\right)}{\Delta t}\chi(r) = \left(\frac{\Delta a}{\Delta t}\right)\sin^{-1}(r).$$

In each case, we can use $\Sigma = a(t)\chi(r)$ to convert the expressions into a more interesting format:

$$v = \frac{1}{a}\left(\frac{\Delta a}{\Delta t}\right)\Sigma.$$

Physically this is very important: the proper motion is a result of the scale factor $a(t)$. The co-moving and co-ordinate distances are not changing. As a result, it is wrong to think of expansion as being due to objects receding from each other *through space*. The proper distance is getting bigger as space is *expanding*, or even better would be to say that space is *rescaling* (equally the objects could be getting closer if $a(t)$ is reducing with time). Even though this expansion is continually happening, we don't see galaxies, stars or buildings getting bigger as the local gravitation and the electrical forces that bind atoms together are sufficient to resist the underlying stretching of space. It is only in the empty distances between objects on a large scale that the effect becomes evident (e.g. between galaxies).

Figure 12.4 shows a representation of how the expansion becomes evident. Observers in each of the three galaxies in the diagram would see the others receding from them, although none of the galaxies are moving in their local space.

This means that the proper velocity can exceed the speed of light without violating relativity, as the velocity is not a rate of change of co-ordinates. Of course, objects can move through their local space, but this is a separate motion to the general flow of matter brought about by the change in scale: *Hubble flow*.

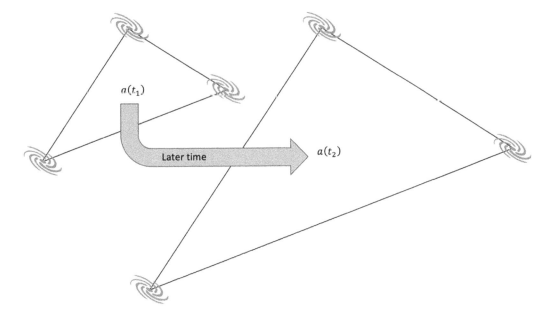

FIGURE 12.4 The expansion of the universe as a change in scale factor.

12.3.2 Hubble's Law

The universe seemed a very different place back before the 1920s. Pale patches of light observed in powerful telescopes, but not fully resolved, were thought to be glowing gas clouds and the population of stars that we see in the night sky was presumed to extend right through the universe. All this was changed by two momentous discoveries made as telescope technology improved.

Some of the glowing patches turned out to be swirling groups of stars at extraordinary distances from us. Evidently the universe did not contain a uniform distribution of stars extending deep into space, but was rather divided up into galaxies, where billions of stars grouped together under gravity, separated by vast reaches of (pretty much) empty intergalactic space. The stars that we see on a clear night are members of our own galaxy, the *Milky Way*.

As various approximations were applied, so the distances to these galaxies became more evident and the study of the light that they emitted revealed their motion. Edwin Hubble[7] systematically measured the distances to galaxies and from their emission spectra confirmed that, aside from the closest to us, they were all redshifted[8] compared to expectations. Using the standard definition of redshift and the Doppler relationship for slow moving light sources:

$$z = \frac{\Delta \lambda}{\lambda} \approx \frac{v}{c}.$$

Hubble was able to plot velocity against distance to reveal the now famous linear relationship called *Hubble's law:*

$$v = H(t)d,$$

where v is the recessional velocity, d is the distance to the galaxy and $H(t)$ is the *Hubble parameter*. This is direct evidence for the expansion of the universe and the recessional velocities revealed by the redshift are the proper velocities calculated in the previous section. However, the redshift is NOT a Doppler effect: that would imply that the galaxies were moving through space. The true source of the redshift is the change of scale parameter acting on the light (Section 12.3.3).

Since Hubble's first announcement, our instrumentation has improved considerably and supernova observations allow us to estimate distances to galaxies hundreds of times further away than Hubble could manage. As a result, Hubble's law is well established (Figure 12.5).

12.3.3 Redshift

To see how the redshift comes about, consider a light wave being emitted from a galaxy at time t_1, and reaching us (as fundamental observers) at time t_2. It will have travelled radially along a null geodesic, so we set $\Delta s = 0$ in the FLRW metric to get:

$$c^2 (\Delta t)^2 = a^2(t) \left[\frac{(\Delta r)^2}{1 - kr^2} \right] \Rightarrow \int_{t_1}^{t_2} \frac{cdt}{a(t)} = \int_0^r \frac{dr}{\sqrt{1 - kr^2}}.$$

To formally carry out the temporal integral, we need to know the exact form of $a(t)$. In practice, this has to be done by numerical approximations using a computer. However, with a little bit of thought we can extract the information that we need without too much difficulty.

If the light ray has a certain frequency, f, on emission, then its period is $1/f$ and a second wave will be emitted $\delta t_1 = 1/f$ after the first. That second wave arrives at us δt_2 after the first wave. The relationship for the second wave is hence:

$$\int_{t_1 + \delta t_1}^{t_2 + \delta t_2} \frac{cdt}{a(t)} = \int_0^r \frac{dr}{\sqrt{1 - kr^2}}.$$

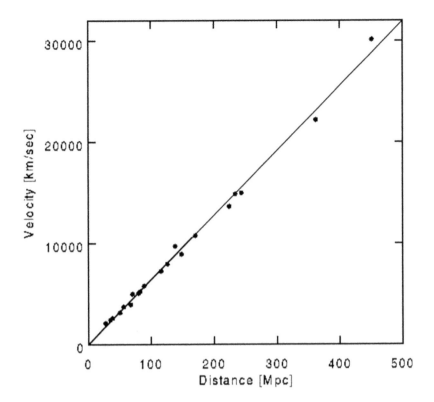

FIGURE 12.5 Modern evidence for Hubble's law. The horizontal scale is mega parsecs and the parsec is an astronomical distance measure roughly equal to 3.26 light years. The total of Hubble's original observations would sit in a small square before the first data points on this graph.

Clearly the right-hand (spatial) integral is the same for both waves, so we can equate the two temporal integrals:

$$\int_{t_1}^{t_2} \frac{cdt}{a(t)} = \int_{t_1+\delta t_1}^{t_2+\delta t_2} \frac{cdt}{a(t)}.$$

One of the basic rules of integration is that the limits can be split up:

$$\int_a^b f(x)dx = \int_a^z f(x)dx + \int_z^b f(x)dx$$

provided $a < z < b$. Craftily applying this rule to the left-hand integral we write:

$$\int_{t_1}^{t_2} \frac{cdt}{a(t)} = \int_{t_1}^{t_1+\delta t_1} \frac{cdt}{a(t)} + \int_{t_1+\delta t_1}^{t_2} \frac{cdt}{a(t)}.$$

Equally, the right-hand integral can be converted into:

$$\int_{t_1+\delta t_1}^{t_2+\delta t_2} \frac{cdt}{a(t)} = \int_{t_1+\delta t_1}^{t_2} \frac{cdt}{a(t)} + \int_{t_2}^{t_2+\delta t_2} \frac{cdt}{a(t)}.$$

Giving on re-assembly:

$$\int\limits_{t_1}^{t_1+\delta t_1} \frac{cdt}{a(t)} + \int\limits_{t_1+\delta t_1}^{t_2} \frac{cdt}{a(t)} = \int\limits_{t_1+\delta t_1}^{t_2} \frac{cdt}{a(t)} + \int\limits_{t_2}^{t_2+\delta t_2} \frac{cdt}{a(t)},$$

so that:

$$\int\limits_{t_1}^{t_1+\delta t_1} \frac{dt}{a(t)} = \int\limits_{t_2}^{t_2+\delta t_2} \frac{dt}{a(t)}.$$

Both of these integrals span small durations of time, so we can approximate the answer by assuming that $a(t)$ does not change significantly during the δts (this is the secret to a result), giving us:

$$\frac{1}{a(t_1)} \int\limits_{t_1}^{t_1+\delta t_1} dt = \frac{1}{a(t_2)} \int\limits_{t_2}^{t_2+\delta t_2} dt.$$

The integral $\int\limits_{a}^{b} dt$ is standard, with result $\int\limits_{a}^{b} dt = [t]_a^b = (b-a)$, so:

$$\frac{\delta t_1}{a(t_1)} = \frac{\delta t_2}{a(t_2)} \text{ rearranging to } \frac{\delta t_1}{\delta t_2} = \frac{a(t_1)}{a(t_2)} = \frac{f_2}{f_1}.$$

Note the similarity between this expression and that of gravitational time dilation, which we will come back to shortly.

The redshift is defined by

$$z = \frac{\Delta\lambda}{\lambda} = \frac{\lambda_2 - \lambda_1}{\lambda_1} = \frac{\lambda_2}{\lambda_1} - 1 = \frac{f_1}{f_2} - 1;$$

so using $f_2 = f_1 \dfrac{a(t_1)}{a(t_2)}$, we get:

$$z = \frac{f_1}{f_1 \dfrac{a(t_1)}{a(t_2)}} - 1 = \frac{a(t_2)}{a(t_1)} - 1 \text{ or } \frac{a(t_2)}{a(t_1)} = 1 + z,$$

which is a useful relationship: cosmologists sometimes label key events or epochs in history by their redshift, rather than the time at which they happened.

Using another of our common approximations, we write:

$$a(t_2) \approx a(t_1) + \frac{\Delta a}{\Delta t}(t_2 - t_1)$$

converting the redshift to:

$$z \approx \frac{a(t_1) + \left(\dfrac{\Delta a}{\Delta t}\right)(t_2 - t_1)}{a(t_1)} - 1 = \frac{\left(\dfrac{\Delta a}{\Delta t}\right)(t_2 - t_1)}{a(t_1)}.$$

The proper distance covered by the light must be $\Sigma = c(t_2 - t_1)$; hence, we have:

$$z \approx \frac{\left(\frac{\Delta a}{\Delta t}\right)(t_2 - t_1)}{a(t_1)} = \frac{1}{a}\left(\frac{\Delta a}{\Delta t}\right) \times \frac{\Sigma}{c}.$$

Comparing this with our earlier expression $v = \frac{1}{a}\left(\frac{\Delta a}{\Delta t}\right)\Sigma$ gives us $z \approx v/c$, making it look just like a Doppler shift, even though the physics is very different. Setting our expression $v = \frac{1}{a}\left(\frac{\Delta a}{\Delta t}\right)\Sigma$ next to the empirical statement of Hubble's law, $v = H(t)d$, reveals the important identity:

$$H(t) = \frac{1}{a}\frac{\Delta a}{\Delta t}.$$

This argument shows that Hubble's law is an approximate relationship for small redshifts. As our observations peer further into the universe (hence further back in time) we would expect to see deviations from linearity.

This argument reveals that the redshift comes about due to a change in scale factor between the time of emission and the time of arrival, rather than a Doppler effect. Occasionally this is portrayed as the expansion of the universe acting to stretch out the wavelength of light while it is in flight. While this is a helpful visualisation, it is not something that can be sustained in a quantum view of photons, where observation of the photon disturbs its properties.

It is also possible to visualise this as a time dilation between the emitting system (galaxy) at scale $a(t_1)$ and the receiving system (us) at $a(t_2)$. An interesting confirmation of this time dilation effect comes from the study of Type 1a supernovae (see Section 12.6.1) which have a characteristic light curve of brightness against time. Data show the width of this curve increasing with the redshift of the supernovae, exactly as we would expect given $\frac{a(t_2)}{a(t_1)} = 1 + z$.

12.4 The Friedmann Equations

If we are going to pin down the exact manner in which the universe evolves, we have to solve the field equations, with an energy density tensor appropriate to the universe.

When Einstein first started his work on cosmology, much of his motivation revolved around trying to smooth over the boundary conditions problem by having a closed universe with no boundary (Section 6.3.3). In order to stop his solutions describing an expanding or contracting universe, he modified his equations to include a *cosmological constant*, Λ:

$$\mathbb{R}_{ab} - \frac{1}{2}\mathcal{R}\left(g^{-1}\right)_{ab} + \Lambda\left(g^{-1}\right)_{ab} = -\frac{8\pi G}{c^4}\mathbb{T}_{ab}$$

which alters the way in which the energy density tensor determines the metric. Subsequently Einstein regarded this as one of his greatest blunders, especially as Hubble later confirmed the expansion of the universe. Nowadays, we have resurrected the cosmological constant in order to model the universe's expansion at an accelerating rate (Section 12.6.1), but in doing so we have re-defined its meaning by moving it to the other side of the equation:

$$\mathbb{R}_{ab} - \frac{1}{2}\mathcal{R}\left(g^{-1}\right)_{ab} = -\frac{8\pi G}{c^4}\mathbb{T}_{ab} - \Lambda\left(g^{-1}\right)_{ab}$$

writing:

$$\mathbb{T}_{ab}^{\Lambda} = \frac{\Lambda c^4 \left(g^{-1}\right)_{ab}}{8\pi G}$$

and interpreting this as a new energy density tensor acting alongside the more traditional matter/energy tensor. The evidence for this step, along with a further discussion of the implications, will come in Section 12.6.1. For the moment, we proceed with the cosmological constant in place and note for future reference that $\mathbb{T}_{00}^{\Lambda} = \rho_{\Lambda} c^2$ in the local Minkowski metric of a fundamental observer, so the slightly mysterious ρ_{Λ} is related to the cosmological constant by $\left(\left(g^{-1}\right)_{00} = \left(\eta^{-1}\right)_{00} = 1\right)$:

$$\rho_{\Lambda} c^2 = \frac{\Lambda c^4}{8\pi G} \text{ or } \rho_{\Lambda} = \frac{\Lambda c^2}{8\pi G}.$$

In order to solve for $a(t)$ and k, we need to use the FLRW metric to calculate the appropriate Christoffel symbols and then the Ricci tensor and scalar. This gives us the left-hand side of the field equations (effectively the Einstein tensor). To get the energy density tensor, we can apply some symmetry arguments, just as we did to build the metric. The starting point is the energy density tensor of dust, as we left it in Chapter 5:

$$\mathbb{T} = nm \begin{pmatrix} \left(\gamma_u\right)^2 c^2 & \boxed{\left(\gamma_u\right)^2 cu_x \quad \left(\gamma_u\right)^2 cu_y \quad \left(\gamma_u\right)^2 cu_z} \\ \boxed{\begin{matrix} \left(\gamma_u\right)^2 cu_x \\ \left(\gamma_u\right)^2 cu_y \\ \left(\gamma_u\right)^2 cu_z \end{matrix}} & \begin{matrix} \left(\gamma_u\right)^2 \left(u_x\right)^2 & \left(\gamma_u\right)^2 u_x u_y & \left(\gamma_u\right)^2 u_x u_z \\ \left(\gamma_u\right)^2 u_y u_x & \left(\gamma_u\right)^2 \left(u_y\right)^2 & \left(\gamma_u\right)^2 u_y u_z \\ \left(\gamma_u\right)^2 u_z u_x & \left(\gamma_u\right)^2 u_z u_y & \left(\gamma_u\right)^2 \left(u_z\right)^2 \end{matrix} \end{pmatrix}.$$

To apply this to the universe, we have to account for the dust being made up of particles of differing mass and velocity. The factor nm has to be taken inside the bracket and then we average over nmu_x, for example. In any isotropic universe, the average value of the 3-momentum along any axis must be zero, $\langle mu_x \rangle = \langle mu_y \rangle = \langle mu_z \rangle = 0$, otherwise there would be a preferred axis. So, as we average the T_{0i} and T_{i0} components (within the boxes), each one must be zero. We can't apply the same argument to $m(u_x)^2$, as $\langle m(u_x)^2 \rangle \neq 0$ with all the values being positive. However, we can argue that $\langle m(u_x)^2 \rangle = \langle m(u_y)^2 \rangle = \langle m(u_z)^2 \rangle$, as these terms act as a pressure, which must be isotropic. When we average the remaining off-diagonal terms in $mu_i u_j$, we can also set them to zero, once again as otherwise there would be an off-axis preferred direction, violating isotropy. So, we have a prototype energy density tensor for the universe, as seen by one of our fundamental observers:

$$\mathbb{T} = \begin{pmatrix} \left(\gamma_u\right)^2 \rho c^2 & 0 & 0 & 0 \\ 0 & \left(\gamma_u\right)^2 P & 0 & 0 \\ 0 & 0 & \left(\gamma_u\right)^2 P & 0 \\ 0 & 0 & 0 & \left(\gamma_u\right)^2 P \end{pmatrix},$$

where ρ is the local average density of matter, and P is the pressure exerted by that matter. In the local Minkowskian space-time of the freely falling fundamental observer, we can write this as:

$$\mathbb{T}_{ab} = \left(\rho + P/c^2 \right) U_a U_b - P \left(\eta^{-1} \right)_{ab}$$

reproducing the matrix form as $U_a = \begin{pmatrix} c & 0 & 0 & 0 \end{pmatrix}$:

$$\mathbb{T}_{00} = \left(\rho + P/c^2 \right) U_0 U_0 - P \left(\eta^{-1} \right)_{00} = \left(\rho + P/c^2 \right) c^2 - P \left(\eta^{-1} \right)_{00} = \rho c^2$$

$$\mathbb{T}_{11} = \left(\rho + P/c^2 \right) U_1 U_1 - P \left(\eta^{-1} \right)_{11} = \left(\rho + P/c^2 \right) \times 0 \times 0 + P = +P,$$

etc. In other co-ordinate systems it would be:

$$\mathbb{T}_{ab} = \left(\rho + P/c^2 \right) U_a U_b - P \left(g^{-1} \right)_{ab}$$

following the normal procedure of switching from η to g in general co-ordinates.

Armed with the energy density tensor, and assuming that we have calculated the Ricci tensor and scalar, we can drop all of these parts into the field equations and plough through the resulting algebra. The end result is the two *Friedmann equations*[9]:

$$\left(\frac{1}{a} \left(\frac{\Delta a}{\Delta t} \right) \right)^2 = H^2(t) = \frac{8\pi G}{3} \rho - \frac{kc^2}{a^2} + \frac{\Lambda c^2}{3}$$

$$\frac{1}{a} \left(\frac{\Delta}{\Delta t} \left(\frac{\Delta a}{\Delta t} \right) \right) = -\frac{4\pi G}{3} \left(\rho + 3\frac{P}{c^2} \right) + \frac{\Lambda c^2}{3}$$

that can be solved (often numerically rather than analytically) to model the universe's evolution, once we insert appropriate expressions for the density and pressure.

12.4.1 Phases of Evolution

Unsurprisingly, the density and pressure appropriate to the universe depends on the nature of the constituent matter. Perhaps surprisingly, this changes as the universe evolves.

Radiation Dominated Phase

In the earliest epochs of the universe's history, it was full of fundamental particles moving at highly relativistic speeds, so their rest mass can be neglected compared to their kinetic energy. To a good approximation, we can treat all the particles in the universe as if they were photons and give them an effective mass density $\rho_r = \varepsilon/c^2$, with ε being their energy density. Photons exert a gentle pressure, $P_r = \frac{1}{3}\rho_r c^2$, when they are absorbed or reflected from objects. On a daily scale this is negligible, but in the early universe this is a significant effect.

The energy density of the radiation will change as the universe evolves. The volume occupied will scale by $a^3(t)$ and the wavelength scales by $a(t)$. Wavelength is inversely proportional to energy, so the energy density of a radiation dominated universe will go down by $1/a^3 (\text{volume}) \times 1/a (\text{energy}) = 1/a^4$.

It is now convenient to rescale the co-ordinates in the FLRW metric. If we make the changes $a = \alpha A$, $r = r/\alpha$ and $k = \alpha^2 k$, where α is any constant, the metric becomes:

$$(\Delta s)^2 = c^2 (\Delta t)^2 - \alpha^2 A^2(t) \left[\frac{(\Delta r)^2/\alpha^2}{1 - \alpha^2 k \dfrac{r^2}{\alpha^2}} + \frac{r^2}{\alpha^2} (\Delta \theta)^2 + \frac{r^2}{\alpha^2} \sin^2(\theta)(\Delta \phi)^2 \right]$$

so that:

$$(\Delta s)^2 = c^2(\Delta t)^2 - A^2(t)\left[\frac{(\Delta r)^2}{1-kr^2} + r^2(\Delta\theta)^2 + r^2\sin^2(\theta)(\Delta\phi)^2\right].$$

Using this freedom, we pick $\alpha = a(t_0)$, taking t_0 to signify the current epoch, and then set the scale so that $\alpha = a(t_0) = 1$. This makes $A(t) \le 1$ and consequently the ratio of scale factors between the current epoch and the time in question. All physical quantities, such as density and pressure, can now be specified in terms of their current values and $A(t)$. So, for example:

$$\rho_r(t) = \frac{\rho_r(t_0)}{A^4(t)} \quad \text{and} \quad P_r(t) = \frac{1}{3}\rho_r c^2 = \frac{1}{3}\frac{\rho_r(t_0)}{A^4(t)}c^2.$$

Matter Dominated Phase

As the universe expands, so the radiation energy density falls until it becomes less than that of (non-relativistic) matter, which has also been falling, but at a slower rate:

$$\rho_M(t) = \frac{\rho_M(t_0)}{A^3(t)}$$

as the expansion does not affect the rest mass energy of slower moving particles, and that energy is the dominant component of their total energy. Hence, the only impact of expansion is on the volume the particles occupy.[10] The manner in which the particle's energy density relates to their pressure depends on the exact species of particle, but can be characterised as:

$$P_M(t) = w\rho_M(t)c^2 = \frac{w\rho_M(t_0)c^2}{A^3(t)},$$

where w is a constant dependent on the specific species.

Dark Energy Dominated Phase

Assuming that there is a cosmological constant at work (and we will review the evidence for this shortly), we can compare the energy density tensor derived from it to the standard form:

$$\mathbb{T}_{ab} = \left(\rho + P/c^2\right)U_aU_b - P\left(g^{-1}\right)_{ab} \qquad\qquad \mathbb{T}_{ab}^\Lambda = \frac{\Lambda c^4\left(g^{-1}\right)_{ab}}{8\pi G}.$$

In the co-moving local system of a fundamental observer, the appropriate metric is η, so:

$$\mathbb{T}_{ab}^\Lambda = \frac{\Lambda c^4\left(\eta^{-1}\right)_{ab}}{8\pi G} = \left(\rho + P/c^2\right)U_aU_b - P\left(\eta^{-1}\right)_{ab}$$

making:

$$\mathbb{T}_{00}^\Lambda = \frac{\Lambda c^4\left(\eta^{-1}\right)_{00}}{8\pi G} = \rho_\Lambda c^2 \text{ or } \rho_\Lambda = \frac{\Lambda c^2}{8\pi G}$$

$$\mathbb{T}_{11}^\Lambda = \frac{\Lambda c^4\left(\eta^{-1}\right)_{11}}{8\pi G} = -\frac{\Lambda c^4}{8\pi G} = -P\left(\eta^{-1}\right)_{11} = +P \text{ or } P = -\frac{\Lambda c^4}{8\pi G} = -\rho_\Lambda c^2.$$

Two things now become clear. Firstly, the energy density ρ_Λ does not scale with expansion. Whatever this energy density is, the value is constant throughout the history of the universe. Hence, as the radiation and matter energy densities decline as $A(t)$ increases, at some stage ρ_Λ must take over as the dominant factor determining how the universe evolves.

Secondly, if we interpret the action of the cosmological constant in these terms, then we could think of it as representing a kind of 'fluid', but a very strange fluid that exerts a negative pressure! This is what brings about the 'repulsive' effect of gravitation (although we should not really talk as if it were a force) triggering an accelerating rate of expansion as the universe becomes dominated by this energy density. Whatever is responsible for this energy density, it has been named *dark energy* and although there is one theoretical frontrunner for its nature, we don't know what it is for sure (Chapter 13).

12.4.2 Friedmann Equations with Density

Putting all this together, we can write the Friedmann equations as:

$$\left(\frac{1}{A}\left(\frac{\Delta A}{\Delta t}\right)\right)^2 = H^2(t) = \frac{8\pi G}{3}\left(\frac{\rho_r(t_0)}{A^4(t)} + \frac{\rho_M(t_0)}{A^3(t)} + \rho_\Lambda\right) - \frac{kc^2}{A^2(t)}$$

$$\frac{1}{A}\left(\frac{\Delta}{\Delta t}\left(\frac{\Delta A}{\Delta t}\right)\right) = -\frac{4\pi G}{3}\left(\frac{\rho_r(t_0)}{A^4(t)} + \frac{\rho_M(t_0)}{A^3(t)} + \rho_\Lambda + \frac{3}{c^2}\left(\frac{1}{3}\frac{\rho_r(t_0)}{A^4(t)}c^2 + \frac{w\rho_M(t_0)c^2}{A^3(t)} - \rho_\Lambda c^2\right)\right).$$

The second equation simplifies by gathering terms together:

$$\frac{1}{A}\left(\frac{\Delta}{\Delta t}\left(\frac{\Delta A}{\Delta t}\right)\right) = -\frac{4\pi G}{3}\left(\frac{2\rho_r(t_0)}{A^4(t)} + \frac{(1+3w)\rho_M(t_0)}{A^3(t)} - 2\rho_\Lambda\right).$$

We pick the value of w to correspond to the matter species that is most prevalent in the universe at the time, or alternatively, we sum the pressure terms for each species with their separate w values.

12.5 The Recipe for the Universe

Supposing for the moment that the universe has a flat spatial geometry, then $k = 0$ and the first Friedmann equation is:

$$H^2(t) = \frac{8\pi G}{3}\left(\rho_r(t) + \rho_M(t) + \rho_\Lambda\right) = \frac{8\pi G}{3}\rho(t)$$

incorporating $\rho(t)$ as a convenient (temporary) shorthand for the total density.

If we introduce a reference density, $D(t)$, and then write the various density components as fractions of that reference, the so-called *density parameters* become:

$$\Omega_r(t) = \frac{\rho_r(t)}{D(t)} \quad \Omega_M(t) = \frac{\rho_M(t)}{D(t)} \quad \Omega_\Lambda(t) = \frac{\rho_\Lambda}{D(t)}$$

allowing us to transform the first Friedmann equation into:

$$H^2(t) = \frac{8\pi G}{3}\left(\rho_r(t) + \rho_M(t) + \rho_\Lambda\right) = \frac{8\pi G}{3}D(t)\left(\Omega_r(t) + \Omega_M(t) + \Omega_\Lambda(t)\right).$$

It is then very tempting to define the reference density as:

$$D(t) = \rho_{\text{crit}}(t) = \frac{3H^2(t)}{8\pi G}$$

and call it the *critical density*. This has the neat effect of establishing:

$$\cancel{H^2(t)} = \frac{8\pi\cancel{G}}{\cancel{3}} \times \frac{\cancel{3}}{\cancel{8\pi G}} \times \cancel{H^2(t)} \times \left(\Omega_r(t) + \Omega_M(t) + \Omega_\Lambda(t)\right),$$

so that:

$$\Omega_r(t) + \Omega_M(t) + \Omega_\Lambda(t) = 1$$

becomes the condition for a spatially flat universe with $k = 0$.

Note that the critical density changes with time: its name can give people the mistaken impression that it is a fixed constant. However, if we are in a spatially flat universe, then the condition $\Omega_r(t) + \Omega_M(t) + \Omega_\Lambda(t) = 1$ is unaffected by the passage of time: i.e. if the universe is launched with a total density equal to the critical density, then it will remain exactly equal to that through history.

Given the established value of the Hubble parameter (see Table 12.1), the critical density in the current epoch is:

$$\rho_C(t) = \frac{3 \times \left(67.8 \text{ km/sMpc}\right)^2}{8\pi G} \approx 8.6 \times 10^{-27} \text{ kg/m}^3$$

or, to give a better 'feel' for the value, ~ 5 protons per cubic metre. We will see how the universe stacks up against this shortly.

If the universe is not spatially flat, then the first Friedmann equation becomes:

$$H^2(t) = \frac{8\pi G}{3} \rho_{\text{crit}}(t)\left(\Omega_r(t) + \Omega_M(t) + \Omega_\Lambda(t)\right) - \frac{kc^2}{A^2(t)}.$$

Substituting in the value of ρ_{crit}, writing $\Omega_r(t) + \Omega_M(t) + \Omega_\Lambda(t) = \Omega$ and cancelling:

$$1 = \Omega - \frac{kc^2}{A^2(t)H^2(t)},$$

which rearranges to:

$$\frac{kc^2}{A^2(t)H^2(t)} = \Omega - 1.$$

Now we can draw the following important conclusions:

- If $\Omega = 1$, then $k = 0$ and the universe is spatially flat
- If $\Omega > 1$, then $k > 0$ and the universe is spatially a 3-sphere
- If $\Omega < 1$, then $k < 0$ and the universe is spatially an (infinite) Pringle.

We normally refer to the case $k > 0$ as being *spatially closed*, $k < 0$ as *spatially open* and $k = 0$ as *spatially flat*. Interestingly, a spatially closed universe is temporally closed as well. Such a universe expands to a maximum scale and then re-contracts again into a *Big Crunch*.[11] Flat and open universes are temporally open: they have no limit to their scale factors or their age.

Although the density parameters change with time, the spatial geometry of the universe does not. If it was launched with $\Omega > 1$, for example, then that remains true throughout history. A closed universe can't evolve into an open one, or vice versa.

As our data only give us values for the various density parameters in the current epoch, we scale them back for earlier periods. For example:

$$\Omega_r(t) = \frac{\rho_r(t)}{\rho_{\text{crit}}(t)} = \frac{\rho_r(t_0)}{A^4(t)\rho_{\text{crit}}(t)} = \frac{\rho_r(t_0)}{A^4(t)\rho_{\text{crit}}(t)} \times \frac{\rho_{\text{crit}}(t_0)}{\rho_{\text{crit}}(t)} = \frac{\Omega_r(t_0)}{A^4(t)} \times \left(\frac{H(t_0)}{H(t)}\right)^2.$$

Converting all the density parameters in a similar way gives:

$$1 = \left(\frac{H(t_0)}{H(t)}\right)^2 \left\{\frac{\Omega_r(t_0)}{A^4(t)} + \frac{\Omega_M(t_0)}{A^3(t)} + \Omega_\Lambda(t_0)\right\} - \frac{kc^2}{A^2(t)H^2(t)}$$

or

$$H^2(t) = H^2(t_0)\left\{\frac{\Omega_r(t_0)}{A^4(t)} + \frac{\Omega_M(t_0)}{A^3(t)} + \Omega_\Lambda(t_0)\right\} - \frac{kc^2}{A^2(t)}.$$

In some quarters, it has also become conventional to incorporate the curvature term $kc^2/A^2(t)$ by writing a *curvature density parameter*, $\Omega_k(t_0) = -\dfrac{kc^2}{H^2(t_0)}$, so that the Friedmann equation becomes:

$$H^2(t) = H^2(t_0)\left\{\frac{\Omega_r(t_0)}{A^4(t)} + \frac{\Omega_M(t_0)}{A^3(t)} + \Omega_\Lambda(t_0) + \frac{\Omega_k(t_0)}{A^2(t)}\right\}.$$

Of course, this is just a convenient notational adjustment. Physically, the curvature parameter is not a density in the same way that the others are.

Tables of cosmological parameters can easily be found on the internet, either directly from experiential datasets or via summarising websites. In many instances, the Hubble parameter is specified in terms of the *reduced Hubble parameter*, h, where:

$$H(t_0) = h \times 100\,\text{km/sMpc}.$$

It is also common to drop the explicit reference to the current epoch by using $\Omega_M = \Omega_M(t_0)$, etc. Hence, we have:

$$H^2(t) = h^2 \left\{\frac{\Omega_r}{A^4(t)} + \frac{\Omega_M}{A^3(t)} + \Omega_\Lambda + \frac{\Omega_k}{A^2(t)}\right\} \times 100\,\text{km/sMpc}.$$

The density parameters are then specified as values of $h^2\Omega_M$, etc.

The current best estimates for these parameters are in shown in Table 12.1.

The data strongly favour $\Omega = 1$, and hence that the density of the universe is exactly equal to the critical density, making it spatially flat with $k = 0$. It is such a singular aspect of these data, that the universe seems to have been launched with a density precisely equal to the exact value that would ensure spatial flatness, that many scientists have been pondering physical mechanisms that could have driven the universe to this density value very early in history. The current frontrunner for this is the *inflationary scenario*, which we will discuss in Chapter 13.

From now on, we will only consider models of the universe with $k = 0$.

TABLE 12.1 The Current (2018) Best Estimate Values of the Cosmological Parameters

Parameter	Value
$\Omega_r(t_0)$	5.37×10^{-5}
$\Omega_M(t_0)$	0.308 ± 0.012
$\Omega_\Lambda(t_0)$	0.692 ± 0.012
$\Omega_T(t_0)$	1.0002 ± 0.0026 (separate fit)
$H(t_0)$	$(0.678 \pm 0.009) \times 100$ km/sMpc
Age of universe	$(13.80 \pm 0.04) \times 10^9$ Years

Source: Taken from http://pdg.lbl.gov/2018/reviews/rpp2018-rev-cosmological-Parameters.pdf. These figures are obtained by combining data from a range of sources, including the *Planck Sky Map* discussed later in this chapter.

12.5.1 The Age of the Universe

To calculate the age of the universe, we start from the relationship:

$$\frac{1}{A}\frac{\Delta A}{\Delta t} = H(t)$$

and consider H to be a function of A, $H = H(A)$. This is valid as $A = A(t)$ and we already have the relationship:

$$H^2(t) = H^2(t_0)\left\{\frac{\Omega_r}{A^4(t)} + \frac{\Omega_M}{A^3(t)} + \Omega_\Lambda(t_0) + \frac{\Omega_k}{A^2(t)}\right\}.$$

The solution to the differential equation is:

$$\int_0^{t_0} dt = \int_0^1 \frac{dA}{AH(A)} = \int_0^1 \frac{AdA}{A^2H(A)} = \int_0^1 \frac{AdA}{H(t_0)A^2\sqrt{\frac{\Omega_r}{A^4} + \frac{\Omega_M}{A^3} + \Omega_\Lambda + \frac{\Omega_k}{A^2}}}$$

making the age of the universe:

$$t_0 = \frac{1}{H(t_0)}\int_0^1 \frac{AdA}{\sqrt{\Omega_r + \Omega_M A + \Omega_\Lambda A^4 + A^2\Omega_k}}.$$

Figure 12.6 shows the results of a numerical calculation (using a spreadsheet[12]) of this integral, based on the values in Table 12.1, the outcome of which is an estimated age of the universe of 13.85 billion years. More sophisticated integrating methods would give an even closer value to the accepted figure from Table 12.1.

12.5.2 The Temperature of the Universe

Any object that is in equilibrium with its thermal radiation emits a characteristic spectrum that only depends on temperature, not the nature of the material in the object. As a perfectly black object would naturally be in equilibrium with its thermal radiation (i.e. at any wavelength it is emitting and absorbing the same amount of energy per second), so this spectrum has been named *black body radiation,*[13] although many examples are not visibly black as they are glowing.

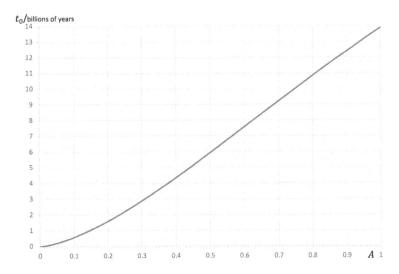

FIGURE 12.6 The age of the universe as a function of A calculated using the cosmological parameters given in Table 12.1. This numerical integration using a spreadsheet gives an age of 13.85 billion years against an accepted value of 13.80 ± 0.04.

In the case of the early universe, the photons emitted and absorbed by the high-energy charged particles, maintained an equilibrium between the constituents and the radiation. Hence, the radiation has a precise black body spectrum, allowing us to define a temperature for the universe. This is done via *Wien's law*, which relates the wavelength at which the most energy is exchanged in the equilibrium to the temperature (on the absolute scale) $\lambda_{max} \propto 1/T$. As the wavelength scales by A, the temperature will scale as A^{-1}. This is the basis for saying that the universe cools as it expands.

Approximately 378,000 years into history (well into the matter dominated epoch), the energy of the radiation dropped below the threshold needed to strip electrons from protons.[14] At this time, the free protons quickly mopped up electrons and became neutral hydrogen, without the photons blasting them apart again. Other charged particles had also dropped out of the mix by then, for related reasons,[15] so the universe became broadly neutral[16] and the bulk interaction between matter and radiation ceased. This is the epoch of *recombination*: a curious name as the charges had never combined before that[17]!

Although the equilibrium between matter and radiation was broken at recombination, the radiation retained its black body spectrum as the photons no longer interacted with matter in any way that would distort that. Hence, their spectrum continues to be a sensible label for the temperature of the universe, scaling as A^{-1}. As they have nowhere else to go (literally!) these photons still fill the universe, and their discovery marked a very significant step in our understanding of cosmology (Section 12.6).

Before recombination the universe was effectively *opaque*. Photons could not cross any significant distance before being scattered or absorbed by a charged particle. It was somewhat like the universe being filled with a dense fog. If we want any data about events in the universe before recombination, then we are not going to get it electromagnetically as the multiple scattering effectively scrambles the information. This is why the onset of gravitational wave astronomy (Section 11.7) is so exciting. Detection of gravitational waves from the universe before recombination will provide valuable data to pin down theoretical models.

12.5.3 The Evolution of the Universe

As we mentioned before, we expect that the early universe was radiation dominated, both as the energy density of radiation $\sim A^{-4}$ whereas for matter the energy density $\sim A^{-3}$, and also as highly relativistic particles act very similarly to massless photons. Under these conditions, with $k = 0$, the first Friedmann equation becomes:

$$\left(\frac{1}{a}\left(\frac{\Delta a}{\Delta t}\right)\right)^2 = \left(\frac{1}{A}\left(\frac{\Delta A}{\Delta t}\right)\right)^2 = \frac{8\pi G}{3}\rho_r(t)$$

making:

$$\frac{\Delta A}{\Delta t} = A\sqrt{\frac{8\pi G}{3}\rho_r(t)} = A\sqrt{\frac{8\pi G\rho_r(t_0)}{3A^4}} = \frac{A}{A^2}\sqrt{\frac{8\pi G\rho_r(t_0)}{3}} = \frac{1}{A}\sqrt{\frac{8\pi G\rho_r(t_0)}{3}}.$$

The solution to this is:

$$\int_0^A A\, dA = \int_0^t \sqrt{\frac{8\pi G\rho_r(t_0)}{3}}\, dt,$$

which is another standard integral with the result:

$$\frac{1}{2}A^2 = \sqrt{\frac{8\pi G\rho_r(t_0)}{3}}\,t \text{ giving } A \sim t^{1/2}.$$

This, in turn, allows us to find the evolution of the Hubble parameter as:

$$H(t) = \frac{1}{a}\left(\frac{\Delta a}{\Delta t}\right) = \frac{1}{A}\left(\frac{\Delta A}{\Delta t}\right) = \frac{t^{-1/2}}{2t^{1/2}} \sim t^{-1}$$

having used the standard result $\frac{\Delta}{\Delta t}\left(t^{1/2}\right) = \frac{1}{2}t^{-1/2}$.

While the universe is radiation dominated, the energy density $\sim A^{-4}$ or $\sim t^{-2}$. Meanwhile, the critical density is $\rho_C(t) = \frac{3H^2(t)}{8\pi G} \sim t^{-2}$. Hence, both the critical density and the radiation density, which is the dominant component of the total density, are varying with time in the same manner, making $\Omega = \frac{\rho(t)}{\rho_C(t)} \sim \frac{t^{-2}}{t^{-2}}$ unchanging over time. As, in this model, we are assuming that the universe was launched with $\Omega = 1$, we see that this value does not change as the universe evolves through the radiation dominated phase.

Eventually, of course, as we run forwards in time, the matter density will overcome the radiation density (Figure 12.7), especially as various species of matter become non-relativistic as they cool.

Our current models suggest that the transition between radiation and matter domination happened about 47,000 years into history (significantly earlier than recombination).

In a *matter dominated universe* with $k = 0$, the Friedmann equation is:

$$\left(\frac{1}{A}\frac{\Delta A}{\Delta t}\right)^2 = \frac{8\pi G}{3}\rho_M = \frac{8\pi G}{3}\frac{\rho_M(t_0)}{A^3(t)}.$$

This can be solved in essentially the same fashion:

$$\frac{\Delta A}{\Delta t} = A\sqrt{\frac{8\pi G}{3}\frac{\rho_M(t_0)}{A^3(t)}} = \frac{A}{A^{3/2}}\sqrt{\frac{8\pi G\rho_M(t_0)}{3}} = A^{-1/2}\sqrt{\frac{8\pi G\rho_M(t_0)}{3}}$$

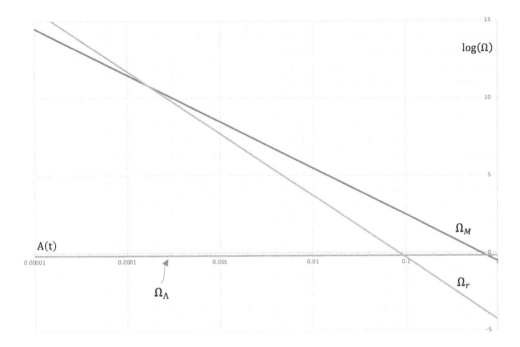

FIGURE 12.7 The variation of Ω components with $A(t)$.

giving an integral:

$$\int_0^A A^{1/2}\, dA = \int_0^t \sqrt{\frac{8\pi G\rho_M(t_0)}{3}}\, dt.$$

The standard result is $\int_0^A A^{1/2}\, dA = \frac{2}{3} A^{3/2}$, so we have:

$$\frac{2}{3} A^{3/2} = \sqrt{\frac{8\pi G\rho_M(t_0)}{3}}\, t$$

meaning that $A \sim t^{2/3}$. From this, we can deduce that $H(t) \sim \dfrac{t^{-1/3}}{t^{2/3}} = t^{-1}$, exactly as before. In fact, we come to the same conclusion about Ω: it remains constant through the matter dominated phase.

When we talk about the matter content of the universe, and the matter dominated epoch, we include both the ordinary matter (which is often, and not wholly correctly, called *baryonic matter*[18]) and the *dark matter*. Back in Section 9.4, we suggested that dark matter could be detected by its gravitational lensing effect. There is plenty of other evidence for the existence of this unknown component of the universe, including the need to include a density of dark matter in order to model the universe's evolution correctly.

The total matter density Ω_M is generally split into two components: the ordinary (baryonic) matter, Ω_b and the *cold dark matter*, Ω_c so that $\Omega_M = \Omega_b + \Omega_c$. The distinction between cold dark matter (sub-relativistic particles) and *hot dark matter* (relativistic particles) has been drawn due to their differing impacts on the process of galaxy formation. It is now generally agreed that evidence and computer models have pretty much ruled out hot dark matter as being an important component of the universe. The proportions of Ω_b and Ω_c can be deduced from the cosmic microwave background (Section 12.6).

Careful inspection of Figure 12.7 shows that about 9.8 billion years into history $(A(t) \approx 0.76)$, we moved into an epoch where the cosmological constant became the largest single component of the energy density. This is backed up by the figures in Table 12.1. It is not reasonable to say that the cosmological component dominates, but it is interesting to see what happens to expansion in an epoch when that would be true (this will pay off later in Chapter 13).

The Friedmann equation is:

$$\left(\frac{1}{A} \frac{\Delta A}{\Delta t} \right)^2 = \frac{8\pi G}{3} \rho_\Lambda \quad \text{or} \quad \frac{\Delta A}{\Delta t} = A \sqrt{\frac{8\pi G \rho_\Lambda}{3}}$$

as ρ_Λ does not scale with the expansion. Remember that ρ_Λ is a *density*, so the energy content of the universe due to this component is *rising* as the universe expands. The solution to the equation takes the form:

$$\int_0^A \frac{dA}{A} = \int_0^t \sqrt{\frac{8\pi G \rho_\Lambda}{3}} \, dt.$$

Once again, the left-hand integral happens to be a standard form (good that the universe co-operates on this...) $\int_0^A \frac{dA}{A} = \log A$, so we have:

$$\log A = \sqrt{\frac{8\pi G \rho_\Lambda}{3}} t \quad \text{or} \quad A \sim e^{\sqrt{\rho_\Lambda} t}.$$

One of the defining properties of the e^x function is that $\frac{\Delta}{\Delta x}\left(e^{kx}\right) = ke^{kx}$. Applying that in our case tells us that $\frac{\Delta A}{\Delta t} \sim \sqrt{\rho_\Lambda} e^{\sqrt{\rho_\Lambda} t}$ and hence that $H(t) \sim \frac{1}{A}\left(\frac{\Delta A}{\Delta t} \right) = \frac{\sqrt{\rho_\Lambda} e^{\sqrt{\rho_\Lambda} t}}{e^{\sqrt{\rho_\Lambda} t}} = \sqrt{\rho_\Lambda}$. In an epoch dominated by the cosmological constant, the Hubble parameter is constant. We are far from that dominant position at our current period in history, but many cosmologists believe that there was a very early phase of the universe when a different form of cosmological constant did dominate and lead to very rapid expansion. This is known as the *inflationary scenario*, to be discussed in Section 13.4.

12.6 Precision Cosmology

In 1965, Robert Wilson and Arno Penzias discovered the photons lingering in the universe since recombination.[19] For a couple of years they had been using a radio antenna (built by Bell Laboratories in America for satellite communication) to study radio emissions from space and had found a low intensity signal in the microwave region of the electromagnetic spectrum. Strangely, the signal was coming equally strongly from all directions and at all times, whereas radio sources are normally localised to a particular part of the sky, even if that moves over time with the rotation of the Earth and the motion of the source. The planet Jupiter[20] in our own solar system would be a good example.

In the early 1990s the Cosmic Background Explorer (COBE) satellite measured the *cosmic microwave background* (the relic photons) over a wide range of wavelengths and confirmed that it followed the expected black body spectrum. The temperature is now measured as 2.72548 ± 0.00057 K.

Further satellite probes, WMAP (2001) and Planck Surveyor (2009), backed up by ground and balloon-based observations (e.g. CBI (1999), BOOMERanG (1997), and MAXIMA (1998)), have mapped tiny temperature variations of the radiation between different points across the sky. The Planck whole sky map is shown in Figure 12.8.

This map is rich in detail regarding the distribution of matter in the universe at the time of recombination. It provides a means of extracting values for a variety of cosmological parameters (such as the density parameters) to a level of precision that was not previously possible. One technique involves averaging the temperature differences squared, $\left\langle (\Delta T)^2 \right\rangle$, between points on the map separated by angle θ. This is then repeated for all angles between $0°$ and $90°$ generating a distribution of temperature fluctuations, such as that shown in Figure 12.9, which is derived from the Planck whole sky map.

The relationship between this plot and the conditions appertaining to the matter distribution at the time of recombination is full of interesting physics. However, we have only the opportunity for a brief skim and the reader is referred to other resources for more detail.[21]

Prior to recombination, the intimate interaction between radiation and matter restricts the ability of matter to clump into regions of higher density. At least, that is true of matter that interacts with

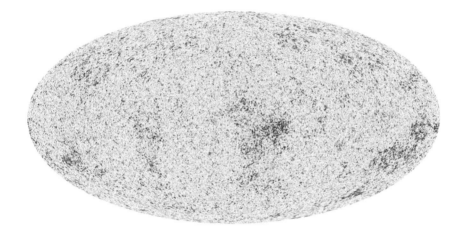

FIGURE 12.8 The whole sky CMB survey from the Planck satellite (2013 data) showing fluctuations in CMB temperature between different portions of the sky. The resolution of this images corresponds to $\sim 0.17°$. The darkest patches are -300μK, the lightest patches $+300\mu$K compared to an average temperature across the whole sky of 2.72548 ± 0.00057 K. (Figure copyright ESA and the Planck Collaboration, with permission.)

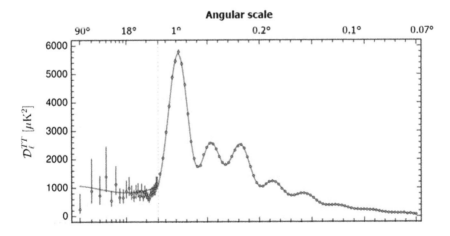

FIGURE 12.9 The cosmic microwave background power spectrum as extracted from the Planck whole sky map. The vertical scale is essentially the $(\Delta T(\theta))^2$ averaged over points on the Planck map separated by angle θ. (Figure copyright ESA and the Planck Collaboration, with permission.)

electromagnetic radiation. Dark matter is not subject to the same constraints, so it had already started to clump well before recombination. In essence, regions with a small difference in density from the average density will tend to diverge further from the average, either by having matter drawn out of them to regions of higher density, or by being a region of higher density hoovering up the surroundings. Dark matter clumps then act as seeds for the ordinary matter to be gravitationally drawn towards them. As this happens, so the radiation interacting within the clumping matter exerts an increasing outwards pressure, which eventually reverses the flow of matter. As matter flows towards or away from one region, it must equally be flowing away from or towards another. The end result is a sequence of *compressions* and *rarefactions* not unlike those found in an acoustic wave travelling through air. Indeed, the photon-matter fluid of the early universe is capable of supporting acoustic waves, albeit on a much grander scale and not related to a 'sound' as such. We expect that the early universe would have many such waves propagating through it, with a range of frequencies (hence periods).

As recombination completes, so these waves will be 'caught' at a compression (high-density region) or rarefaction (low-density region). With the radiation breaking free of the matter at that time (Figure 12.10), its pressure on the matter is released, damping down the acoustic waves. As the radiation breaks free, the matter density in that region will impact on the equivalent temperature of the radiation. In essence, local gravitating matter gives rise to local time dilation, so recombination occurs at slightly differing local times relative to now. As a result, the change in scale factor, and hence redshift, of the radiation varies depending on the local matter density leaving an imprint on the temperature distribution of the background radiation which can still be detected today. Hence, the temperature variations across the whole sky map are a fingerprint of the density variations at recombination, which are in turn a window on the acoustic waves propagating through the universe at that time. We believe that these density variations are the foundations of galaxy formation.

The dominant peak in Figure 12.9 is down to the longest period acoustic waves. They have been caught at the first (and only) compression phase they managed to fit in before the end of recombination. If matter had been streaming into the compression from its surroundings, then a rarefaction has to have formed in another nearby region. The greatest ΔT will be between the crest of the compression and the trough of the next rarefaction, which is exactly what is generating that large peak in the power

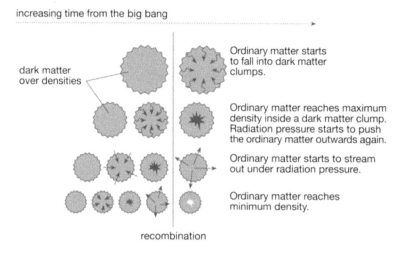

FIGURE 12.10 Setting up acoustic oscillations in the universe before recombination. The process is triggered when ordinary matter is pulled into clumps of dark matter that already exist. As the matter density increases, so the radiation pressure builds up, reversing the inward flow and turning a compression into a rarefaction. This has the effect of feeding a nearby compression and setting up a cycle. The period of a cycle is smallest for the physically smaller length scales. They may go through several cycles leading up to recombination. Note that this diagram suggests that the different size perturbations are separated. In reality, they often lie one inside the other.

spectrum. All things being equal, this peak should be composed of regions about $1°$ across at recombination. However, if the photons have travelled to us from recombination through a curved space-time this will be rather like looking at the background radiation through a lens, distorting the width and the angle of this peak. In the case of a flat universe, there is no dominant curvature of space-time so we would still expect to see the large peak at $1°$ scales. Hence, this peak is a sensitive measure of Ω, with the data strongly favouring $\Omega = 1$.

The second peak in Figure 12.9 is the first rarefaction peak (the fourth row in Figure 12.10). The height of this peak is clearly much lower than that of the first. If we model the acoustic waves in the photon-matter fluid as a mass/spring system, then the pressure exerted by the photons plays the role of the spring and the matter is the mass which gives inertia to the system. The maximum compression achievable is determined by the mass and the pressure. The maximum rarefaction (which is emptying out a region) is determined by the pressure alone. If we could increase the density of ordinary matter compared to the number of photons, then the heights of the first, third and fifth peaks (compression peaks) would increase relative to the second, fourth, etc. (the rarefaction peaks). So, the relative size of the first and second peaks is highly sensitive to the value of Ω_b (ordinary matter). From the Planck power spectrum, $\Omega_b = 0.049$.

Dark matter does not couple to photons, so the density of dark matter tends to damp down the oscillations by its gravitational influence, which makes all the peaks smaller than they would otherwise be. Hence, measuring the relative sizes of the second and third peaks compared to the first, will give an estimation of Ω_c. The Planck data give $\Omega_c = 0.265$.

There is something striking about these figures that should not be overlooked. The total contribution of the ordinary matter that we have known and loved throughout our science education, never mind our everyday life, corresponds to $\Omega_b = 0.049$ out of a total $\Omega = 1$. This stuff is only 4.9% (by mass) of the total universe! Cold dark matter outstrips it by a factor of more than 5:1. While there are many theoretical candidates, we still do not have much idea of what cold dark matter is.

The Planck data tell us that $\Omega = 1$ and independently that $\Omega_M = \Omega_b + \Omega_c = 0.049 + 0.265 = 0.314$: there is still ~0.7 to be accounted for. This is the *dark energy*, which we must turn to next.

12.6.1 Accelerating Expansion

In 1998, there was another dramatic twist in the development of our understanding of the universe. Two independent research groups, the High-Z Supernova Search team and the Supernova Cosmology Project, completed their pioneering studies and reported evidence for a cosmological constant.[22] Improvements in instrumentation allowed the teams to scan portions of the sky the size of the full moon every 5 minutes. With each image containing some 50,000 galaxies, tens of supernova candidates could be discovered every night.[23] Once candidates had been found, follow-up observations confirmed their status and obtained detailed spectra. The teams were looking for a particular class of supernovae known as Type 1a. As we discussed back in Section 10.1.1, there is a maximum mass limit of $1.4M_s$ for white dwarfs, before electron degeneracy pressure can't resist the gravitational collapse. In a Type 1a supernova, a white dwarf star is pulling material from a companion, which adds to the white dwarf's mass. Eventually this accretion of material takes the mass of the white dwarf up to $1.4M_s$, at which time collapse is triggered leading to a supernovae.[24] These events are useful to cosmologists as *standard candles*. As we know, the mass of the star as it explodes (it must be $1.4M_s$), we know the luminosity of the explosion. Measuring the received brightness tells us how far away the explosion was. Inspection of the spectra gives the redshift, hence another distance estimation which can then be compared with that from the brightness. If Hubble's law is correct, the distance measurement from the brightness should be proportional to the redshift. Deviations from Hubble's law will mean that brightness does not correlate correctly with the redshift, which will show in the data. A compilation of results from both teams is shown in Figure 12.11, as well as some theoretical models.

The horizontal scale in Figure 12.11 is the redshift, z, of the light from a supernova. The vertical scale of the top graph will be familiar to astronomers: it is showing the difference between the *magnitude, m*

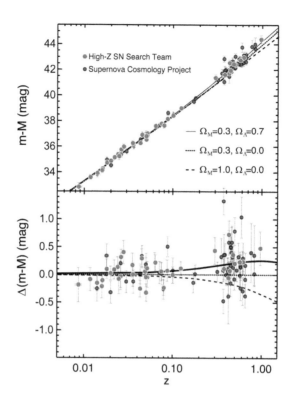

FIGURE 12.11 Supernova data showing the redshift, z, against $(m-M)$, which is a distance estimate (see text) for supernovae seen in distant galaxies (top graph). The data clearly favour the theoretical model with $\Omega_M = 0.3, \Omega_\Lambda = 0.7$. The lower graph is redshift against difference between expected and actual distance. (Image credit: the High-Z Supernova Search team and the Supernova Cosmology Project.)

(the brightness), and the *absolute magnitude M* (which is the brightness the object would have if it were a fixed reference distance away from us); $(m - M)$ is therefore a measure of how far away the object actually is. Three theoretical curves are shown on the graph:

1. A spatially flat universe with $\Omega_M = 0.3, \Omega_\Lambda = 0.7$ leading to $\Omega = 1$.
2. A spatially open universe, with $\Omega_M = 0.3$, which is consistent with the estimations of baryonic and cold dark matter contributions, but no cosmological constant, $\Omega_\Lambda = 0$;
3. A spatially flat universe with no cosmological constant: $\Omega_M = 1, \Omega_\Lambda = 0$;

The data clearly favour option 1.

The lower graph also has z on the horizontal axis, but the vertical axis is now showing:

$$\Delta(m - M) = (m - M)_{\text{measured}} - (m - M)_{\text{expected for }\Omega_M = 0.3, \; \Omega_\Lambda = 0.}$$

It is clear that up to $z \sim 0.1$, the linear redshift/distance relationship expressed by Hubble's law is a reasonable fit to the data. It is only at higher redshifts, and hence further distances and ages, that the data diverge. Once again the $\Omega_M = 0.3, \Omega_\Lambda = 0.7$ model is clearly favoured.

Bear in mind that the supernova research is independent of the data extracted from the cosmic microwave background, yet both have come up with the same essential recipe for the universe: $\Omega_M = 0.3, \Omega_\Lambda = 0.7$ In energy density terms, the cosmological constant, or dark energy, is the greatest single component of the universe. Yet, its nature is, if anything, even more mysterious than that of dark matter. It is possible that it can be explained by the vacuum energy density predicted by quantum field theory. However, that explanation is still fraught with technical difficulties, as we will see in the next chapter.

Notes

1. From *The Inflationary Universe: The Quest for a New Theory of Cosmic Origins* by Alan Guth, Vintage New Ed edition (1998), ISBN-10: 9780099959502.
2. Stephen Hawking has put this well by asking what location lies north of the North Pole?
3. Wienberg, S., 1972. *Gravitation and Cosmology, Principles and Applications of the General Theory of Relativity*. Wiley. ISBN 0471925675, p. 408.
4. Or possibly, at a future Big Crunch.
5. A simpler example is a uniform magnetic field: it is homogeneous, but certainly not isotropic as the alignment of the field picks out a preferred direction.
6. This metric is variously named by taking combinations from Alexander Friedmann, Georges Lemaître, Howard P. Robertson and Arthur Geoffrey Walker who independently contributed to its development and use in cosmological models.
7. Edwin Powell Hubble (1889–1953), University of Chicago and Mount Wilson Observatory. Hubble discovered the existence of galaxies external to our own and used the properties of some variable stars to estimate distances to galaxies. The linear relationship between redshift and distance is attributed to him, although it had also been proposed earlier by Georges Lemaître. The Hubble Space Telescope is named after him.
8. This does mean that the galaxies looked more red in colour. If you lengthen the wavelength of their light, then red becomes infrared and not visible to humans. Equally, ultraviolet becomes violet and so the overall colour stays broadly the same. We can only tell that there is redshift as the spectra of light from stars is crossed by absorption lines where elements in the outer layers of the star have net absorbed some of the wavelengths. The pattern of lines is distinctive to each possible element, and so we can tell if they crop up with the 'wrong' wavelengths.
9. Alexander Friedmann (1888–1925) derived these equations in 1922, before Hubble confirmed the expansion of the universe. They did not get general recognition until they were independently derived by the Belgian cleric Georges Lemaître (1894–1966) in 1927. Cosmological models based on these equations and the FLRW metric are often referred to as Fridemann–Lemaître–Robinson–Walker, models or FLRW models for short.
10. To be clear, expansion will affect their kinetic energy, but as they drop below being relativistic, the kinetic energy is a much smaller part of the total energy than the mass energy.
11. For the proof of this, see the online Appendix.
12. There is something satisfying in using a spreadsheet to calculate the age of the universe and getting an acceptable value while sitting on a train to London…
13. Many objects that look black to the eye are not perfectly black. Any object that looks black absorbs all visible light, but it may not absorb IR or UV radiation – a perfectly black object would absorb all wavelengths equally. Of course, in the case of the cosmic microwave radiation we are talking about an equilibrium between the universe of matter and radiation in the earliest epochs – there is nowhere outside the universe to emit and absorb radiation from!
14. The ionisation energy $\sim 2 \times 10^{-18}$ J.
15. Any particle species, along with its antiparticle, can be created from a high enough energy photon. They can also annihilate into photons. These two processes are in equilibrium maintaining the presence of that species of particle in the universe, until the photon energy drops below the value required to produce the particle/antiparticle. After that, the equilibrium is broken and they steadily annihilate without being replaced at a fast enough rate.
16. It had always been neutral overall (containing the same number of positives as negatives), with particles of both charges wandering freely around. Now all the free charges have vanished, or bonded together to make neutral atoms.
17. The name dates back to earlier theories of the universe, before the Big Bang became the dominant paradigm.

18. Our colleagues in the particle physics branch refer to particles such as protons and neutrons as *baryons*. The early universe certainly contained such particles, but other species as well. Hence calling ordinary matter baryonic is somewhat unjust to the remaining types.
19. They won the 1978 Nobel Prize in Physics for their discovery.
20. Jupiter is a very strong source of radio waves due to electrons spiraling in the planet's very strong magnetic field.
21. My own book, Quarks, Leptons and the Big Bang includes a detailed look at this aspect.
22. The 2011 Noble Prize in Physics was awarded to Saul Perlmutter, Brian P. Schmidt and Adam G. Riess, the leaders of the two research teams, "*for the discovery of the accelerating expansion of the Universe through observations of distant supernovae*".
23. Such supernovae occur ~ once per galaxy per century.
24. Type la can be distinguished from ordinary supernovae by their light curve – how the luminosity changes with time over ~40 days.

13

Quantum Considerations

13.1 Quantum Gravity

Theorists working on quantum gravity are seeking to reconcile general relativity with the principles of quantum theory. These two theoretical frameworks are spectacularly successful in their own areas, but appear to be founded on principles that are incompatible.

As one illustrative example of the tensions that exist between the two theories, consider the problem of time.

In quantum theory, time is an independent background through which the state of a system evolves. Although special relativity has been successfully incorporated into quantum theory, the notion of space-time within the theory still retains a Newtonian tinge, as it is the framework within which events take place. In the general theory, space-time is an aspect of dynamics and interacts with the mass/energy of a system. Merging the two would require some notion of a state evolving through space-time which was itself being influenced by that evolution.

Also, within the framework of quantum theory applied to fundamental forces, energy and momentum exchanges take place between the relevant fields, leading to the interactions that we observe. The granularity of a field gives it particle-like aspects and that wave/particle tension is pacified within the context of a wider *quantum field theory*. If gravitation has field characteristics, and a field is simply a property of space-time that has physical effects, then it must have granularity as well, leading to particles of space-time (we call them *gravitons*). However, these particles need a background space-time through which to propagate, generating another strange conceptual loop for us to resolve.

Finally, quantum theory is *fundamentally probabilistic*. Its calculations reveal the relative probabilities for different possible outcomes, and only under special circumstances tell us exactly what is going to happen. How this probabilistic set of potential outcomes manages to collapse into the definite instantiations that we see in particular experiments, is a fundamental and unresolved issue in quantum theory. This is called the *measurement problem*.

Presumably in a successful quantum gravity, that probabilistic outlook would be retained, leading, via the quantisation of space-time geometry, to a theory in which different geometries would be available to the universe on some probability scale. If we wish to apply a quantum gravity theory to the Big Bang, we are faced with a measurement problem of (literally) universal proportions.

Given the technical and conceptual difficulties that surround the subject, there have to be some compelling reasons to try and tackle quantum gravity in the first place. The motivating factors are broadly as follows:

1. Addressing the fundamental conceptual incompatibilities between the two primary working paradigms in theoretical physics: their conflict being something of an embarrassment;
2. The aesthetic appeal of merging gravity with the other fundamental forces in an over-arching scheme (sometimes called a *Theory of Everything* or TOE);
3. Resolving physical issues that the general theory itself appears inadequate to address; in essence the existence of singularities, either within a black hole or at the start of the universe.

Often in the history of physics, conceptual revolutions are prompted by experimental results which the current theoretical ideas can't incorporate. That was certainly the case with quantum theory and arguably special relativity. In the case of quantum gravity, there are no experimental results to guide us and to help select between competing approaches. Unfortunately, the prospects for such guidance do not seem hopeful, as the situations in which quantum gravity would have a decisive role are at such tiny length scales or vast energy scales, they would appear to be inaccessible to experimental probing. Our best hope lies in some predictive aspect of quantum gravity as applied to the early universe being accessible to observational checking; perhaps via the detection of gravitational waves. In the meantime, various avenues are being explored:

- *Loop quantum gravity* (LQG) as a means of quantising space-time.
- *String theory* as a means of unifying gravity with the other fundamental forces.
- Quantum field theory in curved space-time, which attempts to develop techniques and results by generalising successful quantum field theories in Minkowski space-time. Conspicuous results have been achieved in this way, most notably the establishment of *Hawking radiation* (Section 13.3).
- *Effective field theory*, in which linear gravity theory is quantised and applied to low-energy phenomena where valid calculations can be made, although the effects are too small to be experimentally accessible at the moment.

Much of this active research lies well outside the scope of what we can discuss, but we can touch on various aspects in order to give a brief survey of the state of the game. The plan for this chapter is to first define the sort of scale on which we can expect quantum gravity effects to be significant, then explore the broad features of quantum field theory, briefly survey the two leading contenders for a quantum theory of gravity and then explore in more depth specific contexts where some headway has been made with quantum considerations of gravity: *Hawking radiation* and *inflation*.

13.1.1 The Domain of Quantum Gravity

Back in Section 11.2.2, we suggested that the strength of gravitational effects should be judged by the ratio between the Newtonian gravitational energy and the relativistic rest mass energy:

$$\varepsilon = \frac{GMm/r}{mc^2} = \frac{GM}{c^2r} = \frac{GE}{c^4r}.$$

Now, to get some 'quantum nature' into this, consider a photon with a wavelength equal to r, giving it an energy $E = hf = hc/\lambda = hc/r$. Putting that energy into the strength scale:

$$\varepsilon = \frac{GE}{c^4r} = \frac{Ghc}{c^4r^2} = \frac{Gh}{c^3r^2}$$

and calling for the strong field limit, $\varepsilon = 1$, gives:

$$r^2 = \left(l_p\right)^2 = \frac{Gh}{c^3} \quad \text{or} \quad l_p = \sqrt{\frac{Gh}{c^3}}.$$

At this wavelength, l_p, a photon's energy would generate strong field gravitation. We can use that as a guide to the sort of regime where we might expect quantum effects to become important in gravitation.

Max Planck was the first person to draw attention to the significance of this length scale, l_p, which is now called the *Planck length* in his honour. His argument was based on looking for combinations of the 'most fundamental' constants that produced lengths, times, energies and masses. He proposed:

$$\text{length, } l_p = \sqrt{\frac{Gh}{2\pi c^3}} = 1.62 \times 10^{-35} \text{m}$$

$$\text{time, } t_p = \frac{\text{length scale}}{c} = \sqrt{\frac{Gh}{2\pi c^5}} = 5.39 \times 10^{-44} \text{s}$$

$$\text{energy, } E_p = \frac{hf}{2\pi} = \frac{h}{2\pi t_p} = \frac{h}{2\pi} \times \sqrt{\frac{2\pi c^5}{Gh}} = \sqrt{\frac{2\pi h^2 c^5}{4\pi^2 Gh}} = \sqrt{\frac{hc^5}{2\pi G}} = 1.96 \times 10^9 \text{J}$$

$$\text{mass, } m_p = \frac{E_p}{c^2} = \sqrt{\frac{hc}{2\pi G}} = 2.18 \times 10^{-8} \text{kg}$$

and that these values form scales that Nature would be 'aware of', rather than our human choices of metres, seconds, Joules and kilograms.[1]

The significance of the Planck length is illustrated by the following example. Our weak field arguments from Chapter 11 showed us that:

$$g_{00} \sim 1 + \frac{2\phi}{c^2} \quad \text{and} \quad g_{11} \sim 1 - \frac{2\phi}{c^2}.$$

If we think in experimental terms for the moment, we can only determine the value of ϕ within some degree of uncertainty $\delta\phi$. So, we might expect a corresponding uncertainty in our knowledge of the metric, $\delta g \sim \delta\phi/c^2$. Of course, $\phi = GM/r = GE/c^2 r$, making $\delta g \sim G\delta E/rc^4$.

One of the most famous results in quantum theory is the *Heisenberg uncertainty principle*, which we will fully introduce in Section 13.2.1. For the moment we apply an aspect of that principle which relates our ability to narrow down the energy range over which an interaction takes place to the duration of that interaction: $\delta E \delta \tau \geq h/4\pi$. This could be an experimental interaction measuring ϕ in order to determine the space-time metric. Consequently, $(\delta E)_{min} = h/4\pi\delta\tau$ is the smallest range of energy possible given a duration of $\delta\tau$. Applying this to the uncertainty in the metric $\delta g \sim G(\delta E)_{min}/rc^4$ gives $\delta g \sim Gh/4\pi(\delta\tau)rc^4$. To obtain some duration to work with, consider $\delta\tau \sim \sigma/c$ where σ is the proper size of the region in which we are trying to determine the metric. As $\sigma = \sqrt{g_{11}}r$, we find:

$$\delta g \sim \frac{\sqrt{g_{11}}Ghc}{4\pi\sigma^2 c^4} = \frac{\sqrt{g_{11}}}{2}\left(\frac{l_p}{\sigma}\right)^2 \quad \text{or} \quad \frac{\delta g}{\sqrt{g_{11}}} \sim \left(\frac{l_p}{\sigma}\right)^2. \tag{13.1}$$

We have couched this argument in terms of our ability to measure within degrees of precision, but the uncertainty principle is deeper than that (Section 13.2.1). It relates to reality itself and the applicability of classical concepts. An energy uncertainty such as $(\delta E)_{min}$ delineates the extent to which a system has a definite energy. In this context, a lack of definiteness in energy translates into a lack of definiteness in the metric. The geometry would be 'smeared-out' over a range of metrics within δg.

As long as the spatial dimensions, σ, are much greater than the Planck length, l_p, the resulting δg in Equation 13.1 will be small and the quantum smearing of the metric is insignificant. For example, with an apple ~10 cm, $\delta g/\sqrt{g} \sim 6.6 \times 10^{-69}$! However, at length scales comparable to the Planck length, the quantum smearing in the metric become comparable to the metric itself, and the geometry of the space-time becomes fluid. This suggests that general relativity alone will cease to be an adequate guide to the evolution of the universe when its spatial dimensions are comparable to the Planck length, and that as gravitational collapse drives stellar remnants to this order of size, we might hope that quantum gravity will rescue us from the singularity.

While this calculation tells us the rough order of scale on which we expect quantum gravity to show its head, it is also interesting to know the scale at which quantum field theory becomes appropriate in the context of gravitation. In Section 13.5.5, we will carry out another rough calculation to show that quantum field theory becomes important over length scales $\sim h/mc$.

13.2 Quantum Field Theory

As the name implies, quantum field theory (QFT) melds ideas from the special theory of relativity, orthodox quantum theory and the classical theory of fields.

In quantum theory, mathematical quantities called *amplitudes* determine the *probability* for particular events to take place. For technical reasons, amplitudes are *complex numbers*, i.e. numbers of the form $a + ib$, where $i = \sqrt{-1}$. The *complex conjugate* of a complex number, $z = a + ib$, is defined as:

$$z^* = a - ib.$$

As probabilities have to be real numbers, they are related to the amplitude by

$$\text{probability} = \left|\text{amplitude}\right|^2 = \text{amplitude} \times \text{amplitude}^* = (a + ib)(a - ib) = a^2 - b^2.$$

Suitable events can be interactions between particles, or they could be the outcomes of an experimental measurement made on an individual particle (which is an interaction between particle and experimental device). An important aspect of this probabilistic approach is that the sum total of the probabilities for all the possible outcomes is 1 (something has to happen!). The equations of quantum theory respect *unitarity*. Essentially as a system evolves, its quantum description must retain this sum of probabilities being equal to 1. Here we have one of the most evident clashes between quantum theory and gravity: the world line of any object cannot be extended beyond a singularity (that's one of the definitions of a singularity); hence, if a quantum system evolves so that one or more of its future possibilities involve impacting with a singularity, then unitarity is broken. The possibilities where the system does not encounter a singularity will continue to evolve, but now there will be gaps in the sum. At present, we do not know how to resolve this issue.

In QFT there is a unique *quantum field* for every *flavour*[2] of particle (e.g. the electron field is different to the neutrino field). As the electron field, for example, is responsible for all electron manifestations across the entire universe, this neatly explains why some properties, most evidently the electrical charge, are exactly the same for each electron. The *relativistic quantum field* has another surprise however. Each flavour of field (e.g. the electron field) is also responsible for manifesting an *antiparticle* (e.g. the electron's antiparticle partner, *the positron*), if there is one.

The same powerful theorem that enabled us to write solutions to the wave equation as sums over $y(k) = A\sin(kx - \omega t + \varphi)$ (Section 11.3.1) allows us to expand the field in (roughly) the following manner (Equation 13.2):

$$\psi(X) \sim \sum_K \mathcal{U} A^-(K)\left\{\cos\left(\sum_u K_{\bar{u}} X_u\right) + i\sin\left(\sum_u K_{\bar{u}} X_u\right)\right\} + \sum_{-K} \mathcal{V} A^+(K)\left\{\cos\left(\sum_u K_{\bar{u}} X_u\right) + i\sin\left(\sum_u K_{\bar{u}} X_u\right)\right\}, \quad (13.2)$$

where $\psi(X)$ is the amplitude to find a particle at X. Due to the dictates of quantum theory, the expansion has to be over terms in i as well, and due to the dictates of the Lorentz transformation, *negative frequencies* have to be included. When this field is quantised, the $A^-(K)$ become responsible for melting a particle of momentum K into the field, and the $\mathcal{A}^+(K)$ for making an antiparticle of momentum K arise out of the field. In theoretical parlance they are known, somewhat apocalyptically, as *creation* and *annihilation* operators, respectively. The terms $\mathcal{U}$ and $\mathcal{V}$ relate to the spin of the particles, and I have simplified by leaving out various other factors.

Another expansion of the field gives (Equation 13.3)

$$\psi^*(X) \sim \sum_{-K} \mathcal{U}^* A^+(K)\left\{\cos\left(\sum_u K_{\bar{u}} X_u\right) + i\sin\left(\sum_u K_{\bar{u}} X_u\right)\right\} + \sum_K \mathcal{V}^* \mathcal{A}^-(K)\left\{\cos\left(\sum_u K_{\bar{u}} X_u\right) + i\sin\left(\sum_u K_{\bar{u}} X_u\right)\right\},$$

$$(13.3)$$

where you can see that particles are created and antiparticles are melted.

Alternatively, the same field can be expressed in terms of finding a particle with a given 4-momentum, in which case the expansion proceeds over states which have a definite spatial location.

The Lorentz transformation properties also force us to consider two broad classes of quantum field: *fermion fields*, of which the electron field is one example, and *boson fields*. Fermion fields represent traditional matter and their properties ensured that we came across the particle aspects first in the history of physics, only later discovering that they had wave/field aspects. With any fermion field, only one manifestation can have a specific set of properties at any one time. In Chapter 10, we discussed how this brought about electron degeneracy pressure. On the other hand, boson fields are not so constrained: there is no limit to the number of manifestations that have exactly the same properties. In fact, the more bosons there are with a given set of properties, the greater the amplitude for another identical boson to manifest. This herding aspect of boson fields is exploited in the generation of laser beams and explains the curious properties of *superfluids*. Boson quantum fields give rise to the force fields that we study in classical physics.

13.2.1 Uncertainty

In orthodox quantum theory, the *Heisenberg uncertainty principle* demonstrates the limited applicability of our rigid classical concepts in the more correct quantum description of nature. Some of these concepts turn out to be *conjugate* to each other, for example, position and momentum are *conjugate observables*. Classically, a particle is either at one place or another, and certainly not in more than one place at the same time. That sort of trick is restricted to wave forms, which are legitimately spread across regions.

No experiment ever pins down the location of a particle to exact infinitesimal proportions, the best it can do is point to a region, however small, and say 'its somewhere in there'. Classically, it would follow that the particle has an exact location within this region, but that we are ignorant of that location. However, nature is more slippery than this and we have good experimental grounds to believe that the particle's location is *genuinely blurred across the region*. The experiment has *localised* the particle's position to the region, not *revealed* it to some level of precision. Conjugation comes in because the greater the extent to which the position is localised, the more we blur the particle's momentum, and vice versa.

Mathematically, we express this via the relationship $\delta x \delta p \geq h/4\pi$ where δx is the spatial spread of the particle and δp is the spread across momentum options.

In field theory terms, Equation 13.2 is an expansion over momentum states, localising the position of the particle but surrendering any notion of a precise momentum. On the other hand, the field can manifest a particle with a fixed momentum by summing over position states.

Position and momentum are not the only conjugate classical concepts. We have already seen that energy and duration are conjugate, leading to the relationship $\delta E \delta \tau \geq h/4\pi$.

13.2.2 The Vacuum

In certain situations it is necessary to calculate amplitudes by summing over different field configurations. When this happens, the resulting state can have an indeterminate number of particles, just as a localised particle has no definite momentum state. This sort of situation is when the wave/field aspect is more evident.[3]

Among the field configurations containing a definite number of particles, *the vacuum* is an important example and is defined as the configuration with no particles. However, the absence of particles does not mean the absence of field. This is an important distinction with applications in cosmology, as we will see in Section 13.4.

One of the most interesting results of QFT is that a field contains energy even if no particles are present. Indeed, the sum total of energies from all the different flavours of field may well be the source of the dark energy which is driving the accelerating expansion of the universe (Section 12.6.1). However, current estimates of that total place it some 10^{120} orders of magnitude larger than the necessary background for forming the dark energy. The hope is that some deeper understanding will show how various field energies cancel, in order to bring the total to more manageable proportions. Indeed a popular adjunct to QFT involves supersymmetry (see Section 13.2.6) which suggests that fermion and boson energies cancel. Whatever principle is at work here, it has to balance the various contributions to remarkable precision.

Whatever the value of this vacuum field energy, the uncertainty principle shows that it is imprecise in a manner governed by $\delta E \delta \tau \geq h/4\pi$. This flexibility allows *quantum fluctuations* to take place. In essence particle/antiparticle pairs pop out of the vacuum, exist for an unmeasurably short duration and annihilate each other again. Bizarre though this sounds, the indirect effects of vacuum fluctuations can be detected and measured. It is possible for these fluctuation particles to interact with other particles in the vicinity. Indeed, vacuum fluctuations in the background electromagnetic field destabilise electrons in their atomic energy levels leading them to make transitions into lower energy states and emit photons in the process.

When the particle/antiparticle pairs are created, one has a negative energy to balance the positive energy of the other. If the negative energy particle can collect sufficient energy from an interaction, it can be promoted into a positive energy state and then the pair no longer annihilate and are turned into 'real' particles.

Vacuum fluctuations have a key role in the generation of Hawking radiation.

13.2.3 Tunnelling

If a particle is known to be in the region of location A at a certain time, and then somewhere near B some time later, then classically we would deduce that it had crossed from A to B in-between. In QFT, this is no longer a legitimate conclusion. At the very least, we have no way to determine if it is the same electron at A and then later at B. It would be more correct to think of the particle arising out of the field at A, melting away and then arising at B sometime later.

This leads to one of the most curious quantum mechanical effects, whereby a particle can 'tunnel' from one place to another across a region where, classically, it would be forbidden to stray. One of the most conspicuous example of this comes about in radioactivity where an unstable nucleus becomes more stable by emitting an alpha particle (two protons and two neutrons bound together). Inside the nucleus, the alpha particle temporarily forms by the chance binding of the protons and neutrons. As a result, the particle is raised to a higher energy state within the nucleus. However, this energy is not sufficient to escape from the binding forces holding the nucleus together. It is sufficient to give the alpha particle an independent existence outside the nucleus, but to get there it would have to cross a region where a potential barrier would give it a negative kinetic energy (Figure 13.1).

From time to time, an alpha particle manages to tunnel through the barrier by melting into the field on one side and emerging on the other. The probability for this to happen depends on the width of the

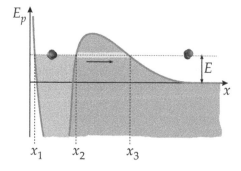

FIGURE 13.1 Alpha particle decay. Between x_1 and x_2, the alpha particle is inside the nucleus with an energy short of the value required to vault the barrier between x_2 and x_3. Beyond x_3 the alpha particle would be free to escape to infinity. The decay takes place when the particle tunnels through this barrier.

barrier, and is sufficiently small that few of the alpha particles formed makes it out. The same effect will turn out to be crucially important in the physics of Hawking radiation (Section 13.3).

13.2.4 Quantum Field Theories of Interactions

In the current *standard model* of particle physics, we recognise the existence of four fundamental forces:

- Electromagnetism between charged particles;
- The weak force which arises at very short range and can cause fermions and bosons to change from one flavour to another, this force is responsible for beta radioactivity;
- The strong force that exists between quarks binding them together into composite particles, such as the proton and neutron;
- Gravitation which exists whenever there is energy.

QFT has been successfully applied to each of the first three. In the process, we have been able to develop an electroweak theory that encompasses the electromagnetic and weak forces. Allowing the various quantum fields to exchange energy and momentum gives rise to interactions and various calculational schemes have to be applied in order to get definite results.

In many situations involving comparatively low energies (on the particle scale), the calculations can be done via a series of approximations known as *perturbation theory*. The famous Feynman diagrams, such as Figure 13.2, arose as part of Richard Feynman's inventive scheme for translating the complex mathematical terms in a perturbation theory expansion into diagrammatic form, enabling them to be classified and studied more effectively. Unfortunately, these Feynman doodles are sometimes regarded as having more physically relevant descriptive powers than they were ever intended to possess.

During the development of perturbation theory, many of the calculations were shown to produce absurd infinite results. A major conceptual breakthrough took place with the development of *renormalisation* which is a systematic technique for sweeping up infinite answers and eliminating them via counter terms generated by quantum corrections to the fundamental properties (e.g. mass, charge, spin) of the particles involved. This works fine as long as the infinite results can be brought together into a number of classifications that equals the number of tuneable fundamental properties available. A non-renormalisable theory would contain more varieties of infinite result than the number of physical properties available.

Outside of the regime where perturbation theory can work, and the interactions of quarks inside a proton would be a good example, theorists have to work a lot harder to get results. Even a simple sounding thing like calculating the mass of a proton accurately is a considerable achievement. Techniques have been developed to allow the calculations to be put in a form amenable to computer processing. With the level of processing power now available, these calculations are increasingly precise and fruitful.

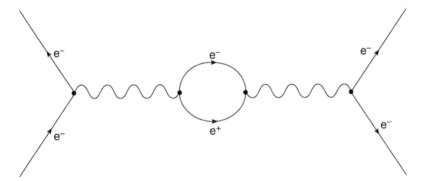

FIGURE 13.2 A Feynman diagram for the interaction between two electrons. The wavy line represents a photon and the loop in the middle is a quantum fluctuation in the electromagnetic field.

13.2.5 QFT Meets Gravitation

All our quantum field theories assume a background space-time that is Minkowskian. While, in Weinberg's terms, this is exactly where we need to start, near a source of gravitation we should adapt our QFT of non-gravitational interactions via a curved metric. However, if we start with a curved metric, developing a QFT is mathematically challenging. For example, in the expansion of the field, it becomes difficult to define positive and negative frequencies consistently between co-ordinate systems when there is no fixed notion of temporal co-ordinate.[4] This has a particular (pun sort of intended) application when it comes to the vacuum state, making it impossible to define the vacuum in the same manner for each observer. This opens up the possibility that the vacuum in one system is not the same as the vacuum in another, which is another aspect of Hawking radiation (Section 13.3).

The situation is bleak, but not hopeless. It turns out that the equations that form the basis of a certain type of boson field in Minkowskian space-time are exactly the linear vacuum field equations that we derived in Chapter 11. This is the *graviton field* which can be used in some limited, low-energy situations to make sensible calculations and hence predictions. In linear gravity theory, we split the dynamic part of the metric from the Minkowski background when we write $g_{\overline{ab}} = \eta_{\overline{ab}} + h_{\overline{ab}}$, which gets around the conceptual loop of space-time being a necessary background for a field theory in which space-time is a propagating particle, but we know that this is an approximation with limited validity. If we try and extend the calculations to higher energies, then various approximation terms 'blow up', and we start to get infinite answers. The theory is *non-renormalisable*. Never the less, if we assume that the infinite terms arise at higher energies and that a more complete quantum gravity will rescue us from their irritating presence, then the resulting *effective field theory of gravitation* is capable of limited but useful results.

13.2.6 Gravitation Meets QFT

> Loopy people go to loopy conferences. Stringy people go to stringy conferences. They don't even go to 'physics' conferences anymore. I think it's unfortunate that it developed this way.
>
> **J Pullin**[5]

Given the clash between the conceptual structures of the general theory and QFT, and as there are no new concepts queuing up to rescue us, then the best thing to do is to treat one theory as primary and develop the other to fit. In broad terms, this corresponds to the approaches taken by the two most popular candidate theories.

Loop Quantum Gravity

LGQ started in 1986 with the work of Abhay Ashtekar[6] who introduced a new way of looking at the general theory. Focussing on how representations of spin are parallel transported, allows various structures, such as the metric, to be recast in a fashion that makes them more like traditional structures found in particle physics. Using this formalism, Lee Smolin[7] and Ted Jacobson[8] found that they could obtain exact solutions to equations of quantum gravity based on the general theory. With additional input from Carlo Rovelli,[9] this then developed into LQG: a quantisation of space-time starting from general relativistic principles.

The theory reveals a microstructure to space-time geometry which makes space look like a network with discrete edges. In essence, space is quantised into distinct volumes $\sim l_p^3$ in size. The surfaces dividing one volume from the next are also quantised $\sim l_p^2$.

This approach has had some notable successes, not the least of which has been calculating the Hawking temperature of a black hole, via quantisation of the surface area of the event horizon. It has also given quantum corrections to the Hawking radiation formula, which should be observable and so help to verify/falsify the theory. Applying it to the Big Bang has produced the intriguing suggestion that the singularity can be avoided by changes in the quantum structure at high densities. This might allow fundamental constants, including the speed of light, to change giving the universe the ability to 'bounce'.

As with all work in this field, LQG has its problems. For example, LQG appears to be in conflict with the special theory. Firstly, if there is a minimum length scale $\sim l_p$ arising from the quantisation of area and volume, that would contradict the basis of the Lorentz transformations.[10] Also, the speed of light acquires an energy dependence, with high energy photons travelling slightly slower than their low-energy counterparts (an effect that is potentially observable in gamma ray bursts from space). Recent work by Simolin and others has shown that the assumptions of special relativity can be modified to accommodate this. The effect also has important consequences for cosmology, potentially allowing the speed of light to be faster in the hot, dense early universe, achieving the aims of the inflationary scenario (see Section 13.4) without what is seen by some as the baggage that comes with it. At this stage of development, it is unclear if LQG actually reproduces the general theory at macroscopic scales, but as an approach, its popularity is starting to rise as some see frustration growing with *string theory*.

String Theory

String theory approaches the problem from a different direction, by trying to develop quantum theory to incorporate gravity. Its history goes deep into the 1940s with origins in the early theory behind the strong force. Its development has been the work of many physicists, principally among them Ed Witten,[11] John Henry Schwarz,[12] David Gross,[13] Leonard Susskind[14] and Michael Green.[15]

The fundamental basis of the theory is the suggestion that all the point-like particles that we observe are in fact different vibrational modes of extended objects. By providing a smallest size scale for particles, the theory automatically side-steps some issues in QFT related to infinites that arise when you allow a particle to be point-like. The strings that give the theory its name are extended one-dimensional objects that can close on themselves to form vibrating loops. It is assumed that the size of the strings will be $\sim l_p$.

In order to work fully, string theory needs a connection between fermions and bosons, which is called *supersymmetry*. Many physicists believe that supersymmetric effects should exists, on aesthetic grounds if nothing else, as supersymmetrical theories entail deep connection between bosons and fermions that otherwise enter a theory independently. Many attractive theoretical consequences follow from this. However, supersymmetry should also imply the existence of a class of supersymmetric particles, which so far have not been seen in experiments and it is becoming hard to explain why they have not been found as yet.

Even more extraordinarily, string theory requires six or seven (depending on the variant) other spatial dimensions for its mathematical consistency. These dimensions are assumed to be unobservable at our scales as they are *compactified* (rolled-up) on themselves on scales $\sim l_p$ so that they form closed structures.

String theory successfully incorporates the fundamental forces into one framework as the graviton arises as one of the string vibrations. As with LQG, the theory presents formidable mathematical challenges and struggles to produce experimentally testable predictions. The best hope again lies in supersymmetric effects in the early universe being magnified by inflation to become observable.

13.3 Hawking Radiation

In my opinion, Hawking's remarkable calculation of the entropy and temperature of a black hole… is the only reasonably reliable conclusion that has been obtained, to date, from any quantum-gravity theory.

R Penrose[16]

In 1974, Stephen Hawking[17] discovered that Schwarzschild black holes radiate particles from the region just outside their event horizons, and as a result loose mass and will eventually 'evaporate'. This effect, known as *Hawking radiation*, overturns our traditional view that black holes swallow matter in a one-way fashion and once formed are eternal, at least until the universe itself comes to an end. His derivation is not a quantum gravity result as such, but rather comes from applying traditional field theory in a curved space-time. It transpires that Hawking radiation is closely linked with three pieces of physics that we have already discussed, albeit briefly: the Penrose process for extracting energy from spinning black holes, quantum vacuum fluctuations and quantum tunnelling.

13.3.1 Killing Vectors

If we look at the Schwarzschild metric in standard co-ordinates:

$$(\Delta s)^2 = \boxed{\left(1 - \frac{2GM}{c^2 r}\right)} c^2 (\Delta t)^2 - \frac{(\Delta r)^2}{\boxed{\left(1 - \frac{2GM}{c^2 r}\right)}} - \boxed{r^2}(\Delta\theta)^2 - \boxed{r^2 \sin^2(\theta)}(\Delta\varphi)^2,$$

we see that none of the metric's components explicitly depend on t or φ. As a result, the form of the metric is preserved if we make a change to either of these co-ordinates: $t' = t + \varepsilon K$ or $\varphi' = \varphi + \varepsilon K$, where $\varepsilon \ll 1$. This allows us to define two *Killing vectors*:[18]

$$\boldsymbol{K}^1 = \begin{pmatrix} 1 & 0 & 0 & 0 \end{pmatrix} \text{ or } K_u^1 = \delta_{0u} \qquad \boldsymbol{K}^2 = \begin{pmatrix} 0 & 0 & 0 & 1 \end{pmatrix} \text{ or } K_u^2 = \delta_{3u}$$

with transformations of co-ordinates along these vectors leaving the metric unchanged. They are quoted above in their contravariant form; covariantly they are:

$$K_{\bar{u}}^1 = \sum_v g_{\overline{uv}} \delta_{0v} \quad \overline{\boldsymbol{K}}^1 = \begin{pmatrix} \left(1 - \dfrac{R_s}{r}\right) & 0 & 0 & 0 \end{pmatrix}$$

$$K_{\bar{u}}^2 = \sum_v g_{\overline{uv}} \delta_{3v} \quad \overline{\boldsymbol{K}}^2 = \begin{pmatrix} 0 & 0 & 0 & -r^2 \sin^2(\theta) \end{pmatrix}.$$

We can generalise this approach, so that it can be applied to any metric, in the following way. The change in a metric due to a co-ordinate transformation is the same as that for any tensor, so under the transformation $x_i \to x_i'$ the metric is:

$$g_{ab}(x) = \sum_{u,v} \left\{ \frac{\Delta x_u'}{\Delta x_a} \frac{\Delta x_v'}{\Delta x_b} g_{uv}(x') \right\}.$$

Specifically, if the transformation is of the form $x_i' = x_i + \varepsilon K_i$, then:

$$\frac{\Delta x_u'}{\Delta x_a} = \frac{\Delta}{\Delta x_a}(x_u + \varepsilon K_u) = \delta_{au} + \varepsilon \frac{\Delta K_u}{\Delta x_a}$$

and so:

$$g_{ab}(x) = \sum_{u,v} \left\{ \left(\delta_{au} + \varepsilon \frac{\Delta K_u}{\Delta x_a} \right) \left(\delta_{bv} + \varepsilon \frac{\Delta K_v}{\Delta x_b} \right) g_{uv}(x') \right\}$$

$$= \sum_{u,v} \left\{ \delta_{au} \delta_{bu} g_{uv}(x') + \delta_{au} \varepsilon \frac{\Delta K_v}{\Delta x_b} g_{uv}(x') + \delta_{bv} \varepsilon \frac{\Delta K_u}{\Delta x_a} g_{uv}(x') \right\}$$

$$= g_{ab}(x') + \varepsilon \sum_v \frac{\Delta K_v}{\Delta x_b} g_{av}(x') + \varepsilon \sum_u \frac{\Delta K_u}{\Delta x_a} g_{ub}(x')$$

neglecting terms in ε^2. With a little more dexterous algebraic manipulation, which is done in the online Appendix for interested readers, we find that general Killing vectors satisfy the condition:

$$\mathbb{D}_a\left(K_{\bar{b}}\right) + \mathbb{D}_b\left(K_{\bar{a}}\right) = 0.$$

13.3.2 Killing Vectors and Conserved Quantities

The utility of the Killing vectors derives from being able to use them to generate *conserved quantities*. Given an object with 4-momentum components $P_a = mU_a$ the quantity $\sum_a mK_{\bar{a}}U_a$ *is conserved along a geodesic*. To show this, we parameterise both $K_{\bar{a}}$ and U_a using τ so that:

$$\frac{D}{\Delta \tau}\left(\sum_a mK_{\bar{a}}U_a\right) = \sum_a \left\{ m\frac{DK_{\bar{a}}}{\Delta \tau}U_a + mK_{\bar{a}}\frac{DU_a}{\Delta \tau} \right\}.$$

Now:

$$\frac{DK_{\bar{a}}}{\Delta \tau} = \sum_b \left(\frac{DK_{\bar{a}}}{\Delta x_b}\right)\left(\frac{\Delta x_b}{\Delta \tau}\right) = \sum_b \left(\frac{DK_{\bar{a}}}{\Delta x_b}\right) U_b,$$

so that:

$$\frac{D}{\Delta \tau}\left(\sum_a mK_{\bar{a}}U_a\right) = \sum_a \left\{ m\sum_b \left(\frac{DK_{\bar{a}}}{\Delta x_b}\right) U_b U_a + mK_{\bar{a}}\frac{DU_a}{\Delta \tau} \right\}.$$

As we are following a geodesic, by definition $DU_a/\Delta \tau = 0$, hence we are reduced to:

$$\frac{D}{\Delta \tau}\left(\sum_a mK_{\bar{a}}U_a\right) = m\sum_{a,b} \left(\frac{DK_{\bar{a}}}{\Delta x_b}\right) U_b U_a.$$

Working on the summation:

$$\sum_{a,b} \left(\frac{DK_{\bar{a}}}{\Delta x_b}\right) = \frac{1}{2}\sum_{a,b} \left\{ \frac{DK_{\bar{a}}}{\Delta x_b} + \frac{DK_{\bar{b}}}{\Delta x_a} \right\} = 0$$

by the definition of the Killing vector. Hence, we have established:

$$\frac{D}{\Delta \tau}\left(\sum_a mK_{\bar{a}}U_a\right) = 0$$

and so that the quantity $\sum_a mK_{\bar{a}}U_a$ is conserved.

For the Schwarzschild metric, $\overline{\boldsymbol{K}}^1 = \begin{pmatrix} \left(1-\dfrac{R_s}{r}\right) & 0 & 0 & 0 \end{pmatrix}$ giving us the conserved quantity:

$$\sum_a mK_{\bar{a}}^1 U_a = mK_{\bar{0}}^1 U_0 = m\left(1-\frac{R_s}{r}\right)c\frac{\Delta t}{\Delta \tau} = mc\left(1-\frac{R_s}{r}\right)\frac{\Delta t}{\Delta \tau}.$$

If we calculate this a great distance from the gravitating mass, so that $r \gg R_s$, then it becomes:

$$\sum_a mK_{\bar{a}}U_a \approx mc\frac{\Delta t}{\Delta \tau} = \gamma mc.$$

If we define the *Killing energy* $E_K = c\sum_a mK_{\bar{a}}U_a$, we see that this becomes the energy at a great distance from the gravitating mass, and so is the conserved energy along a geodesic. Hence, we have:

$$E_K = mc^2\left(1-\frac{R_s}{r}\right)\frac{\Delta t}{\Delta \tau}.$$

To check that this is giving us something sensible, we consider a stationary observer sitting at r, θ, ϕ in the Schwarzschild metric. The interval becomes:

$$(\Delta s)^2 = (1-R_s/r)c^2(\Delta t)^2 \quad \text{so that} \quad \Delta \tau_s = \sqrt{1-R_s/r}\,\Delta t_s.$$

Hence, the observer has a 4-velocity given by:

$$U^{\text{stat}} = \frac{\Delta}{\Delta \tau}\begin{pmatrix} ct & r & \theta & \phi \end{pmatrix} = \begin{pmatrix} c\dfrac{\Delta t}{\Delta \tau} & 0 & 0 & 0 \end{pmatrix} = \begin{pmatrix} \dfrac{c}{\sqrt{1-R_s/r}} & 0 & 0 & 0 \end{pmatrix}.$$

To calculate the energy this observer sees when a particle falls past them along a geodesic, we make use of a neat relationship. In general, if $\boldsymbol{P}$ is the 4-momentum of an object which is being studied by an observer with 4-velocity $\boldsymbol{U}$, then the energy of the object as seen by the observer is:

$$E' = \sum_u P_{\bar{u}}U_u.$$

To prove this, consider the rest system of the observer which has $U_u = \begin{pmatrix} c & 0 & 0 & 0 \end{pmatrix}$, hence:

$$E' = \sum_u P_{\bar{u}}U_u = \frac{E}{c}c = E.$$

As with all invariants, if this is the result in one system, it must be the result in all of them.

Armed with this dodge, the energy observed by our static observer as a particle falls past on a geodesic must be:

$$E_{\text{stat}} = \sum_u P_{\bar{u}}U_u^{\text{stat}} = \frac{cE_K/c}{\sqrt{1-R_s/r}} \quad \text{or} \quad E_K = E_{\text{stat}}\sqrt{1-R_s/r}.$$

For $r \gg R_s$ this becomes:

$$E_K = E_{\text{stat}}\sqrt{1-R_s/r} = E_{\text{stat}}(1-R_s/2r).$$

If the velocity, v, of the falling object is small then we can take $E_{\text{stat}} \approx mc^2 + \frac{1}{2}mv^2$ so that:

$$E_K \approx \left(mc^2 + \frac{1}{2}mv^2 \right)\left(1 - R_s/2r\right) = mc^2 - \frac{R_s mc^2}{2r} + \frac{1}{2}mv^2 - \frac{R_s mv^2}{4r}$$

substituting for R_s reveals:

$$E_K \approx mc^2 + \frac{1}{2}mv^2 - \frac{GMm}{r} - \frac{GMmv^2}{2rc^2} = mc^2 + \frac{1}{2}mv^2 - \frac{GMm}{r}$$

dropping the term in v^2/c^2 as being too small.

13.3.3 Positive Energy, Negative Energy and the Flow of Time

Provided $r > R_s$, the Killing vector $\overline{K}^1 = \left(\begin{array}{cccc} \left(1 - \dfrac{R_s}{r}\right) & 0 & 0 & 0 \end{array} \right)$ is time-like so that $\sum_u K_u^1 K_u^1 > 0$ and the Killing energy:

$$E_K = mc^2 \left(1 - \frac{R_s}{r}\right)\frac{\Delta t}{\Delta \tau}$$

is positive if $\Delta t/\Delta \tau > 0$, i.e. time is 'flowing' forwards.[19] This seems so much like common sense, it hardly seems worthy of comment. However, once we get inside the Schwarzschild radius, the Killing vector becomes space-like and more possibilities open up.

To understand this, it helps to start by considering an object hovering (aided by propulsion of some form) near the event horizon of a Schwarzschild black hole. If $r = R_s$, the Killing energy is zero. However, an object of the same mass at rest well outside the gravitational influence of the black hole would have a Killing energy of mc^2. Clearly, we can't get from one to the other along a geodesic, as the Killing energy is conserved along such paths. However, if we did wish to lift the object from the event horizon (or more correctly just outside the event horizon) then we have to provide energy mc^2 to get it to infinity.

If we have a slightly different scenario, whereby we had to provide *more* energy than mc^2 in order to get the object to rest at infinity, that would imply that the initial Killing energy was *negative*. At first the existence of negative energies seems counter-intuitive, but it is the energy referenced to an observer at infinity. Another observer, placed near the Schwarzschild radius would see the negative energy objects as having a positive energy. However, just having negative energies on their own is not sufficient for this to be a physically realisable situation. The Killing vector that is derived from the time independence of the appropriate metric has to be space-like in the same region.

The real trick would be to find a region of space-time where the Killing energy could be negative, but the 4-momentum of the object was time-like. After all, real objects have to have time-like 4-momenta:

$$E^2 - p^2c^2 = m^2c^4 \geq 0$$

for real masses (i.e. masses that are not complex numbers). We would then have the situation:

$$E_K = c \sum_a \left(\text{Space-like Killing vector}\right)_{\bar{a}} \left(\text{time-like 4 momentum}\right)_a$$

which can be negative.

Such a region, with $E_K < 0$, lies within the Schwarzschild radius on a non-rotating black hole, but also in the ergosphere of a rotating black hole. The more practically interesting second option is the one that we will look at first.

13.3.4 The Penrose Process Again

The Kerr metric for a spinning black hole gives an interval:

$$(\Delta s)^2 = \left(1 - \frac{R_s r}{\rho^2}\right)c^2(\Delta t)^2 + \frac{2R_s r\alpha \sin^2 \vartheta}{\rho^2}c\Delta t\Delta\phi - \frac{\rho^2(\Delta r)^2}{D} - \rho^2(\Delta\vartheta)^2 - \left\{\left(r^2 + \alpha^2\right)\sin^2 \vartheta + \frac{R_s r\alpha^2 \sin^4 \vartheta}{\rho^2}\right\}(\Delta\phi)^2$$

with $D = r^2 - R_s r + \alpha^2$ and $\rho^2 = r^2 + \alpha^2 \cos^2 \vartheta$. As none of the metric's components depend on t or ϕ, we have the Killing vectors:

$$\mathbf{K}^1 = \begin{pmatrix} 1 & 0 & 0 & 0 \end{pmatrix} \quad \text{or} \quad K_i^1 = \delta_{0i} \qquad \mathbf{K}^2 = \begin{pmatrix} 0 & 0 & 0 & 1 \end{pmatrix} \quad \text{or} \quad K_i^2 = \delta_{3i}.$$

We are only interested in $\mathbf{K}^1$, which in covariant form is:

$$K_{\bar{u}}^1 = \sum_v g_{\overline{uv}} K_v^1 = \begin{pmatrix} \left(1 - \dfrac{R_s r}{\rho^2}\right) & 0 & 0 & \dfrac{R_s r\alpha \sin^2 \vartheta}{\rho^2} \end{pmatrix}.$$

We can calculate the invariant for the first Killing vector:

$$\left|\mathbf{K}^1\right|^2 = \sum_u K_{\bar{u}}^1 K_u^1 = 1 - \frac{R_s r}{\rho^2}.$$

For this to be null, $\dfrac{R_s r}{\rho^2} = 1$ or $r = \rho^2/R_s$ which is exactly the definition of the ergosphere from Section 10.5.3. It follows that inside the ergosphere, the Killing vector is space-like, $\left|\mathbf{K}^1\right|^2 < 0$.

Calculating the Killing energy we obtain:

$$E_K = c\sum_a mK_{\bar{u}}^1 U_u = mcK_{\bar{0}}^1 U_0 + mcK_{\bar{3}}^1 U_3.$$

As the displacement 4-vector in these co-ordinates is $X = \begin{pmatrix} ct & r & \vartheta & \phi \end{pmatrix}$, the 4-velocity is $U = \begin{pmatrix} \dfrac{c\Delta t}{\Delta\tau} & \dfrac{\Delta r}{\Delta\tau} & \dfrac{\Delta\vartheta}{\Delta\tau} & \dfrac{\Delta\phi}{\Delta\tau} \end{pmatrix}$, making:

$$E_K = mc^2\left(1 - \frac{R_s r}{\rho^2}\right)\left(\frac{\Delta t}{\Delta\tau}\right) + mc\frac{R_s r\alpha \sin^2 \vartheta}{\rho^2}\left(\frac{\Delta\phi}{\Delta\tau}\right).$$

If we confine ourselves (for simplicity) to the equatorial plane $\vartheta = \pi/2$ then $\rho^2 = r^2$ and we get:

$$E_K = mc^2\left(1 - \frac{R_s}{r}\right)\left(\frac{\Delta t}{\Delta\tau}\right) + mc\frac{R_s\alpha}{r}\left(\frac{\Delta\phi}{\Delta\tau}\right)$$

which will be negative if $r < R_s$ and:

$$mc^2\left|1 - \frac{R_s}{r}\right|\left(\frac{\Delta t}{\Delta\tau}\right) > mc\frac{R_s\alpha}{r}\left(\frac{\Delta\phi}{\Delta\tau}\right).$$

In 1969, Roger Penrose indicated how it would be possible to use a negative energy state to extract rotational energy from a black hole. As we briefly discussed in Section 10.5.5, the process requires a little delicate manoeuvring (Figure 13.3).

A spacecraft with 4-momentum $\mathbf{P}^0$ flies into the ergosphere of a rotating black hole and then divides into two stages. If the impulses controlling the staging are correctly arranged, one of the two pieces can

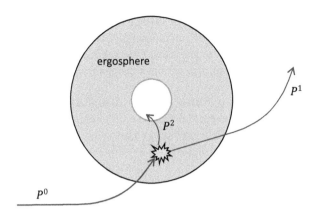

FIGURE 13.3 The Penrose process for extracting energy from a rotating black hole.

be placed onto a negative energy geodesic and the other onto a positive energy geodesic which exits the ergosphere and escapes to infinity. Within the ergosphere negative energy states are physically possible as the Killing vector is space-like. That stage will be trapped within the ergosphere and will eventually cross the event horizon on the way to the singularity.

Conservation of 4-momentum demands $\boldsymbol{P}^0 = \boldsymbol{P}^1 + \boldsymbol{P}^2$ (Figure 13.3) which when contracted with the Killing vector gives $E_0 = E_1 + E_2$. So, if $E_1 < 0$, $E_2 > E_0$ and the stage that departs the ergosphere has more energy than the whole spacecraft on entry, at least as measured at infinity. This is balanced by a reduction in the rotation rate of the black hole caused by absorbing a negative energy stage with negative angular momentum. Interestingly, this also places a limit on the process, as the black hole's rotation can be reduced to zero by repetitive applications, however the mass is not reduced to zero in the same way. Eventually the black hole would stop spinning, the ergosphere would vanish and the Penrose process would no longer work.

Although no mention is made in the script, it is very tempting to suppose that the spacecraft manoeuvring that takes place at the end of *Interstellar* is designed to make use of the Penrose process to boost one of the astronauts towards a habitable planet.

13.3.5 And Finally, Hawking Radiation

Just outside the event horizon there will be virtual pairs of particles, one with negative energy and one with positive energy. The negative particle is in a region which is classically forbidden but it can tunnel through the event horizon to the region inside the black hole where the Killing vector which represents time translations is spacelike. *In this region the particle can exist as a real particle with a timelike momentum vector even though its energy relative to infinity as measured by the time translation Killing vector is negative.* The other particle of the pair, having a positive energy, can escape to infinity where it constitutes a part of the thermal emission described above.

S Hawking[20] (my emphasis)

This quote from Stephen Hawking's paper pretty much explains how Hawking radiation comes about. Vacuum fluctuations are taking place everywhere and all the time. In the process, pairs of particles are created and annihilated in the continual dance which is part of the quantum world.

Fluctuations that take place away from the vicinity of a black hole's event horizon appear and disappear as they are classically non-physical states. However, if this pair production takes place within striking distance of the event horizon, then it is possible for one of the particles to fall into the black hole, potentially leaving the other to escape to an independent life. This is only physically possible if the particle that falls in has negative energy, so given that $E < 0$ is not classically allowed in this space-time, the particle would have to tunnel through the non-physical region to get inside the event horizon, where such states are physically sustainable.

A useful measure of the distance a particle would have to tunnel comes from the uncertainty principle and the scale at which we might expect QFT to become an important factor in our calculations. If we wish to resolve the location of a particle to within an uncertainty δx, the uncertainty principle tells us that the uncertainty in its momentum, δp, is tied in by $\delta x \delta p \geq h/4\pi$. At the same time, momentum is related to the energy and mass via $E^2 - p^2 c^2 = m^2 c^4$ (in flat space-time) so:

$$2E\delta E - 2c^2 p \delta p = 0 \quad \text{making} \quad \delta E = \frac{c^2 p \delta p}{E}.$$

Substituting from the uncertainty principle for $\delta p \geq h/4\pi\delta x$:

$$E = \frac{c^2 p \Delta p}{E} \geq \frac{hc^2 p}{4\pi E \delta x} \quad \text{or} \quad \delta E_{\min} = \frac{hc^2 p}{4\pi E \delta x},$$

where $\delta E_{\min}$ is the smallest energy uncertainty arising from a spatial uncertainty of δx. This is very significant: at some level of spatial resolution, the uncertainty in a particle's energy will be larger than the mass energy of a particle-antiparticle pair. In that case, *we can no longer be sure how many particles there are occupying that region* as there is enough energy uncertainty for particle/antiparticle pairs to pop in and out of existence. On such scales, QFT needs to be deployed in any calculations.

In that case:

$$\delta E_{\min} = 2mc^2 = \frac{hc^2 p}{4\pi E \delta x} \quad \text{or} \quad \delta x = \frac{hp}{8\pi E m}.$$

Now if $E^2 - p^2 c^2 = m^2 c^4$ and the 'kinetic energy' term $p^2 c^2 \sim 4m^2 c^4$, we have:

$$E^2 - p^2 c^2 = p^2 c^2/4 \quad \text{or} \quad E = \frac{\sqrt{5}}{2} pc \sim pc.$$

Finally this gives us:

$$\delta x \sim \frac{hp}{8\pi pcm} = \frac{h}{8\pi mc} = \frac{\lambda_c}{4}$$

which is a quarter of the *Compton wavelength* of the particle, $\lambda_c = h/2\pi mc$. The Compton wavelength itself can be defined as the wavelength of a photon which carries the same energy as the mass of the particle, hence the name Compton *wavelength*, although in this application it is more of a significant length scale. To a rough order of magnitude, we will say that QFT needs to be applied to distances below the Compton wavelength.

So, within a Compton wavelength of the event horizon, we can expect vacuum fluctuations to be important. This also gives us a scale over which the negative energy particle would have to tunnel in order to get to the physically viable region within the event horizon. We now have different, but equivalent, ways to think about Hawking radiation.

One possibility is to imagine a virtual particle pair emerging near to the event horizon, then the negative energy state tunnelling through to within the Schwarzschild radius, where it is a physically possible

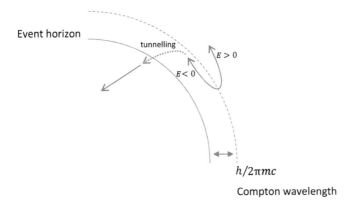

Event horizon

tunnelling

$E > 0$

$E < 0$

$h/2\pi mc$

Compton wavelength

FIGURE 13.4 Hawking radiation: a virtual particle pair pops into existence near the event horizon and the negative energy state tunnels to where it is physically sustainable while the positive energy state escapes to infinity.

state (Figure 13.4). This negative energy state is absorbed by the black hole, reducing its net mass and allowing the positive energy particle to escape to infinity. To the outside observer, it seems that the positive energy particle is emitted from the region of the black hole.

Another way of explaining Hawking radiation is shown in Figure 13.5. Here a particle within the event horizon tunnels through and annihilates with the negative energy particle conjured by the vacuum fluctuation. Once again the mass/energy of the black hole is down by the mass/energy of the particle that escapes to infinity.

A third way of thinking about Hawking radiation is related to the problems of defining a quantum vacuum in a curved space-time. As I mentioned before, the standard techniques of QFT rely on the ability to expand the field over positive and negative frequencies. In a curved space-time it is impossible to do this in a consistent way in all co-ordinate systems. If we add in the constraint of a close-by event horizon, providing a physical cut-off, then the vacuum becomes equivalent to a collection of thermally excited particles, which is the Hawking radiation.

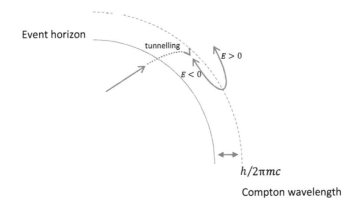

Event horizon

tunnelling

$E > 0$

$E < 0$

$h/2\pi mc$

Compton wavelength

FIGURE 13.5 Hawking radiation: a particle from inside the event horizon tunnels through and annihilates with the negative energy of the virtual pair.

13.3.6 Black Hole Evaporation

Hawking's detailed calculations confirm that the black hole is emitting particles like a black body with an equivalent temperature, T_H related to its mass, M:

$$T_H = \frac{hc^3}{16\pi^2 GMk}$$

as seen at infinity (k being Boltzmann's constant, $1.38 \times 10^{-23}\,\text{m}^2\,\text{kg/s}^2\text{K}$). So, as the mass of the black hole reduces it gets 'hotter'.

Standard black body radiation theory relates the total radiated power to the temperature of the body and its surface area, $P = \sigma A T^4$, with σ being Stefan's constant ($5.670 \times 10^{-8}\,\text{W/m}^2\text{K}^4$). As the black hole's mass reduces, the increased temperature will cause the rate of radiation to increase as well, so the mass reduces at a greater rate, and we have an accelerating 'evaporation' of the black hole. The increasing temperature also shortens the wavelength at which the most energy is being radiated (Wien's law, Section 12.5.2), ending with a burst of gamma rays.

If we take the emitting area as the surface of the event horizon $A = 4\pi (R_s)^2 = 16\pi G^2 M^2 / c^4$, then:

$$P = \frac{16\pi\sigma G^2 M^2}{c^4}\left(\frac{hc^3}{16\pi^2 GMk}\right)^4 = \frac{\sigma c^8 h^4}{16^3 G^2 \pi^7 k^4 M^2}.$$

This radiant power is reducing the mass of the black hole at the rate $c^2 \Delta M / \Delta t = -P$, giving:

$$c^2 \frac{\Delta M}{\Delta t} = -\left(\frac{\sigma c^8 h^4}{16^3 G^2 \pi^7 k^4}\right)M^{-2} = CM^{-2}$$

gathering all the constants into C for convenience. The solution to this equation is:

$$c^2 \int_{M_0}^{0} \frac{M^2 dM}{C} = -\int_{0}^{T} dt,$$

where T is the time required for the black hole to evaporate completely, via Hawking radiation. The standard integral $\int M^2\,dm = \frac{1}{3}M^3$, so the result is:

$$T = \frac{c^2 M_0^3}{3C} = 8.41 \times 10^{-17} M_0^3.$$

Our reference black hole has been Sgr A*, which is a supermassive black hole of mass $\sim 4 \times 10^6\,M_s$, this monster would take:

$$T = 4.24 \times 10^{94}\,\text{s} \sim 1.3 \times 10^{87}\,\text{years}$$

to evaporate. Even a much smaller candidate, like Cygnus X-1, with $M_0 \sim 15\,M_s$ would take $\sim 10^{71}$ years. Working backwards from the age of the universe, the most massive black holes that could be evaporating right now would have $M_0 \lesssim 1.7 \times 10^{11}$ kg which is in the mini black hole category. Astronomers are certainly aware of this possibility and are on the lookout for the characteristic radiation signature of such black hole evaporations.

Another consideration is the cosmic microwave background. If the temperature of the black hole matches that of the CMB, then it will be absorbing energy at the same rate that it is radiating. Such a black hole has a mass today $\sim 0.8\%$ of the Earth or ~ 0.6 of the mass of the Moon. However, as the CMB temperature gets colder over time, this stability will not last.

13.4 Inflation

As I mentioned in Chapter 12, it is very striking that Ω is so close to being 1. We seem to have ended up with the *one unique value* of Ω that stays constant over time. It is hard to believe that this has come about by chance. There is a name for this issue in cosmological circles: *the flatness problem*. Physicists don't believe in co-incidences; if Ω is 1 there has to have been some physical process that drove the universe to this density at an early stage.[21] If so, then this might also be a solution to a related problem, that of the horizon.

13.4.1 The Horizon Problem

In Section 12.2.2, I suggested that the map of the cosmic microwave background provided powerful justification for assuming isotropy in our universe. However, there is also an argument to suggest that it is too isotropic. The map displays variations in the equivalent black body temperature for the cosmic microwave background from point to point across the sky. Broadly, these represent variations ~10^{-5} K in an average temperature across the sky of ~2.7 K.

The confirmation of these temperature variations was a milestone in cosmology as we expected regions of the universe where the density of matter at recombination was slightly greater than average would have slightly higher CMB temperatures and vice versa. With density variations being the key to galaxy formation, not finding any temperature fluctuations would have been problematical. However, the measured scale of these fluctuations presents us with a problem.

Physically two objects come to the same temperature when energy flows from one to the other. The fastest way to exchange energy is via photons. However, in the universe at large, very few photons interact with matter after recombination. So, any mechanism to exchange energy and bring the universe to a uniform temperature had better be complete by recombination, as it will become hopelessly inefficient afterwards.

If the age of the universe at recombination is t_r, then the size of the region brought to equilibrium by this mechanism is $x_r \leq ct_r$. However, that will not represent the visible size now, as it has to be scaled up by expansion. If microwaves from the opposite edges of the region have only just reached us, they have been travelling for a time $t_0 - t_r$. As $t_0 \sim 13.8 \times 10^9$ years and $t_r \sim 380{,}000$ years, $t_0 - t_r \approx t_0$. As a result, the edges of the region must be $d = ct_0$ from us. The next step is to use basic geometry to find the angular size of this region on the sky. In Figure 13.6, two points on the sky (drawn as galaxies) are both a distance $d = ct_0$ from us. Assuming that the angle they subtend is small (and this anticipates the answer), they are separated from each other by a distance $x_0 = d\vartheta_0 = ct_0\vartheta_0$, where ϑ_0 is their angular

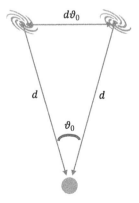

FIGURE 13.6 Two points in the sky that are a distance d form us and separated by $d\vartheta_0$ from each other would subtend an angle ϑ_0 on our view of the sky.

separation. In order to link this back to the size of the region, x_r, at recombination, we have to scale d down (or scale x_r up). Accordingly we have:

$$\frac{x_0}{x_r} = \frac{A_0}{A_r}.$$

In Chapter 12, we set the scale so that $A_0 = 1$, giving $x_r = A_r x_0$. To find a value for A_r, we consider the temperature of the CMB. Recombination was triggered by the photon energy dropping below that of the ionisation energy of hydrogen, which corresponds to a black body temperature of ~3,000 K. In Chapter 12, we showed that the CMB temperature scaled as A^{-1}, so if the black body temperature now is ~2.7 K, we have:

$$\frac{T_0}{T_r} = \frac{A_r}{A_0} = \frac{A_r}{1} = \frac{2.7}{3000} \quad \text{making} \quad A_r = 9 \times 10^{-4}.$$

Putting this all together:

$$x_r = A_r x_0 \quad \text{so} \quad ct_r = A_r ct_0 \vartheta_0$$

giving:

$$\vartheta_0 = \frac{t_r}{A_r t_0} = \frac{378000}{9 \times 10^{-4} \times 13.8 \times 10^9} = 0.03 \,\text{rad}$$

which corresponds to ~2°. For comparison, the full Moon covers an angle of about 0.5°. Even if these elementary considerations were wrong by several orders of magnitude, it is clear that the degree of isotropy observed in the CMB sky map could never have been established by a conventional physical mechanism, at least not before recombination took place. In the community, this is known as the *horizon problem* – how can two places in the universe be the same temperature if one is 'over the horizon' from the other?

13.4.2 Expanding Our Horizons

On 7 December 1979, Alan Guth[22] (then a post-PhD researcher at SLAC) cycled into his office carrying a notebook with *spectacular realization* scribbled across the top of one of its pages. His calculations, concluded late the previous evening, had shown him a way of solving the horizon and flatness problems in a manner that arose very naturally from particle physics. Since Guth's ground-breaking work, physicists have refined the model and its predictions, but the central tenant remains.

According to Guths' insight, at a very early phase of the universe's evolution it would have been kicked into exponential expansion by a change of state in the fields present. This period of *inflation* probably started at ~10^{-36} s into history and stopped at ~10^{-32} s. However, during this microscopic period of time the scale of the universe increased by a factor of 10^{26} or more: equivalent to a proton swelling to 10^{11} m across... There is also the more ordinary expansion that has taken place since inflation completed.

If you reverse the argument and scale the visible universe back using ordinary (Friedmann) expansion (radiation dominated, then matter dominated then dark energy), you find that it must have been ~3 m across when inflation stopped. One thing is for sure, if this scenario did happen, the total universe was considerably bigger than 3 m by the time inflation had done its work. This would imply that the current visible universe is only a microscopic speck in the totality of what is out there (Figure 13.7).

Although this is clearly an extraordinary process for the early universe to have gone through, it does rather elegantly solve the flatness problem. As it transpires, the field causing inflation swells with the expansion so that a tiny patch of field pre-inflation becomes responsible for the matter density of a vast region post inflation.

13.4.3 The Inflaton Field

In Guth's early calculations, the root cause of the inflationary expansion was a version of the Higgs field (Chapter 5). That model (now called *old inflation*) was subsequently abandoned as it proved incapable of

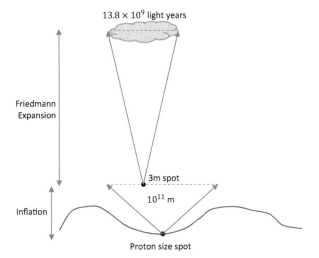

FIGURE 13.7 During inflation, regions of the early universe the size of a proton expanded until they were 10^{11}. Inside one such region, a patch roughly 3 m in size would then expand over the millennia since inflation to become the size of the current visible universe. Every such patch would have done the same, making our current visible universe a tiny fraction of what is out there.

generating a sufficiently homogeneous universe after inflation ended. A new model (*new inflation*) was proposed by Andre Linde[23] in 1982. In this scenario, the cause of inflation is an *inflaton field* which, like the Higgs field, is a *scalar field* (under the Lorentz transformations) with very distinctive properties.

As we discussed earlier, it is possible for a field to contain a great deal of energy even if there are no particles present. In an outline way, this is like the surface of a very deep and smooth lake. Without any ripples crossing the surface of the lake, we are hard pressed to 'see' that the water is present. The field cannot be 'seen' if its strength is zero, as no particles are being excited into existence. However, pushing the analogy, as the lake has great depth, so the field can still contain a great deal of energy, even though its presence cannot be directly detected. Generally we don't worry about such things as we only experience and measure energy *changes*. However Guth realised that such energy densities should be included in the Friedmann equations.

The energy density tensor for a scalar (non-quantised) field with field strength φ is[24]:

$$\mathbb{T}_{ab} = \sum_{u,v}\left(g^{-1}\right)_{au}\left(g^{-1}\right)_{bv}\frac{\Delta\varphi}{\Delta x_u}\frac{\Delta\varphi}{\Delta x_v} - \left(\frac{1}{2}\sum_{u,v}\left\{\left(g^{-1}\right)_{uv}\frac{\Delta\varphi}{\Delta x_u}\frac{\Delta\varphi}{\Delta x_v}\right\} - V(\varphi)\right)\left(g^{-1}\right)_{ab},$$

where $V(\varphi)$ is the potential energy density of the field with different choices for $V(\varphi)$ underling the different models of inflation that exist. By comparing this to the standard energy tensor:

$$\mathbb{T}_{ab} = \left(\rho + P/c^2\right)U_aU_b - P\left(g^{-1}\right)_{ab},$$

we can extract the equivalent 'pressure' and 'density', as if this field were an exotic form of fluid:

$$\rho_\varphi c^2 = \frac{1}{2}\left(\frac{\Delta\varphi}{c\Delta t}\right)^2 + V(\varphi)$$

$$P_\varphi = \frac{1}{2}\left(\frac{\Delta\varphi}{c\Delta t}\right)^2 - V(\varphi)$$

noting that $\varphi = \varphi(t)$ only, for consistency with the homogeneity in the FLRW metric. As the expression $\frac{1}{2}\left(\frac{\Delta\varphi}{c\Delta x_t}\right)^2$ looks somewhat like the formula for classical kinetic energy, this is known as the *kinetic energy density*.

The first Friedmann equation then becomes:

$$H^2(t) = \frac{8\pi G}{3c^2}\left(\frac{1}{2}\left(\frac{\Delta\varphi}{c\Delta t}\right)^2 + V(\varphi)\right)$$

assuming a flat space-time ($k = 0$) with this new energy density dominating over all other forms (including the dark energy). If we then apply the standard condition to the energy density tensor:

$$\sum_a \frac{D\mathbb{T}_{ab}}{\Delta x_a} = 0,$$

we obtain an equation that determines the rate of change of the field:

$$\frac{\Delta}{c\Delta t}\left(\frac{\Delta\varphi}{c\Delta t}\right) + 3H\frac{\Delta\varphi}{c\Delta t} + \frac{\Delta V(\varphi)}{\Delta\varphi} = 0 \tag{13.4}$$

which can be solved (remembering that H is a function of time) to find the time evolution of φ. Physicists have noted that Equation 13.4 has a passing resemblance to the formula for an object sliding down a slope (Figure 13.8), so the coefficient of $\Delta\varphi/c\Delta t$ has been named the *Hubble friction*, although physically it is a radically different sort of phenomenon.

Inflation comes into play if the kinetic term is small compared with the potential energy, i.e. $\frac{1}{2}\left(\frac{\Delta\varphi}{c\Delta x_t}\right)^2 \ll V(\varphi)$, in which case we have:

$$\rho_\varphi c^2 \approx V(\varphi) \quad \text{and} \quad P_\varphi \approx -V(\varphi)$$

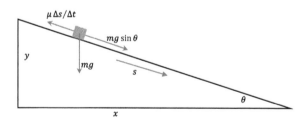

Net force down slope $= ma = m\frac{\Delta}{\Delta t}\left(\frac{\Delta s}{\Delta t}\right) = mg\sin\theta - \mu\frac{\Delta s}{\Delta t}$

$s = x\cos\theta \quad$ so $\quad \frac{\Delta s}{\Delta t} = \frac{\Delta x}{\Delta t}\cos\theta \quad$ and $\quad \frac{\Delta}{\Delta t}\left(\frac{\Delta s}{\Delta t}\right) = \frac{\Delta}{\Delta t}\left(\frac{\Delta x}{\Delta t}\right)\cos\theta$

So that: $\quad m\frac{\Delta}{\Delta t}\left(\frac{\Delta x}{\Delta t}\right)\cos\theta + \mu\frac{\Delta x}{\Delta t}\cos\theta - mg\sin\theta = 0$

Hence: $\quad m\frac{\Delta}{\Delta t}\left(\frac{\Delta x}{\Delta t}\right) + \mu\frac{\Delta x}{\Delta t} - mg\tan\theta = 0 \quad$ and $\quad \tan\theta = \frac{\Delta y}{\Delta x}$

Finally: $\quad \frac{\Delta}{\Delta t}\left(\frac{\Delta x}{\Delta t}\right) + \frac{\mu}{m}\frac{\Delta x}{\Delta t} - g\frac{\Delta y}{\Delta x} = 0$

FIGURE 13.8 Obtaining the equation that determines how an object slides down a slope.

with the Friedmann equation:

$$H^2(t) = \frac{8\pi G V(\varphi)}{3c^2}.$$

Note that we now have a negative pressure, just as we discussed in relation to dark energy in Chapter 12. In that scenario, the Friedmann equation was:

$$\left(\frac{1}{A} \frac{\Delta A}{\Delta t} \right)^2 = H^2(t) = \frac{8\pi G}{3} \rho_\Lambda$$

leading to $A \sim e^{\sqrt{\rho_\Lambda} t}$. In this case, we obtain $A \sim e^{\sqrt{V(\varphi)} t}$ (if $V(\varphi)$ is hardly changing with time) which is the exponential expansion characteristic of inflation. The problem now becomes stopping it, which was also the issue with Guth's old inflation – it stopped 'raggedly' leading to different densities in different regions. This is where the selection of $V(\varphi)$ comes in.

With a typical choice of $V(\varphi)$ (Figure 13.9) the potential evolves with φ over a broad flat 'shoulder' which turns down into a pronounced minimum. While φ takes values across the shoulder, the potential is reasonably constant and larger than the kinetic term, so inflation is happening. Note also that in these circumstances $\Delta V / \Delta \varphi$ is small and can be neglected, making Equation 13.4:

$$\frac{\Delta}{c \Delta t} \left(\frac{\Delta \varphi}{c \Delta t} \right) + 3H \frac{\Delta \varphi}{c \Delta t} \approx 0.$$

Once φ reaches the elbow, the energy drops substantially to the minimum value. The kinetic term now becomes larger than the potential, bringing the inflationary period to a close. What happens next depends to a degree on how the scalar field couples to matter fields.

Around the minimum of $V(\varphi)$, particles which are the granularity of this field can be excited into existence. As a loose analogy to explain why this can happen, consider the shape in Figure 13.9 being assembled from a track so that a ball could roll along; $V(\varphi)$ is then the height of the track above the ground. A ball moving along the track from the left will reach the 'bowl' at the bottom and settle into rolling back and forth around the minimum (lowest point). Either side of the minimum, the slope points down and so provides a restoring force against the motion of the ball. The oscillation of the ball can be thought of as the field excitations, which in quantum theory correspond to particles.

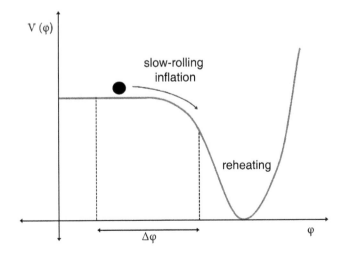

FIGURE 13.9 Slow roll inflationary potential.

13.4.4 Refilling the Universe

In cosmological terms, once the field has evolved into the potential energy minimum, floods of field particles are excited into existence. The energy that was potential in the field is now manifest in these particles. The inflaton field has not vanished, but it has had a negligible energy density since inflation ended. Remarkably, the energy that is ultimately the source of these particles comes from the expansion itself. Simplistically, as inflation swells the universe with the inflaton field its *energy density* stays roughly constant, so the *total energy* within a region must increase.[25] At the same time, negative gravitational energy is also appearing within the same region, cancelling the energy gained by the inflaton field. This energetic slight-of-hand may well indicate that the energy in the universe is exactly zero, when you include gravitational energy as well. This is certainly an attractive prospect, as otherwise we need a source of energy to create the universe in the first place.

If the φ field couples to matter fields (and there would really be no point if it did not), then the φ particles will decay into ordinary matter particles, filling up the inflated universe. As the extraordinary change of scale brought about by inflation has scattered the pre-inflationary matter to far distances, the density has dramatically fallen. However, now the universe is re-filled with matter from the decaying φ, which has the effect of driving the density close to the critical value: Ω is pushed towards one. One rough way of thinking about how this happens is to picture the hyper-expansion brought about by inflation as 'flattening' the spatial curvature (just as the surface of a very large ball looks flatter than that of a smaller one). As this curvature links to the matter density via the field equations, we can see that the value of Ω before inflation becomes irrelevant. That matter density has been scaled to zero. Now the flattened universe fills with the particles generated by φ, so the new density will be that required to have the flattened spatial curvature.

The inflation of every subatomic speck in the early universe was pushed by the inflaton field within that speck. As it grows, it creates more field with the same energy density. Before inflation the inflaton field would have had different values in different parts of the universe, as there is no physics that can force it to exactly the same value everywhere, that's the horizon problem. But, at subatomic scales the differences would have been extremely tiny. Consequently as inflation proceeds, each speck grows to a vast size with an inflaton field inside it that is *virtually identical across the whole of the region*. Once that smooth field has transferred energy into ordinary matter via decay of the φ particles, the density of matter is remarkably constant across the whole vast region: exactly what we see in the CMB results.

Local interactions between these new particles quickly set up a thermal equilibrium, and the fact that they have all come from the energy density of the inflaton field, means that they are all being generated with similar energies. Hence, a broad isotropy and homogeneity is produced, without the need for a separate physical process.

Although smoothing out density variations in the universe is a great virtue of the inflationary scenario, it is vital that they are not *completely* eliminated or galaxy formation has nothing to get a gravitational grip on. Fortunately, quantum fluctuations in the φ field means that inflation does not stop at exactly the same time across the universe as a whole, or equivalently the number of φ particles produced varies slightly from point to point. As a result, although the overall density of the universe is driven close to the critical value, density fluctuations are built in. These quantum fluctuations give rise to a very specific pattern of density variations, matching what has been seen in the CMB sky map and also in the distribution of galaxies recorded in the Sloan Digital Sky Survey. This observational confirmation has given physicists confidence that the inflationary scenario is not simply a mathematical construct to explain flatness and the horizon problem: it is capable of making independent scientific predictions that are confirmed by observations. The detection of gravitational waves from inflation would be another exciting confirmation.

Inflation is currently part of the standard model of Big Bang cosmology, but it is not without its critics. The nature of the inflaton field itself is still something of a mystery, although the recent discovery of the Higgs Boson, confirming that the Higgs field does exist, has re-kindled interest in seeing if the inflaton field is a cousin of Higgs. Perhaps more tellingly, many detractors point to the need for a particular and not naturally 'obvious' $V(\varphi)$ to bring about inflation and suggest that this is as much 'fine tuning' as having $\Omega = 1$ selected by Divine fiat.

13.4.5 Quantum Cosmology

Given that inflation can expand a tiny region of the universe into a vast space, it is natural to wonder if the initial trigger was a quantum fluctuation in some complex vacuum. When one adds the possibility that the total energy of the universe is zero (balancing the energy of the matter against the negative energy of gravity), it becomes plausible that a tiny universe appeared as a fluctuation in some pre-existing quantum reality. Inflation would then expand this proto-universe into the one that we are living in now.

This is a highly speculative extension of our current theories, but it is only by pushing ideas in this manner that we learn what works and what does not. Perhaps all our current approaches will turn out to be dead ends, but the exercise is worthwhile for what we can learn from the effort.

Admittedly I am prejudiced, but I still cannot think of a more exciting subject to be involved with right now.

Notes

1. The significance of these Planck scales is sometimes over-stated, as there is still some element of choice involved. Planck elected to use the constants c, G and $h/2\pi$ as he adjudged them to be the most fundamental (one might challenge $h/2\pi$ rather than h). These scales may be natural, but that does not imply that they are fundamental.
2. When you are grappling with terminology, you have to reach for any foothold you can see. In particle physics, different species are often referred to as favours. So, the electron field is one flavour of quantum field, the muon field another flavour. Then, we can talk about particle and antiparticle varieties of the same flavour.
3. This is also expressed via conjugation: the number of particles in a state, n, and the phase of the wave aspect, φ, are conjugate observables.
4. In terms of the development of Section 13.3.1, a global time-like Killing vector is needed.
5. Jorge Pullin (1963–), Horace Hearne Chair in theoretical Physics, Louisiana State University. As quoted in https://www.quantamagazine.org/string-theory-meets-loop-quantum-gravity-20160112/.
6. Abhay Vasant Ashtekar (1949), Director of the Institute for Gravitational Physics and Geometry at Pennsylvania State University.
7. Lee Smolin (1955–), Professor of Physics at the University of Waterloo.
8. Ted Jacobson (1954–), Professor of Physics at the University of Maryland.
9. Carlo Rovelli (1956–), Centre de Physique Théorique de Luminy of Aix-Marseille University.
10. If we start with a photon wavelength equal to the Planck length, we can always boost into a system that is moving towards the wave source, Doppler shifting the measured wavelength to a smaller value. If that were not possible, then there would have to be a co-ordinate system in which the photon had a minimum possible wavelength, singling that system out as being rather special among otherwise Lorentz-equivalent systems.
11. Edward Witten (1951–), Professor of Mathematical Physics at the Institute for Advanced Study in Princeton, New Jersey.
12. John Henry Schwarz (1941–), Harold Brown Professor of Theoretical Physics at Caltech.
13. David Gross (1941–), Noble Prize 2004, Professor of Theoretical Physics at the Kavli Institute for Theoretical Physics of the University of California.
14. Leonard Susskind (1940–), Professor of Theoretical Physics at Stanford University.

15. Michael Green (1946–), Professor of Applied Mathematics and Theoretical Physics at Cambridge, Lucasian Professor of Mathematics.

16. Penrose, R., 2005. *The Road to Reality*. Vintage; New Ed edition, p. 825. ISBN-10: 0099440687.

17. Stephen Hawking (1942–2018), Director of Research at the Centre for Theoretical Cosmology at the University of Cambridge, Lucasian Professor of Mathematics.

18. So named after Wilhelm Karl Joseph Killing, (1847–1923), Professor of Mathematics at the University of Münster.

19. Always providing that the mass is positive, but we have no indications that negative masses exist, nor how we would interpret them if they did.

20. Hawking, S.W., 1975. Particle creation by black holes. *Communications in Mathematical Physics*, **43**, pp. 119–220.

21. Or some other constraint that makes this value the only possible one.

22. Alan Guth (1947–), Professor of Physics at MIT.

23. Linde, A.D., 1982. A new inflationary universe scenario: A possible solution of the horizon, flatness, homogeneity, isotropy and primordial monopole problems. *A Linde Physics Letters B*, **108** (6), pp. 389–393.

24. This is another result that can be obtained from the calculus of variation, which can also be used to derive the field equations. Unfortunately, such arguments lie well beyond what we can achieve here. Note that for a scalar, there is no dependence on axis vectors; hence, an ordinary rate of change is the same as a covariant rate of change…

25. As discussed in Chapter 11, we struggle with a total energy for the universe, but we can look into regions within the universe.

Bibliography

While researching this book, I made use of a range of excellent resources on the Internet as well as text books and articles by a variety of authors. In my attempts to find the clearest blends of mathematical development or conceptual discussion, I can't attribute lines of thought directly to separate sources. I am more than happy, however, to acknowledge all of them here. I apologise to any that I inadvertently omitted.

Alexander, G.P., 2017. *General Relativity 2017*. Warwick, UK: University of Warwick Centre for Complexity Science. https://warwick.ac.uk/fac/sci/physics/staff/academic/galexander/general_relativity_px436.pdf.

Balbus, S., 2018. *Notes on General Relativity and Cosmology*. Oxford, UK: Oxford Physics Department. https://www2.physics.ox.ac.uk/sites/default/files/profiles/balbus/ht18-copy-42841.pdf.

Baumann, D., 2014. *Cosmology. Part III: Mathematical Tripos*. Cambridge, UK: Cambridge University. https://www.damtp.cam.ac.uk/user/db275/Cosmology/Lectures.pdf.

Bini, D., Chicone, C. and Mashhoon, B., 2012. Spacetime splitting, admissible coordinates, and causality. *Physical Review D*, **85**(10), p. 104020.

Blau, M., 2011. *Lecture Notes on General Relativity*. Bern, Germany: Albert Einstein Center for Fundamental Physics. http://www.blau.itp.unibe.ch/Lecturenotes.html.

Brown, P.M., 2002. *Einstein's Gravitational Field*. Ithaca, NY: Cornell University. *arXiv preprint physics/0204044*.

Carroll, S.M., 1997. *Lecture Notes on General Relativity*. Ithaca, NY: Cornell University. *arXiv preprint gr-qc/9712019*.

Collier, P., 2012. *A Most Incomprehensible Thing*. Harlow, UK: Incomprehensible Books.

Crowell, B., 2008. *General Relativity*. Fullerton, CA: Light and Matter. https://www.lightandmatter.com.

Earman, J. and Glymour, C., 1978. Lost in the tensors: Einstein's struggles with covariance principles 1912–1916. *Studies in History and Philosophy of Science Part A*, **9**(4), pp. 251–278.

Engelhardt, W., 2013. *On the Origin of the Lorentz Transformation*. Ithaca, NY: Cornell University. *arXiv preprint arXiv:1303.5309*.

Flanagan, E.E. and Hughes, S.A., 2005. The basics of gravitational wave theory. *New Journal of Physics*, **7**(1), p. 204.

Hirata, C.M., 2012. *Lecture VIII: Linearized Gravity. University Lecture Notes CALTECH*. Pasadena, CA: California Institute of Technology. https://www.tapir.caltech.edu/~chirata/ph236/lec08.pdf.

Janssen, M., 2008. *'No Success like Failure...': Einstein's Quest for General Relativity, 1907–1920*. Pittsburgh, PA: PhilSci-Archive. http://philsci-archive.pitt.edu/4377/.

Kokkotas, K.D., 2002. Gravitational wave physics. *Encyclopedia of Physical Science and Technology*, 7(3), pp. 67–85. Academic Press.

Lambourne, R.J., 2010. *Relativity, Gravitation and Cosmology*. Cambridge, UK: Cambridge University Press.

Lancaster, T. and Blundell, S.J., 2014. *Quantum Field Theory for the Gifted Amateur.* Oxford, UK: Oxford University Press.

Maurer, S.M., 2001. Idea man. *Beamline (SLAC),* **31**, p. 1.

Misner, C.W., Thorne, K.S., Wheeler, J.A. and Kaiser, D.I., 2017. *Gravitation.* Princeton, NJ: Princeton University Press.

Norton, J.D., 1993. General covariance and the foundations of general relativity: Eight decades of dispute. *Reports on Progress in Physics,* **56**(7), p. 791.

Özer, M. and Taha, M.O., 2000. *On Proper Time in General Relativity.* Ithaca, NY: Cornell University.

Penrose, R., 2004. *The Road to Reality.* London, UK: Jonathan Cape.

Rizzi, G. and Tartaglia, A., 1998. Speed of light on rotating platforms. *Foundations of Physics,* **28**(11), pp. 1663–1683.

Rovelli, C., 2017. *Reality Is Not What It Seems: The Journey to Quantum Gravity.* 1st edn. London: Penguin.

Rugh, S.E. and Zinkernagel, H., 2010. *Weyl's Principle, Cosmic Time and Quantum Fundamentalism.* Ithaca, NY: Cornell University. *arXiv preprint ArXiv:1006.5848.*

Schutz, B., 2009. *A First Course in General Relativity.* Cambridge, UK: Cambridge University Press.

Sperhake, U., 2014. *General Relativity. Part III: Mathematical Tripos.* Cambridge, UK: Cambridge University. https://www.damtp.cam.ac.uk/user/us248/Lectures/Notes/gr.pdf.

Taylor, E.F. and Wheeler, J.A., 2000. *Exploring Black Holes Introduction to General Relativity.* San Francisco, CA: Addison Wesley Longman.

Tolish, A., *General Relativity and the Newtonian Limit.* The University of Chicago. https://www.math.uchicago.edu/~may/VIGRE/VIGRE2010/REUPapers/Tolish.pdf.

Ule, A., 2006. Einstein, Gödel and the disappearance of time. *Synthesis Philosophica,* **21**(2), pp. 223–231.

Visser, M., 2007. *The Kerr Spacetime: A Brief Introduction.* Ithaca, NY: Cornell University. *arXiv preprint arXiv:0706.0622.*

Weinberg, S., 1972. *Gravitation and Cosmology: Principles and Applications of the General Theory of Relativity.* 1st edn. New York: Wiley.

Index

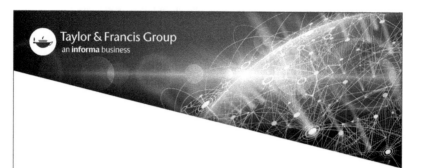

Taylor & Francis eBooks

www.taylorfrancis.com

A single destination for eBooks from Taylor & Francis
with increased functionality and an improved user
experience to meet the needs of our customers.

90,000+ eBooks of award-winning academic content in
Humanities, Social Science, Science, Technology, Engineering,
and Medical written by a global network of editors and authors.

TAYLOR & FRANCIS EBOOKS OFFERS:

A streamlined
experience for
our library
customers

A single point
of discovery
for all of our
eBook content

Improved
search and
discovery of
content at both
book and
chapter level

REQUEST A FREE TRIAL
support@taylorfrancis.com

 Routledge
Taylor & Francis Group

 CRC Press
Taylor & Francis Group

Milton Keynes UK
Ingram Content Group UK Ltd.
UKHW051946071024
449327UK00026B/2182